# Natural Computing Series

Series Editors: G. Rozenberg (Managing)
Th. Bäck  A.E. Eiben  J.N. Kok  H.P. Spaink
Leiden Center for Natural Computing

T0189993

Springer
*Berlin*
*Heidelberg*
*New York*
*Hong Kong*
*London*
*Milan*
*Paris*
*Tokyo*

Hans-Paul Schwefel   Ingo Wegener
Klaus Weinert (Eds.)

# Advances in Computational Intelligence

Theory and Practice

With 91 Figures and 30 Tables

 Springer

*Editors*

Prof. Dr. Hans-Paul Schwefel
Universität Dortmund
Fachbereich Informatik
Lehrstuhl Informatik XI
44221 Dortmund, Germany
hps@udo.edu

Prof. Dr. Ingo Wegener
Universität Dortmund
Lehrstuhl Informatik II
44221 Dortmund, Germany
wegener@cs.uni-dortmund.de

Prof. Dr. Klaus Weinert
Universität Dortmund
Fakultät Maschinenbau
Institut für Spannende Fertigung
44221 Dortmund, Germany
weinert@isf.maschinenbau.uni-dortmund.de

*Series Editors*

G. Rozenberg (Managing Editor)
Th. Bäck, A.E. Eiben, J.N. Kok,
H.P. Spaink

Leiden Center for Natural Computing
Leiden University
Niels Bohrweg 1
2333 CA Leiden
The Netherlands
rozenber@liacs.nl

Library of Congress Cataloging-in-Publication Data

Advances in computational intelligence: theory and practice/Hans-Paul Schwefel, Ingo
Wegener, Klaus Weinert (eds.)
    p. cm. – (Natural computing series)
    Includes bibliographical references and index.

    1. Computational intelligence. I. Schwefel, Hans-Paul. II. Wegener, Ingo. III. Weinert,
Klaus, 1943– IV. Series.
Q342 .A37 2002
006.3–dc21                                                                    2002021693

ACM Computing Classification (1998): F.1, F.2, G.1.6, G.3, G.4, I.2.3, I.2.8, I.5.1, J.2, J.6

ISBN 978-3-642-07758-6

Springer-Verlag Berlin Heidelberg New York,
a member of BertelsmannSpringer Science+Business Media GmbH
http://www.springer.de

© Springer-Verlag Berlin Heidelberg 2010
Printed in Germany

Cover Design: KünkelLopka, Heidelberg

# Preface

Most authors of this book are, only a few were, members of the

| | |
|---|---|
| Collaborative Research Center *Design and Management of Complex Technical Processes and Systems by Means of Computational Intelligence Methods* (SFB 531) | Sonderforschungsbereich *Design und Management komplexer technischer Prozesse und Systeme mit Methoden der Computational Intelligence* (SFB 531) |

This interdisciplinary research center was founded at the University of Dortmund in the beginning of 1997, and it is supported by the Deutsche Forschungsgemeinschaft (DFG). We are very grateful for the support from both the DFG and our university.

The book contains overviews of recent results of the SFB and, at the same time, provides a good introduction to modern trends in computational intelligence (CI). The areas of fuzzy logic, evolutionary algorithms, and machine learning are discussed by researchers from theoretical and applied computer science, as well as from the disciplines of mechanical engineering, electrical engineering, and chemical engineering. The results presented range in nature from theoretical to applied. They clearly illustrate how theoretical results make their way into real-world applications and how applications guide theoretical research. Altogether, the book reflects our interdisciplinary approach to CI techniques.

This is also the place to thank all those people who have helped behind the scenes, especially when different manuscripts had to be put together into a coherent whole for this publication. We cannot name everyone who helped, but at least Mrs. Gundel Jankord and Maria Bayer working as secretaries for the SFB should be thanked here as well as Lutz Schönemann, the present chief executive officer of the SFB. Last, but not least, we want to thank the editors of this book series and Mrs. Ingeborg Mayer, for their excellent collaboration, and Kenneth De Jong for "degermanizing" this preface.

April 2002

The board of directors of the SFB 531
H.-P. Schwefel, I. Wegener, and K. Weinert

# Contents

# 1 Introduction

Ulrich Hammel[1], Boris Naujoks[2], and Hans–Paul Schwefel[2]

[1] NuTech Solutions GmbH, Martin-Schmeißer-Weg 15,
44227 Dortmund, Germany
ulrich.hammel@nutechsolutions.de
[2] University of Dortmund, Department of Computer Science, Informatik XI,
44221 Dortmund, Germany
{naujoks, schwefel}@Ls11.cs.uni-dortmund.de

**Summary.** This introduction aims at helping newcomers to the field of computational intelligence (CI) to understand the scope of CI, to learn about its brief history, and to glimpse neighboring fields, be they well established already or still emerging from within the living system called science.

## 1.1 Computational Intelligence Methods

Fuzzy systems, artificial neural networks, and evolutionary computation – referred to as CI methods – have many features in common:

- Their basic concepts have been known for a long time and have been surrounded by controversy at times. One reason, among others, for the hesitant acceptance of CI methods in the beginning was the lack of sufficient computational power. Moreover, some resistance within established scientific communities had to be overcome. And finally, as in the case of other new fields, a critical mass of people had to accumulate in order to initiate research at a visible scale.
- They become more and more successful as applications spread into ever broader areas. Similar to other scientific domains, the critical mass in the field of CI first accumulated in applied research. Concrete practical problems exist, and they have to be solved, whether basic research is aware of them or not, and even if somebody proves that the problem at hand is generally unsolvable under certain premises. New methods are often applied to such problems tentatively at first and demonstrate their practical utility without a priori existing formal proof of their applicability. The observation that it works is sometimes all that is required to stimulate investigations to clarify the why and under which conditions a new method works.
- Their theoretical foundations are not complete. Basic research tends to keep an expectant attitude before it assimilates new methods, the operability of which is difficult if not impossible to prove. In parts of CI, especially in the domain of fuzzy logic, substantial progress in laying its theoretical foundation has already been achieved; a comprehensive picture, however, is still missing.

- Repeatedly, new and specialized variants of CI methods are created in a trial and error fashion. The main reason for this very inefficient procedure of reinventing the wheel again and again is the lack of an appropriate theoretical basis.

- The areas of applications are overlapping. To apply CI methods is reasonable only when conventional concepts fail or when exact solutions gained with such methods are by far too expensive. In these cases, CI methods may lead to good approximations or reasonable probability of success. Furthermore, they can often be applied alternatively. Typical applications are found in nonlinear control, optimization, identification, classification, image processing, pattern recognition, and decision support, especially when there are multiple, conflicting objectives.

- They supplement each other excellently. Sometimes, the application of one CI method results in a new problem, which is also not solvable in a canonical way, but by means of a second CI method. Examples are the combined structure and weight optimization in neural networks, the parameterization and genetic operator selection in evolutionary algorithms, as well as the determination of fuzzy rule sets and appropriate membership functions. Such tasks can often be tackled by combining different CI methods. At present, such hybridization is typically done in a sequential manner (method A as a pre-processing step for method B). One may speculate that coupling methods with one another more intelligently will increase the overall performance. This holds in particular as soon as the foundations of the different methods are better understood in detail.

- They are intermediate steps on a long march. The general goal is to push problem solving algorithms forward into problem classes of increasing complexity. There are many other methods aiming in the same direction, e.g., cellular automata, immune networks, agent systems, possibilistic approaches, multivalued logic as well as competitive, cooperative, and self-organizing structures. Ideas for uses of CI methods in these areas are expected to continue growing in number.

- They were not even mentioned in the 1988 edition of the well-known German encyclopedia *Duden Informatik*. The second edition of 1993 contained already paragraphs on fuzzy logic, neural networks, and genetic algorithms. The latter concerns at least one of the two American variants of the broader class of evolutionary algorithms. Only the third edition of 2001 [7] briefly mentions further variants like evolution strategies and genetic programming. Research establishments have begun to recognize the power of CI methods. In industry and economy they have been in use since the early 1970s, and they are just entering worldwide the curricula of computer science.

## 1.2    Parallel Problem Solving from Nature

The common features of CI methods reach even deeper than described in Section 1.1. Apart from the fact that they mainly use a subsymbolic (numerical) representation of information – which is in stark contrast to the classic artificial intelligence (AI) and its emphasis on symbolic representations – CI methods are intended to mimic the information processing within natural systems algorithmically. Aside from the fascination from observing efficacious and efficient processes and systems in nature, there are two main reasons for dealing with them: the urge of scientists to understand real-world phenomena, and the eagerness of engineers to make use of nature's patents.

Computational intelligence can be looked at as a continuation of cybernetics and bionics, two older movements in the history of science. Meanwhile, computers have entered the scene, both as simulation tools for systems analyses and as target instruments for implementing bio-inspired problem solving methods. Single-processor personal computers, multiprocessor supercomputers, and networks of workstations are serving those purposes. A precondition for this connection was and still is the transition from analytical to algorithmic models of the processes observed and from purely sequential to immanently parallel procedures. Parallel processing does not mean merely making use of the divide-and-conquer principle; it means exploiting the power of diverse collective actions.

Dennett [8] argued vividly that Darwin's main merit lies in his recognition of organic evolution as an iterative and parallel process. Similar arguments hold for other CI paradigms as well. These open the door to fiddling with simulation models of synthetic living systems as has been done in the realm of so-called artificial life (AL). They also offer the opportunity to use the basics of computer science for analyzing and thus gaining deep insight into the phenomena. The collaborative research center Design and Management of Complex Technical Processes and Systems by Means of CI Methods (SFB 531) follows the second road that, hopefully, is leading to a methodology creating problem solving tools via analyzing algorithmic models of natural processes.

## 1.3    On the Development of Computational Intelligence

The international interest in research related to CI and its practical significance is witnessed by corresponding national grant initiatives, especially in Japan and the U.S.A., as well as by the numbers of annual publications and conferences since 1985 [33]. At present, artificial neural networks (NN) [9], fuzzy logic (FL) [30], and evolutionary algorithms (EA) [2] form the most important constituent parts of CI. The strong methodological relationships among these three approaches were publicly honored for the first time during the first IEEE world congress on computational intelligence (WCCI) in 1994

[37], an assembly of three concurrent and colocated conferences on NN, FL, and EA. The second WCCI took place in 1998, the third one during May 2002. These congresses form milestones in establishing the field of computational intelligence.

The term computational intelligence, as used here, was coined by Bezdek [3, 4] and since then has been accepted as a generic term for the three sets of methods mentioned above. In related areas, some other ideas are seeking attention under such designations as autonomous agents, from animals to animats (or simulation of adaptive behavior), swarm intelligence, evolvable hardware, and the like. Although CI is seen as an extension of AI in some regions, especially in America, the representation and processing of information within CI methods differs fundamentally from the traditional AI approach as used in connection with expert systems. A short review of the history of CI may serve to distinguish between the different solution paths.

An old dream of creating artificially intelligent machines seemed to be realized when the first digital computers became available and their universal (or Turing) computability was discovered. In those early days, the interest focussed on universally adaptive and learning systems operating on the basis of simple numerical representations. Many different attempts were made, one of them being based on the perceptron model of Rosenblatt [28] in the mid 1950s. It became the precursor of today's neural networks. The first evolutionary algorithms showed up in the early 1960s. Independently of one another, Holland [18, 19] devised the class of genetic algorithms (GA) as a model of adaptation and classification processes in Ann Arbor, Michigan, and Fogel [11, 12] began to train finite automata for time series prediction by means of simulated evolution in San Diego, California. This second variant of EA is better known as evolutionary programming (EP) [10]. Around the same time, evolution strategies (ES) were used as experimental optimization heuristics by Rechenberg [26, 27] and Schwefel [31, 32] in Berlin, Germany. Also in the 1960s, Zadeh [36] initiated the transition to fuzzy sets as a new means for qualitative model building, thus creating the foundations of fuzzy logic.

The breakthrough of the new proposals did not happen immediately, the main reasons for that being the low information processing power of early computers as well as the insufficient theory-based understanding of the new concepts. It can be seen as a consequence of the second deficiency, that Minsky and Papert [25] could disprove the universal computability of the perceptron model in the late 1960s. This led to a shift of focus from connectionist models to symbolic knowledge processing, later called crisp in contrast to soft computing by Zadeh. Indeed, the construction of expert systems and software development tools led to spectacular successes in the 1970s and 1980s [6, 16, 24]. Concepts, such as evolutionary algorithms, were nearly forgotten during that period.

In the mid 1980s, the limits of pure symbolic knowledge processing became apparent. Many technological problems have inherently numerical, or subsymbolic representations, on which symbolic operators are not suitable for solving the task at hand. This observation, together with the still exponential increase of computational power summoned a revival of some CI methods. The work of Hopfield [21], Rumelhart, McClelland, and the PDP Research Group [29] brought neural networks back to the research laboratories. Some remarkable successes of systems based on either NN, EA, or FL techniques (see Sejnowski and Rosenberg [34], Kohonen et al. [23], Goldberg [13, 14], Holmblad and Østergaard [20]) pushed research activities worldwide and resulted in many successful applications. These successes called attention to the impressive potential of the three CI paradigms for problem solving (see Alander's [1] bibliography containing more than 3000 references on EA topics, Klimaskausas [22] for NN topics, as well as Hirota [17], Sugeno [35], and Gupta and Yamakawa [15] for FL).

Bezdek [3] characterized the methods of computational intelligence as having the following properties:

- adaptivity,
- fault tolerance,
- computational speed in the same range as achieved in biological neural networks (especially by making use of the inherent parallelism), and
- error rate optimality (balancing training speed versus error frequency).

The attribute *computational* refers to the subsymbolic, or numerical representation of the problem and the resulting subsymbolic knowledge aggregation and information processing. Bezdek rates computational intelligence only as a means carrying necessary but not sufficient conditions for the construction of artificially intelligent systems. The collaborative research center (SFB 531) takes a pragmatic view: it does not primarily aim at constructing intelligent systems but aims instead at developing CI methods for tackling complex technological challenges.

## 1.4   State of the Art and Trends in Computational Intelligence

The state of CI can be characterized as follows:

- A couple of isolated solutions for special types of problems exist, and their protagonists defend them against any attack from outside (for example, genetic algorithms versus evolutionary programming proponents).
- Out of all existing methods, practitioners tend not to cautiously choose the one best suited to the situation at hand. Due to time pressure and only secondary interest in the methods' working principles they accept each improvement without assessing the reason for its success. Understanding

the cause of the good result and how it was achieved could help to find even better solutions by a properly adapted method.

- The considerable number of success stories provokes euphoria and exaggerated expectations, pushing developers to ever new ad-hoc variants. Scientific findings appear accidentally under such conditions and remain incoherently separated from each other.
- Currently most advanced theory is in the field of fuzzy computation. However an integrative view is rare, though rudiments exist, e.g., in the area of evolutionary computation.
- No methodical superstructure for all CI methods (FL, NN, and EA) is available. That is the main reason why no constructive concept has yet overcome ad-hoc solution path finding.

As a subject that has its origins in applied research, CI has reached a phase that typically occurs in many young disciplines, including computer science. One should remember the early programming languages that existed before a theoretical superstructure was developed (e.g., formal languages, theory of automata, complexity theory, and formal logic); the influence of Codd's algebra of relations on the development of data bases; the impact of formal logic on symbolic AI; and the importance of the theory of data types for software engineering.

Grand challenges in technological disciplines, natural science including molecular biology, and computer science itself, including data bases, image processing, data mining, and visualization need effective and efficient procedures. These procedures should approximate problem solving capacities of living systems and comprise such features as adaptability, fault tolerance, low error rates, and high performance. In addition to using single CI methods, such problems are now more and more often handled by means of coupling different CI methods or hybridizing CI methods with AI techniques, traditional optimization procedures, or statistical data analyses. This may lead to the detection and exploitation of synergetic effects from combining methods developed in different disciplines. Hybrid approaches may yield unexpected, qualitatively new behaviors. Therefore, coupling single CI methods with one another is an objective of the collaborative research center in addition to the analysis and further development of plain methods.

Further paradigms gleaned from organic information-processing procedures, such as immune networks, multicellular tissues, and co-evolving multi-agent systems, to mention only a few of those fascinating visions, are likely to gain significance in the future.

## 1.5   Goals of the Collaborative Research Center

The popularity and the successes of CI methods are mainly based on their generality. The independence from any particular problem setting results in

a wide range of applicability for all methods derived from CI. The majority of practical applications can be found in technical disciplines. The number of applications in business and management is growing continuously, too, as shown by Biethan and Nissen [5]. In both areas problems arise that need subsymbolic solution methods due to their inherent representation; but often no problem-specific analytical or numerical approximation methods are known for these problems. Formally speaking, these problems are not solvable in an exact or sufficiently efficient exact way. Or it may be that the cost for creating a new exact method for solving a particular problem is by far too high in comparison to the merit gained. In such cases, practitioners tend to rest content then with coarse approximations based on a few tests and experience. These are the situations in which huge benefits could be gained by applying CI methods as general heuristics for implementing problem solvers.

The desire and necessity to solve complex tasks in all kinds of operations, at least approximately, will grow in a technology-driven world with ubiquitous information and global free trade. This will require researchers to:

- advance existing methods, to couple them with each other as well as with traditional techniques,
- prove their performance on ever new problems of growing complexity and to compare them with already existing other methods,
- analyze their formal foundations in order to detect their capabilities and limits, and
- look for alternatives and check them, too.

Since 1997, the collaborative research center (SFB 531) has made an effort toward contributing to the first three requirements mentioned above. The stated objectives of the effort are to advance understanding of information-processing networks (neural nets being only one special subgroup here), the processing of fuzzy (i.e., vague and incomplete) information and evolutionary adaptation and amelioration mechanisms – in isolation as well as in combination. At the collaborative research center, we want to explore the computer-assisted problem solving potential of these methodologies, and we want to clarify their limits. Moreover, we want to push the field in Europe as others do abroad by means of prototypical applications using improved CI methods. Our vision is to initiate a development toward a constructive methodologic superstructure and contributing substantially to such movement. Nevertheless, a strategy of small steps will be necessary to proceed toward the goal.

All subtasks of the collaborative research center will first concentrate on the three established CI techniques fuzzy logic, neural networks, and evolutionary algorithms in isolation in order to enhance a deeper understanding of their working principles and their limits. Furthermore, the application domains ought to be classified and distinguished from each other. Next, hybrid methods will be developed and the expected synergetic effects studied. Finally, a redefinition of CI and its borders to other problem solving strategies will become worthwhile.

From the analytic and experimental work described above, we hope to gain deep insight into the working principles of the CI methods, so that in a next step we will be able to create a methodology that synthesizes specific procedures made up of well-understood components, or building blocks, a term used in connection with genetic algorithms. This is the long-term perspective of our SFB 531.

The application-oriented subtasks of the collaborative research center were chosen from engineering disciplines, in particular from electrical, mechanical, and chemical engineering. The problem solving concepts to be established will, however, be transferable to other, non-engineering disciplines, because they cover rather general tasks, such as identification, modeling, classification, pattern recognition, image processing, process control, and optimization.

It should perhaps be mentioned that one or the other team member of the SFB 531 hopes to attain some measure of enlightenment toward understanding the real-world processes that are mimicked somehow by the different CI algorithms, even if the emulations are crude simplifications of the natural models. If this is achieved, the efforts would not only contribute to the often-demanded "mathematization" within the life sciences, which has not yet been as successful as in physics and parts of chemistry, they might even become a decisive component for dealing with more and more complex artificial systems, embedded into a sensitive natural environment. Having passed a long-term survival test, the organic models imitated by CI techniques are without doubt vital examples for robust and often astonishingly efficient processes.

Certainly, many more adaptation and amelioration or, more generally, problem solving concepts exist out there in natural systems that have not yet been, but perhaps should be exploited one day. Beginning with the question of why adaptability is often found at criticality borders (edge of chaos) and whether one could learn something from that fact, up to multi-agent strategies with cooperative or even altruistic behavior of single individuals and subpopulations that might be especially suitable for finding Pareto-optimal solutions between Scylla (efficient, but not that robust method) and Charybdis (effective, but too expensive method), so much know-how is waiting to be discovered!

## 1.6   On the Organization of the Chapters of This Book

The chapters of this book summarize the achievements in the field of CI that have been obtained during the last years by members of the collaborative research center SFB 531 at the University of Dortmund. The working groups of the SFB 531 are organized as shown in the figure below. Three types of subtasks exist: type A are theoretically oriented tasks that aim at fostering the foundations of the CI methods, whereas type C are application-oriented

tasks that make use of the CI techniques, eventually demonstrating their usefulness. Application domains are prototypic tasks from three engineering disciplines, i.e., electrical, mechanical, and chemical engineering. Type B, literally in between A and C subtasks, are the necessary bridges to help the users making the most of the results from theory, e.g., by delivering appropriate software tools, and to help theoreticians recognize the needs of practitioners.

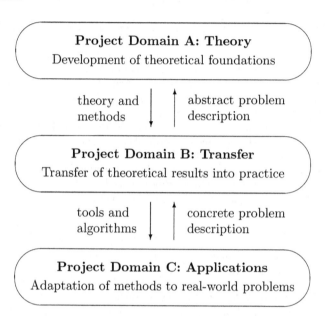

The chapters of this book do not mirror the structure of the subtasks. Instead, they are organized thematically along the lines of the different CI methods, i.e., from fuzzy logic to evolutionary algorithms. Within the chapters, there are paragraphs emphasizing either theory or applications. The fact that there is no chapter on neural networks does not indicate any aversion to them. Simply, there was no major SFB subtask on this topic during the years up to the early 2002, when this book was written.

# References

1. J. T. Alander. *An Indexed Bibliography of Genetic Algorithms: Years 1957–1993*. Art of CAD Ltd., Espoo, Finland, 1994.
2. T. Bäck, D. B. Fogel, and Z. Michalewicz, editors. *Handbook of Evolutionary Computation*. Oxford University Press, New York, and Institute of Physics Publishing, Bristol, 1997.
3. J. C. Bezdek. What is computational intelligence? In Zurada et al. [37], pages 1–12.

4. J. C. Bezdek. Computational intelligence and edge detection. In A. Grauel, W. Becker, and F. Belli, editors, *Fuzzy-Neuro-Systeme '97 - Computational Intelligence, Proceedings in Artificial Intelligence*, pages 1–31. infix-Verlag, Sankt Augustin, 1997.

5. J. Biethahn and V. Nissen. *Evolutionary Algorithms in Management Applications*. Springer, Berlin, 1995.

6. B. G. Buchanan and E. H. Shortcliff. *Rule Based Expert Systems: The MYCIN Experiments of the Stanford Heuristic Programming Project*. Addison-Wesley, Reading, MA, 1985.

7. V. Claus and A. Schwill, editors. *Duden Informatik*. Dudenverlag, Mannheim, 3rd edition, 2001.

8. D. C. Dennett. *Darwin's Dangerous Idea*. Simon & Schuster, New York, 1995.

9. E. Fiesler and R. Beale, editors. *Handbook of Neural Computation*. Oxford University Press, New York, and Institute of Physics Publishing, Bristol, 1997.

10. D. B. Fogel. *Evolutionary Computation: Toward a New Philosophy of Machine Intelligence*. IEEE Press, New York, 1995.

11. L. J. Fogel. Autonomous automata. *Industrial Research*, 4:14–19, 1962.

12. L. J. Fogel, A. J. Owens, and M. J. Walsh. *Artificial Intelligence Through Simulated Evolution*. Wiley, New York, 1966.

13. D. E. Goldberg. Genetic algorithms and rule learning in dynamic system control. In J. J. Grefenstette, editor, *Proceedings First International Conference Genetic Algorithms and Their Applications*, pages 8–15. Lawrence Erlbaum, Hillsdale, NJ, 1985.

14. D. E. Goldberg. *Genetic Algorithms in Search, Optimization and Machine Learning*. Addison-Wesley, Reading, MA, 1989.

15. M. M. Gupta and T. Yamakawa, editors. *Fuzzy Computing: Theory, Hardware and Applications*. Elsevier, Amsterdam, 1988.

16. F. Hayes-Roth, D. A. Waterman, and D. B. Lenat, editors. *Building Expert Systems*. Addison-Wesley, Reading, MA, 1983.

17. K. Hirota. *Industrial Applications of Fuzzy Technology*. Springer, Tokyo, 1993.

18. J. H. Holland. Outline for a logical theory of adaptive systems. *Journal of the ACM*, 3:297–314, 1962.

19. J. H. Holland. *Adaptation in Natural and Artificial Systems*. The University of Michigan Press, Ann Arbor, MI, 1975.

20. L. P. Holmblad and J. J. Østergaard. Control of a cement kiln by fuzzy logic. In M. M. Gupta and E. Sanchez, editors, *Approximate Reasoning in Decision Analysis*, pages 389–400. North-Holland, Amsterdam, 1982.

21. J. J. Hopfield. Neural networks and physical systems with emergent collective computational abilities. *Proceedings National Academy of Science*, 79:2554–2558, 1982.

22. C. C. Klimaskausas, editor. *The 1989 Neuro-Computing Bibliography*. MIT Press, Cambridge, MA, 1989.

23. T. Kohonen, M. Shozakai, J. Kangas, and O. Venta. Microprocessor implementation of a large vocabulary speech recognizer and phonetic typewriter for Finnish and Japanese. In J. A. Laver and M. A. Jack, editors, *Proceedings European Conference on Speech Technology*, pages 377–380. CEP Consultants, Edinburgh, 1987.

24. R. K. Lindsay, B. G. Buchanan, E. A. Feigenbaum, and J. Lederberg. *Applications of Artificial Intelligence for Organic Chemistry: The DENDRAL Project*. McGraw-Hill, New York, 1980.

25. M. Minsky and S. Papert. *Perceptrons*. MIT Press, Cambridge MA, 1969.
26. I. Rechenberg. Cybernetic solution path of an experimental problem. Royal Aircraft Establishment, Library translation No. 1122, Farnborough, Hants., U.K., August 1965.
27. I. Rechenberg. *Evolutionsstrategie: Optimierung technischer Systeme nach Prinzipien der biologischen Evolution*. Dr.-Ing. Dissertation, Technical University of Berlin, 1971.
28. F. Rosenblatt. The perceptron: A probabilistic model for information storage and organization in the brain. *Psychological Review*, 65(6):386–408, 1958.
29. D. E. Rumelhart and J. L. McClelland. *Parallel Distributed Processing – Explorations in the Microstructure of Cognition*, volume 1: Foundations. MIT Press, Cambridge, MA, 1986.
30. E. H. Ruspini, P. P. Bonissone, and W. Pedrycz, editors. *Handbook of Fuzzy Computation*. Institute of Physics Publishing, Bristol, 1998.
31. H.-P. Schwefel. Experimentelle Optimierung einer Zweiphasendüse Teil I. Technical Report No. 35 of the Project MHD–Staustrahlrohr 11.034/68, AEG Research Institute, Berlin, October 1968.
32. H.-P. Schwefel. *Evolutionsstrategie und numerische Optimierung*. Dr.-Ing. Dissertation, Technical University of Berlin, May 1975.
33. H.-P. Schwefel. Parallel problem solving from nature. In A. Kent and J. G. Williams, editors, *Encyclopedia of Computer Science and Technology*, volume 37, supplement 22, pages 225–246. Marcel Dekker, New York, 1997.
34. T. J. Sejnowski and C. R. Rosenberg. Parallel networks that learn to pronounce English text. *Complex Systems*, 1:145–168, 1987.
35. M. Sugeno. *Industrial Applications of Fuzzy Control*. North-Holland, Amsterdam, 1985.
36. L. A. Zadeh. Fuzzy sets. *Information and Control*, 8:338–353, 1965.
37. J. M. Zurada, R. J. Marks II, and C. J. Robinson, editors. *Computational Intelligence: Imitating Life*. IEEE Press, New York, 1994.

Part I

# Fuzzy Logic

# 2 Mathematical Foundations of Fuzzy Inference

Stephan Lehmke, Bernd Reusch, Karl–Heinz Temme, and Helmut Thiele

University of Dortmund, Department of Computer Science, Informatik I,
44221 Dortmund, Germany
{lehmke, reusch, temme, thiele}@Ls1.cs.uni-dortmund.de

**Summary.** The research project reported here has been active since 1997. The goals and results of the whole project are outlined here, but only results obtained since 2000 are reported in detail.

The project is concerned with the theoretical foundations of fuzzy logic, in particular functional analytic foundations of inference with fuzzy IF-THEN rule bases. The goals of this project are as follows:

- Systematic characterization and comparison of fuzzy inference methods, applying methods from functional analysis.
- Axiomatic characterization of fuzzy inference operators.
- Investigation of functional analytic semantics of complex fuzzy inference mechanisms.
- Investigation of extensions and generalizations of fuzzy inference methods.

## 2.1 Introduction

Since the early publications [109, 110, 111, 112, 113, 114] by L. A. Zadeh, a large variety of applications for fuzzy IF-THEN rule bases have been found, mainly for designing fuzzy controllers, but also for formalizing fuzzy expert systems, in decision theory and in approximate reasoning, to name a few applications. Even today, a significant scarcity of basic theoretical investigations of the concept fuzzy IF-THEN rule base and its handling can be perceived. To illustrate this, classical and fuzzy control theorems are compared.

Classical control theory possesses a well-established basis in classical analysis, in particular, the theory of differential and integral equations and the corresponding numerical methods, consisting of accepted definitions, theorems, and tested algorithms. Fuzzy control theory, on the other hand, shows an almost complete lack of such a basis, in particular, precise definitions are frequently replaced by intuitive explanations, and establishing theorems is circumvented by trial-and-error methods.

It is the goal of this project to somewhat remedy this scarcity of theoretical investigations. It has to be pointed out that, for lack of space here, mainly definitions and results are reported here. Proofs can be found in the references. The remainder of this introduction is devoted to defining some fundamental notations and concepts.

For arbitrary sets $M$ and $N$, the cartesian product $M \times N$ is defined, as usual, by $M \times N =_{\text{def}} \{[x,y] \mid x \in M \wedge y \in N\}$, where $[x,y]$ denotes the ordered pair of $x$ and $y$. The $n$-fold cartesian product of the set $M$ is denoted by $M^n$, where $n$ is a natural number with $n \geq 1$. $\alpha : M \to N$ means that $\alpha$ is a unique mapping from $M$ into $N$. The set of all mappings from $M$ into $N$ is denoted by $N^M$.

$\mathbb{R}$ and $\langle 0,1 \rangle$ denote the set of all real numbers and the closed real unit interval, respectively. Fuzzy sets on a set $M$ are mappings $F : M \to \langle 0,1 \rangle$. $\mathfrak{P}M$ and $\mathfrak{F}M$ denote the usual power set and the fuzzy power set of $M$, defined by $\mathfrak{P}M =_{\text{def}} \{X \mid X \subseteq M\}$ and $\mathfrak{F}M =_{\text{def}} \{F \mid F : M \to \langle 0,1 \rangle\}$, respectively. T-norms are mappings $\tau : \langle 0,1 \rangle \times \langle 0,1 \rangle \to \langle 0,1 \rangle$, which are commutative, associative, and nondecreasing in the first place, and which fulfil the boundary condition $\forall s\,(s \in \langle 0,1 \rangle \to \tau(1,s) = s)$.

Sections 2.2 and 2.3 introduce some further indispensable concepts and problems with respect to fuzzy inference with fuzzy IF-THEN rule bases (also known as approximate reasoning). In Sections 2.4, 2.5, and 2.6, investigations of the fundamental concept of solutions for fuzzy IF-THEN rule bases (in particular, existence and uniqueness of solutions) are reported. Section 2.7 contains some new results on functional analytic properties of fuzzy inference operators. Sections 2.8 and 2.9 are devoted to two important related generalisations of approximate reasoning, based on a generalised concept of fuzzy set. Sections 2.10, 2.11, and 2.12 report research on the algebraic and axiomatic foundations of fuzzy inference, in particular with respect to fuzzy approximation, information granulation, and modelling with words.

## 2.2   Structure of Fuzzy Inference Mechanisms

Consider a control circuit in the following simplified form:

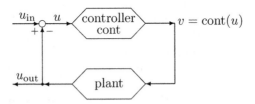

Assume that there exist nonempty sets $U$ (inputs) such that $u \in U$ and $V$ (outputs) such that $v \in V$.

In classical control, there is an important distinction between controllers without memory (at every point in time, the value of $v$ is a function of the value of $u$) and controllers with memory (at every point in time, the value of $v$ can depend on the present value of $u$ as well as all values of $u$ at earlier points in time, for example, a PID controller).

For simplicity here, for fuzzy control we only consider the first case ($v$ is a function of $u$). It is trivial to construct a fuzzy PID controller by replacing

the linear combinations of a classical PID controller by one fuzzy controller with several inputs. It is a long-term goal of this project to achieve a *true* incorporation of temporal aspects into fuzzy inference by using methods from functional analysis and classical temporal logic (see Section 2.8). For more information on technical aspects of the structure and design of fuzzy control circuits, see [26, 85].

When the block

is realized by a fuzzy controller, following Zadeh, we assume that cont can be factored in the form

$$\text{cont} = \Delta \circ \Psi \circ \Phi,$$

with components defined and denoted as follows:

$\Phi : U \to \mathfrak{F}U$   fuzzifier
$\Psi : \mathfrak{F}U \to \mathfrak{F}V$  inference engine
$\Delta : \mathfrak{F}V \to V$   defuzzifier.

This means that for $u \in U$, the equation

$$\text{cont}(u) = \Delta(\Psi(\Phi(u)))$$

holds, illustrated graphically by:

$$u \quad \Phi \quad \Phi(u) = F \quad \Psi \quad \Psi(F) = G \quad \Delta \quad \Delta(G) = v$$

Obviously, $\Psi : \mathfrak{F}U \to \mathfrak{F}V$ is a functional operator. In Sections 2.4 and 2.5, methods for constructing such operators are discussed. One essential prerequisite for such a construction is that a functional operator $\Psi$ is a solution of the interpretation problem of a given fuzzy IF-THEN rule base (see Section 2.3 and [40, 43, 77, 79, 80, 81, 82, 83, 84, 85]).

The defuzzifier $\Delta : \mathfrak{F}V \to V$ in the factorization cont $= \Delta \circ \Psi \circ \Phi$ above can be considered to be a functional which maps a function $G : V \to \langle 0, 1 \rangle$ to a crisp value $v = \Delta(G)$ from the domain of $G$. Accordingly, $\Phi$ is considered to be an inverse functional which maps a crisp value $u \in U$ to a function $F : U \to \langle 0, 1 \rangle$.

While there has been substantial research concerning defuzzifiers [6, 26, 27, 28, 36, 37, 65, 66, 67, 68, 83, 84], only few results concerning fuzzifiers are known [7, 53].

For space restrictions, the concepts fuzzifier and defuzzifier are not considered further here; see the referenced literature.

One goal of the project reported here is to study the semantics of complex fuzzy inference mechanisms, which are composed by arbitrarily combining inference modules in a layout as it is done for digital circuits. Inference modules are fuzzifiers, rule-based inference engines and defuzzifiers, but they are also modifiers and fuzzy arithmetic operators.

Some first steps towards defining the semantics of complex fuzzy inference mechanisms by using concepts of classical switching theory are described in [40]. A software tool for constructing, testing, comparing and automatically transforming complex fuzzy inference mechanisms has been implemented [2]. Research on this topic is ongoing.

## 2.3   Interpretation of Fuzzy IF-THEN Rule Bases Using Concepts of Functional Analysis

Let a fuzzy IF-THEN rule base

$$\text{RB} =_{\text{def}} \begin{array}{c} F_1 \Rightarrow G_1 \\ \vdots \\ F_n \Rightarrow G_n \end{array}$$

be given, where $n \geq 1$, $F_1, \ldots, F_n$ are fuzzy sets on $U$, and $G_1, \ldots, G_n$ are fuzzy sets on $V$.

*Remark 1.* The form of IF-THEN rule base given above is said to be *in normal form*. In applications, a fuzzy IF-THEN rule base is most often composed of finitely many rules of the form

IF $L_1 = T_1$ and $\cdots$ and $L_n = T_n$ THEN $L_{n+1} = T_{n+1}$,

where $L_1, \ldots, L_n, L_{n+1}$ are linguistic variables and $T_1, \ldots, T_n, T_{n+1}$ are linguistic terms assigned to the linguistic variables.

An interpretation assigns to every linguistic variable $L_i$ a non-empty set $U_i$ (universe) and to every term $T_i$ a fuzzy set $F_i$ on $U_i$.

It is easy to see that after interpretation, the expression $L_1 = T_1$ and $\cdots$ and $L_n = T_n$ can be replaced by a fuzzy set $F$ which is a combination of the fuzzy sets $F_1, \ldots, F_n$ (using an appropriate mapping from $\langle 0, 1 \rangle^n$ to $\langle 0, 1 \rangle$ interpreting "and") on a universe $U$, which is a combination of the universes $U_1, \ldots, U_n$ and the expression $L_{n+1} = T_{n+1}$ can be replaced by the fuzzy set $F_{n+1}$ on $U_{n+1}$.

This construction is easily extended to rule bases.

The concrete normal form construction is described in [82].

For the following considerations, this normal form is sufficient and facilitates the notation considerably.

Consider a functional black-box of the form

$$
F \longrightarrow \boxed{\begin{array}{c} F_1 \Rightarrow G_1 \\ \vdots \\ F_n \Rightarrow G_n \end{array}} \longrightarrow G
$$

induced by the rule base RB. We require that for every $i \in \{1, \ldots, n\}$, the input $F = F_i$ is mapped to the output $G = G_i$

A first step towards specifying the interpretation problem for a rule base is to interpret RB as a system of functional equations giving a partial specification of a functional operator $\Psi : \mathfrak{F}U \to \mathfrak{F}V$.

**Definition 1 (Solution for a fuzzy IF-THEN rule base).** $\Psi$ is said to be a solution (of the interpretation problem) for the fuzzy IF-THEN rule base RB iff

$$\Psi(F_1) = G_1$$

$$\vdots \qquad \vdots$$

$$\Psi(F_n) = G_n.$$

The set of all solutions for RB is denoted by SOL(RB).

Based on this definition, we formulate the following three fundamental problems;

I  *The existence problem.* Given a rule base RB, decide whether a solution for RB exists (SOL(RB) $\neq \emptyset$). If a solution exists, construct at least one solution.

II  *The uniqueness problem.* Given a rule base RB, decide whether a solution for RB has to be unique (if SOL(RB) $\neq \emptyset$, then SOL(RB) = $\{\Psi\}$).

III  *The restriction problem.* Given a rule base RB, how should the set of eligible functional operators $\Psi$ be restricted such that SOL(RB) contains only sensible solutions, without compromising existence of a solution?

The remainder of this section is a brief discussion of the restriction problem. The other two problems are discussed in subsequent sections.

First, obviously a rule base can be considered to be a set of rules, as neither reordering nor duplicating rules has an influence on the solution set.

**Observation 1.** *If* RB $\subseteq$ RB$'$, *then* SOL(RB$'$) $\subseteq$ SOL(RB).

This observation means that adding rules to a rule base will reduce the set of solutions.

**Observation 2.** *If a rule base* RB *contains rules* $F \Rightarrow G$ *and* $F \Rightarrow G'$ *with* $G \neq G'$, *then* $\mathrm{SOL}(\mathrm{RB}) = \emptyset$.

One nontrivial restriction which leads to more sensible solutions is to require the continuity of eligible functional operators $\Psi : \mathfrak{F}U \to \mathfrak{F}V$, assuming the existence of corresponding topologies in $\mathfrak{F}U$ and $\mathfrak{F}V$. This restriction is especially interesting in applications of fuzzy IF-THEN rule bases as it expresses the property that a small change of the output $\Psi(F)$ can be controlled by an appropriately small change of the input $F$.

If the topologies in $\mathfrak{F}U$ and $\mathfrak{F}V$ are generated by metrics $\rho$ and $\rho'$, then it is possible to consider the most promising (strong) form of continuity, namely Lipschitz continuity, defined by

$$\exists c (c \in \mathbb{R} \wedge c > 0 \wedge \forall F \forall G (F, G \in \mathfrak{F}U \to \rho'(\Psi(F), \Psi(G)) \leq c \cdot \rho(F, G))).$$

This formulation can be interpreted as a principle of strong causality for fuzzy inference expressed by $\Psi$: a small change of input leads to a small change (up to a multiplicative constant) of output.

For further information see [4, 43, 79, 82]. See also [29], where further restrictions such as Monotonicity are discussed.

## 2.4   The Generalized Modus Ponens and the Compositional Rule of Inference

From theory and applications of logical calculi, the inference rule modus ponens is well known.

$$\mathrm{MP} =_{\mathrm{def}} \boxed{\begin{array}{c} A \to B \\ A \\ \hline B \end{array}} \qquad (2.1)$$

where $A$ and $B$ are expressions of the calculus $\mathfrak{C}$ and $\to$ is a syntactic connective that allows the joining of expressions $A$ and $B$ into the compound expression $A \to B$.

If in $\mathfrak{C}$, a concept Th of theorem is defined, then MP is said to be correct with respect to Th if it holds that

If Th$A \to B$ and Th$A$, then Th$B$. $\qquad (2.2)$

In practice, this means that MP can be applied to derive Th$B$ from Th$A \to B$ and Th$A$.

In analogy with this deduction scheme, Zadeh [112] formulated the generalized modus ponens as:

$$\mathrm{GMP} =_{\mathrm{def}} \boxed{\begin{array}{c} F_1 \Rightarrow G_1 \\ F \\ \hline G \end{array}} \qquad (2.3)$$

where $F_1$, $G_1$, $F$ and $G$ are fuzzy sets on the universe $U$ or $V$.

That for $F = F_1$ and $G = G_1$, the scheme GMP coincides syntactically with MP has been a constant source of confusion in the past. While for the classical modus ponens, the conclusion $B$ is constructed by simply detaching the premise $A$ from the implication $A \to B$ (taking the concept Th into account), in the context of the generalized modus ponens, an analogous procedure as well as the introduction of an analogous concept of theorem are obviously unreasonable.

One further source of confusion is to mistake the symbol $\Rightarrow$ used to denote the IF-THEN connective with the symbol $\to$ for logical implication.

In the remainder of this section, we present the working hypothesis for interpreting the generalized modus ponens and in general, for interpreting fuzzy IF-THEN rule bases as it has been developed and investigated in the past five years [43, 79, 80, 81, 82, 83, 84, 85], namely the consistent application of concepts and methods from functional analysis. It has turned out that the classical functional analysis is a suitable tool for treating the interpretation problem for fuzzy IF-THEN rule bases.

For historical correctness, we mention that this approach is present in principle in the Compositional Rule of Inference developed by Zadeh [108, 109, 110, 111, 112, 113, 114] for interpreting the generalized modus ponens.

Remember that for a crisp relation $R \subseteq U \times V$ and a crisp set $X \subseteq U$, the $R$-image $R(X)$ of $X$ is defined

$$R(X) =_{\text{def}} \{y \,|\, \exists x(x \in X \wedge [x, y] \in R)\} \,. \tag{2.4}$$

Obviously, this definition induces a set operator $\Phi_R : \mathfrak{P}U \to \mathfrak{P}V$.

This definition was fuzzified by Zadeh in [110] under the name compositional rule of inference as follows;

The crisp components above are replaced by a binary fuzzy relation $S : U \times V \to \langle 0, 1 \rangle$ and a fuzzy set $F : U \to \langle 0, 1 \rangle$; furthermore $\wedge$ is interpreted by min, and $\exists$ is interpreted by sup, yielding the $S$-image $S(F)$ of $F$:

$$S(F)(y) =_{\text{def}} \sup \{\min(F(x), S(x, y)) \,|\, x \in U\} \qquad (y \in V) \tag{2.5}$$

Obviously, this definition induces a functional operator $\Psi_S : \mathfrak{F}U \to \mathfrak{F}V$.

This approach to interpreting the generalized modus ponens is completed by defining the fuzzy relation $S$.

For the generalized modus ponens in the form

$$\begin{array}{c} F_1 \Rightarrow G_1 \\ \underline{F} \\ G \end{array} \tag{2.6}$$

Zadeh defined $S : U \times V \to \langle 0, 1 \rangle$ for $x \in U$ and $y \in V$ by

$$S(x, y) =_{\text{def}} \min(F_1(x), G_1(y)) \tag{2.7}$$

and constructed the conclusion $G$ by

$$G =_{\text{def}} S(F). \tag{2.8}$$

The remarks above are summarized by the following definition.

**Definition 2 (Compositional rule of inference in standard form).**
The generalized modus ponens given in (2.6) is interpreted as follows:

1. $S(x, y) =_{\text{def}} \min(F_1(x), G_1(y))$ $\qquad\qquad (x \in U, y \in V)$,
2. $G(y) =_{\text{def}} \sup\{\min(F(x), S(x, y)) \mid x \in U\}$,
3. $\Psi^{F_1, G_1}(F) =_{\text{def}} G$.

With respect to the existence problem posed in Section 2.3, it has to be asked whether the compositional rule of inference yields a solution for the fuzzy IF-THEN rule base RB $=_{\text{def}} F_1 \Rightarrow G_1$, i.e. (compare Def. 1) whether $\Psi^{F_1, G_1}(F_1) = G_1$ holds. This property is also called correctness of the compositional rule of inference with respect to the generalized modus ponens.

Soon after introduction of the concepts mentioned above, it became apparent that, for applications, the compositional rule of inference in standard form is too special. This led to the following generalized definition, replacing min and sup by arbitrary functions [43, 79];

**Definition 3 (Interpretation).** $I = [\pi, \kappa, \mathcal{Q}]$ is said to be an interpretation iff

1. $\pi, \kappa : \langle 0, 1 \rangle \times \langle 0, 1 \rangle \to \langle 0, 1 \rangle$,
2. $\mathcal{Q} : \mathfrak{P}\langle 0, 1 \rangle \to \langle 0, 1 \rangle$.

**Definition 4 (Generalized Compositional Rule of Inference).** Given an interpretation $I = [\pi, \kappa, \mathcal{Q}]$, the generalized modus ponens given in (2.6) is interpreted as follows:

1. $\text{FREL}(\pi, F_1, G_1)(x, y) =_{\text{def}} \pi(F_1(x), G_1(y))$,
2. $G(y) =_{\text{def}} \mathcal{Q}\{\kappa(F(x), \text{FREL}(\pi, F_1, G_1)(x, y)) \mid x \in U\}$,
3. $\Psi_I^{F_1, G_1}(F) =_{\text{def}} G$.

## 2.5 Constructing Solutions for Fuzzy IF-THEN Rule Bases

Let a fuzzy IF-THEN rule base

$$
\text{RB} =_{\text{def}} \quad
\begin{matrix}
F_1 \Rightarrow G_1 \\
\vdots \\
F_n \Rightarrow G_n
\end{matrix}
\tag{2.9}
$$

be given. In this section, we discuss some methods for constructing solutions for RB. Considering Def. 1, this means finding a functional operator $\Psi : \mathfrak{F}U \to \mathfrak{F}V$ such that for every $i \in \{1, \ldots, n\}$, $\Psi(F_i) = G_i$. In the terminology introduced in Section 2.3, this means the current subsection is concerned with problem I – the *existence problem*.

Sections 2.5.1 and 2.5.2 refer to extending the generalized compositional rule of inference to rule bases using the principles FITA (first inference, then aggregation) and FATI (first aggregation, then inference). These principles were introduced (for a special case) in [106]. A discussion of the equivalence of these principles is presented in Section 2.5.3.

The relationship between the construction of solutions for fuzzy IF-THEN rule bases and solutions of fuzzy relational equations is discussed in Section 2.5.4. A method called similarity-based reasoning which is not based on the generalized compositional rule of inference is discussed in Section 2.5.5.

It has to be pointed out that most results concerning the subject of this section predate the year 2000. Due to restrictions, we only refer to the numerous publications on this subject. Nevertheless, a discussion of different methods is indispensable as a foundation for fuzzy inference and as a basis for more recent investigations reported in subsequent sections.

## 2.5.1  The Principle FITA

Let a rule base of the form (2.9) be given. For every $i \in \{1, \ldots, n\}$, fix an interpretation $I_i = [\pi_i, \kappa_i, Q_i]$.

Construct $\Psi_{I_i}^{F_i, G_i}$ as in Def. 4, i.e. for $F \in \mathfrak{F}U$ and $x \in U$,

$$\Psi_{I_i}^{F_i, G_i}(F)(y) =_{\text{def}} Q_i \left\{ \kappa_i(F(x), \pi_i(F_i(x), G_i(y))) \mid x \in U \right\}. \qquad (2.10)$$

Thus, FITA means first inference, then aggregation, i.e. after applying inference $n$ times, the inference results $\Psi_{I_1}^{F_1, G_1}(F), \ldots, \Psi_{I_n}^{F_n, G_n}(F)$ are aggregated.

**Definition 5 (FITA oriented interpretation).** $\text{INT}_{\text{FI}} =_{\text{def}} [I_1, \ldots, I_n, \alpha]$ is said to be an FITA oriented interpretation iff

1. $I_1, \ldots, I_n$ are interpretations in the sense of Def. 4,
2. $\alpha : \langle 0, 1 \rangle^n \to \langle 0, 1 \rangle$.

$\alpha$ is called aggregation function.

Given a FITA oriented interpretation $\text{INT}_{\text{FI}}$, we define a functional operator for interpreting RB.

**Definition 6 (FITA).**
$$\text{FITA}_{\text{INT}_{\text{FI}}}^{\text{RB}}(F)(y) =_{\text{def}} \alpha(\Psi_{I_1}^{F_1, G_1}(F)(y), \ldots, \Psi_{I_n}^{F_n, G_n}(F)(y)).$$

The nontrivial question, under which conditions is the functional operator $\text{FITA}^{\text{RB}}_{\text{INT}_{\text{FI}}}$ a solution of the interpretation problem for RB remains to be answered. A number of results can be found in [43, 79]; it has to be remarked that the results achieved so far do not come close to a final solution of this problem.

### 2.5.2   The Principle FATI

Let a rule base of the form (2.9) be given. When applying the inference method FITA, there are $n$ applications of functional operators of the type (2.10) before the inference result can be determined. If the operator $\Psi$ is computationally expensive, it is desirable to reduce the number of applications.

This leads to the principle FATI, denoting first aggregation, then inference.

**Definition 7 (FATI oriented interpretation).**
$\text{INT}_{\text{FA}} =_{\text{def}} [\pi_1, \ldots, \pi_n, \beta, \kappa, \mathcal{Q}]$ is said to be an FATI oriented interpretation iff

1. $\pi_1, \ldots, \pi_n, \kappa : \langle 0,1 \rangle^2 \rightarrow \langle 0,1 \rangle$,
2. $\beta : \langle 0,1 \rangle^n \rightarrow \langle 0,1 \rangle$,
3. $\mathcal{Q} : \mathfrak{P}\langle 0,1 \rangle \rightarrow \langle 0,1 \rangle$.

Given an FATI oriented interpretation $\text{INT}_{\text{FA}}$, we define a functional operator for interpreting RB (see Def. 4 for the definition of FREL).

**Definition 8 (FATI).**

1. $S(x,y) =_{\text{def}} \beta\,(\text{FREL}(\pi_1, F_1, G_1)(x,y), \ldots, \text{FREL}(\pi_n, F_n, G_n)(x,y))$,
2. $\text{FATI}^{\text{RB}}_{\text{INT}_{\text{FA}}}(F)(y) =_{\text{def}} \mathcal{Q}\,\{\kappa(F(x), S(x,y)) \mid x \in U\}$.

How far the computational complexity is reduced by reducing $n$ applications of $\mathcal{Q}$ to one application has to be discussed.

Again, results concerning the question under which conditions the functional operator $\text{FATI}^{\text{RB}}_{\text{INT}_{\text{FA}}}$ is a solution of the interpretation problem for RB can be found in [43, 79].

### 2.5.3   On the Equivalence of the Principles FITA and FATI

Considering the differences and similarities in the definitions of FITA and FATI (and especially the fact that FATI is less computationally expensive), it is interesting to ask under which conditions both methods yield the same results. To this end, we give a general definition.

**Definition 9 (Equivalence of functional operators).** Two functional operators $\Psi, \Psi' : \mathfrak{F}U \rightarrow \mathfrak{F}V$ are said to be equivalent with respect to $\mathcal{F} \subseteq \mathfrak{F}U$ iff $\forall F(F \in \mathcal{F} \rightarrow \Psi(F) = \Psi'(F))$.

We give some restrictions for interpretations $\mathrm{INT_{FI}}, \mathrm{INT_{FA}}$ and the base set $\mathcal{F}$ such that $\mathrm{FITA}^{\mathrm{RB}}_{\mathrm{INT_{FI}}}$ and $\mathrm{FATI}^{\mathrm{RB}}_{\mathrm{INT_{FA}}}$ are equivalent.

For specifying a special set $\mathcal{F}$, we need the concept of singleton fuzzy set. A singleton fuzzy set $F_u$ on $U$ is determined by an element $u \in U$ as follows: For all $u' \in U$

$$F_u(u') =_{\mathrm{def}} \begin{cases} 1, & \text{if } u' = u \\ 0, & \text{if } u' \neq u \end{cases}$$

Let $\mathcal{F} =_{\mathrm{def}} \{ F_u \mid u \in U \}$.

**Theorem 1 (Equivalence of FITA and FATI).**

1. *If* $\mathrm{INT_{FI}} = [\pi_1, \ldots, \pi_n, \kappa_1, \ldots, \kappa_n, \sup, \ldots, \sup, \alpha]$,

   $\mathrm{INT_{FA}} = [\pi_1, \ldots, \pi_n, \alpha, \kappa, \sup]$
   *and furthermore, every* $\varphi \in \{\kappa_1, \ldots, \kappa_n, \kappa\}$ *fulfills the boundary condition of t-norms, i.e.* $\forall s\, (s \in \langle 0, 1 \rangle \rightarrow \varphi(1, s) = s)$, *then* $\mathrm{FITA}^{\mathrm{RB}}_{\mathrm{INT_{FI}}}$ *and* $\mathrm{FATI}^{\mathrm{RB}}_{\mathrm{INT_{FA}}}$ *are equivalent with respect to* $\mathcal{F}$ *(without further conditions on* $\pi_1, \ldots, \pi_n, \alpha$*).*

2. *If* $\mathrm{INT_{FI}} = [\pi_1, \ldots, \pi_n, \kappa, \ldots, \kappa, \sup, \ldots, \sup, \max]$,

   $\mathrm{INT_{FA}} = [\pi_1, \ldots, \pi_n, \max, \kappa, \sup]$
   *and furthermore,* $\kappa$ *is monotone, then* $\mathrm{FITA}^{\mathrm{RB}}_{\mathrm{INT_{FI}}}$ *and* $\mathrm{FATI}^{\mathrm{RB}}_{\mathrm{INT_{FA}}}$ *are equivalent with respect to* $\mathfrak{F}U$ *(without further conditions on* $\pi_1, \ldots, \pi_n$*).*

In applications, it is still common to use only singleton fuzzy sets as inputs to inference engines. That in this case, FITA and FATI are equivalent for a wide range of interpretations, can be seen as a reason why the distinction between FITA and FATI is not well known in the literature. But as the use of fuzzy inputs gains popularity [7, 50, 53], this distinction becomes more important because in the case of nonsingleton inputs, the range of interpretations has to be restricted severely to assure equivalence.

For more sophisticated results on the equivalence of FITA and FATI for nonsingleton inputs, see [43, 79].

### 2.5.4   Solutions for Fuzzy IF-THEN Rule Bases as Solutions of Relational Equations

The equation (2.5) can be generalized to a definition scheme for a relation-based functional operator as follows.

**Definition 10 (Relation oriented interpretation).** $\mathrm{INT_R} =_{\mathrm{def}} [\kappa, \mathcal{Q}]$ is said to be a relation oriented interpretation iff

1. $\kappa : \langle 0, 1 \rangle^2 \rightarrow \langle 0, 1 \rangle$,
2. $\mathcal{Q} : \mathfrak{P}\langle 0, 1 \rangle \rightarrow \langle 0, 1 \rangle$.

Given a relation oriented interpretation $\mathrm{INT_R}$ and $S : U \times V \to \langle 0, 1 \rangle$, we define a functional operator implementing a generalized $S$-image.

**Definition 11 (Generalized $S$-image).**
$\Psi^S_{\mathrm{INT_R}}(F)(y) =_{\mathrm{def}} \mathcal{Q} \{ \kappa(F(x), S(x, y)) \, | \, x \in U \}.$

Finding a solution for a given rule base of the form (2.9) can be seen as the quest for a fuzzy relation which simultaneously solves the following system of relational equations:

$$\Psi^S_{\mathrm{INT_R}}(F_1) = G_1$$

$$\vdots \qquad\qquad\qquad (2.11)$$

$$\Psi^S_{\mathrm{INT_R}}(F_n) = G_n$$

This problem has been thoroughly investigated in the last two decades [3, 14, 15, 16, 17, 18, 54, 55, 56]. Obviously, there is a strong relationship between the operators $\Psi^S_{\mathrm{INT_R}}$ and $\mathrm{FATI}^{\mathrm{RB}}_{\mathrm{INT_{FA}}}$. The following result is well known; we only adapt it to the terminology used here.

**Theorem 2 (Relationship between $\Psi^S_{\mathrm{INT_R}}$ and $\mathrm{FATI}^{\mathrm{RB}}_{\mathrm{INT_{FA}}}$).**

1. *If $\mathrm{FATI}^{\mathrm{RB}}_{\mathrm{INT_{FA}}}$ is a solution for RB, where $\mathrm{INT_{FA}} = [\pi_1, \dots, \pi_n, \beta, \kappa, \mathcal{Q}]$, then $S$ given by*

    $$S(x, y) =_{\mathrm{def}} \beta \left( \mathrm{FREL}(\pi_1, F_1, G_1)(x, y), \dots, \mathrm{FREL}(\pi_n, F_n, G_n)(x, y) \right)$$

    *is a solution for (2.11), where $\mathrm{INT_R} = [\kappa, \mathcal{Q}]$.*
2. *Assume $\kappa$ is a left-continuous t-norm and $\mathcal{Q} = \sup$. Then (2.11) is solvable for some $S$ if and only if $\mathrm{FATI}^{\mathrm{RB}}_{\mathrm{INT_{FA}}}$ is a solution for RB, where $\mathrm{INT_{FA}} = [\pi, \dots, \pi, \min, \kappa, \mathcal{Q}]$ such that $\pi$ is the residual implication of $\kappa$, defined for $r, s \in \langle 0, 1 \rangle$ by*

    $$\pi(r, s) =_{\mathrm{def}} \sup \{ t \, | \, t \in \langle 0, 1 \rangle \text{ and } \kappa(r, t) \le s \} .$$

It is also possible to define a relational variant of FITA by aggregating the values of relation-based functional operators constructed from relational solutions of single equations. This method is not described in detail here.

## 2.5.5   Similarity-Based Reasoning

We present an method for defining functional operators which is not based on the compositional rule of inference – in particular, no quantification over the whole universe is necessary.

The approach was introduced for a special case in [117, 118]. More detailed investigations and further references are to be published in a forthcoming article.

**Definition 12 (Similarity oriented interpretation).** SI = [SIM, mod] is said to be a similarity oriented interpretation iff

1. SIM : $\mathfrak{F}U \times \mathfrak{F}U \to \langle 0, 1 \rangle$,
2. mod : $\langle 0, 1 \rangle \times \langle 0, 1 \rangle \to \langle 0, 1 \rangle$.

For interpreting the generalized modus ponens in the form (2.6), we define the following functional operator:

**Definition 13 (Similarity-based reasoning operator).**
$\Psi_{\text{SI}}^{F_1, G_1}(F)(y) =_{\text{def}} \text{mod}(\text{SIM}(F, F_1), G_1(y))$ for every $y \in V$.

While the compositional rule of inference is based on the idea of taking the image under a fuzzy relation, similarity-based inference is built on the idea of determining the degree of similarity between the input fuzzy set and the premise of the rule and then modifying the conclusion, based on the similarity degree. Next, this definition has to be extended to rule bases of the form (2.9). It is easy to define a FITA-like method.

For every $i \in \{1, \ldots, n\}$, fix a similarity oriented interpretation $\text{SI}_i =_{\text{def}} [\text{SIM}_i, \text{mod}_i]$ and furthermore, fix an aggregation function $\alpha : \langle 0, 1 \rangle^n \to \langle 0, 1 \rangle$.

**Definition 14 (Similarity-FITA oriented interpretation).**
$\text{SI}_{\text{FI}} =_{\text{def}} [\text{SI}_1, \ldots, \text{SI}_n, \alpha]$ is said to be a Similarity-FITA oriented interpretation.

**Definition 15 (Similarity-FITA).**
$\text{SFITA}_{\text{SI}_{\text{FI}}}^{\text{RB}}(F)(y) =_{\text{def}} \alpha(\Psi_{\text{SI}_1}^{F_1, G_1}(F)(y), \ldots, \Psi_{\text{SI}_n}^{F_n, G_n}(F)(y)).$

We give one condition for $\text{SFITA}_{\text{SI}_{\text{FI}}}^{\text{RB}}$ to be a solution for RB. Further investigations are found in [11, 12].

**Theorem 3.** $\text{SFITA}_{\text{SI}_{\text{FI}}}^{\text{RB}}$ *is a solution for* RB *if*

1. *For every* $i \in \{1, \ldots, n\}$, $\text{SIM}_i(F_i, F_i) = 1$ *and*
2. *For every* $i \in \{1, \ldots, n\}$ *and* $s \in \langle 0, 1 \rangle$, $\text{mod}_i(1, s) = s$ *and*
3. *For all* $i, j \in \{1, \ldots, n\}$ *with* $i \neq j$, $\text{SIM}_i(F_i, F_j) = 0$ *and*
4. *For every* $i \in \{1, \ldots, n\}$ *and* $s \in \langle 0, 1 \rangle$, $\text{mod}_i(0, s) = 0$ *and*
5. *For every* $s \in \langle 0, 1 \rangle$, $\alpha(0, \ldots, 0, s, 0, \ldots, 0) = s$.

We give one more result related to the equivalence between SFITA and FITA.

**Theorem 4.** *If*

$$\text{INT}_{\text{FI}} = [\pi_1, \ldots, \pi_n, \pi_1, \ldots, \pi_n, \sup, \ldots, \sup, \alpha],$$
$$\text{SI}_{\text{FI}} = [[\text{SIM}_1, \text{mod}_1], \ldots, [\text{SIM}_n, \text{mod}_n], \alpha]$$

*such that every $\pi_1, \ldots, \pi_n$ is a left-continuous t-norm and for every $i \in \{1, \ldots, n\}$,*

$$\mathrm{SIM}_i(F, F') = \sup \{\pi_i\left(F(x), F'(x)\right) \mid x \in U\},$$
$$\mathrm{mod}_i(s, G)(y) = \pi_i\left(s, G(y)\right),$$

*then $\mathrm{FITA}^{\mathrm{RB}}_{\mathrm{INT_{FI}}}$ and $\mathrm{SFITA}^{\mathrm{RB}}_{\mathrm{SI_{FI}}}$ are equivalent with respect to $\mathfrak{F}U$.*

Less restrictive conditions are possible when limiting the base set $\mathcal{F}$, for instance to singleton fuzzy sets. FATI-like similarity-based inference operators are possible, but the problem of aggregating rules is nontrivial without the intermediate representation in the form of a fuzzy relation.

The investigations of similarity-based inference are ongoing.

### 2.5.6   Additional Remarks

The presentation in the preceding subsections is a mere sketch of the most popular fuzzy inference methods. One goal of the research project reported here is to give a systematic and comprehensive account of the existing inference methods, their properties, possible equivalences, and conditions for the inference methods that will provide solutions for fuzzy IF-THEN rule bases.

Discussions of further inference methods can be found in [5, 19, 34, 35, 57, 70, 116].

## 2.6   Uniqueness of Solutions for Fuzzy IF-THEN Rule Bases

The uniqueness problem was posed in Section 2.3 as one of the three fundamental problems concerning solutions of fuzzy IF-THEN rule bases. In the theory of classical crisp logical consequence operators, it is a fundamental and useful result that two consequence operators are identical if both agree on the empty set, are compact and monotone, and fulfill the inference and deduction properties. This also means that every operator fulfilling these properties is unique.

Unfortunately, in the theory of fuzzy inference operators, the situation is much more complicated. The attempt to directly translate the results of classical logic to fuzzy inference yields some results, but they involve very strong conditions, and it is hard to estimate the applicability to real-world inference problems [100].

We give one of the results from [100] here; see [100] for further results.

**Theorem 5.** *If*

1. $\Psi, \Psi' : \mathfrak{F}U \to \mathfrak{F}V$
2. $\mathcal{F} \subseteq \mathfrak{F}U$ and $F \in \mathfrak{F}U$

3. $\Psi$ and $\Psi'$ are monotone on $\mathfrak{F}U$, i. e., if for $F, F' \in \mathfrak{F}U$, it holds that $F \subseteq F'$, then $\Psi(F) \subseteq \Psi(F')$ and $\Psi'(F) \subseteq \Psi'(F')$
4. $\Psi$ and $\Psi'$ are $\mathcal{F}$-compact on $\mathfrak{F}U$, i. e.,

$$\forall x \forall F \left( \begin{array}{l} x \in U \wedge F \in \mathfrak{F}U \wedge \bigcup \mathcal{F} \subseteq F \\ \rightarrow \exists F' \exists F'' \left( F' \in \mathcal{F} \wedge F'' \subseteq F' \wedge \Psi(F)(x) \leq \Psi(F'')(x) \right) \end{array} \right)$$

5. $\forall F \left( F \in \mathcal{F} \rightarrow \Psi(F) = \Psi'(F) \right)$
6. $\bigcup \mathcal{F} \subseteq G$,

then $\Psi(F) = \Psi'(F)$.

It is obvious that for a usual fuzzy IF-THEN rule base of the form (2.9) such that $\mathcal{F} = \{F_1, \dots, F_n\}$, the conditions of the theorem above will not yield an equivalence of $\Psi$ and $\Psi'$ for fuzzy sets $F$, which are likely to occur as inputs of the inference engine. On the whole, our investigations of this problem so far yield the insight that the current state of knowledge about functional properties of fuzzy inference operators may be insufficient to find a satisfying result. It is hence sensible to first study the restriction problem much more thoroughly, to find suitable conditions for fuzzy inference operators that restrict the set of all admissible solutions for a rule base sufficiently to make uniqueness conditions that are more easily fulfilled possible.

One important restricting condition is Lipschitz continuity, as mentioned in Section 2.3. Further conditions in the form of axioms are mentioned in [1, 9, 13, 29], but no systematic study of axiomatic characterizations of fuzzy inference operators has emerged so far. It is one mid-term goal of this research project to carry out a systematic investigation of restricting conditions for fuzzy inference operators, hopefully leading to more suitable unique characterizations.

## 2.7 Functional Analytic Properties of Fuzzy Inference Operators

One goal of the interpretation of fuzzy inference methods as functional operators in the research project reported here was to apply the methods of classical functional analysis to give deeper insights into the nature of fuzzy inference. This goal has been pursued throughout the research project [43, 72, 80, 85]; see [69] for a predecessor of this work.

One important part of this research is the investigation of fixed points of functional operators interpreting fuzzy IF-THEN rule bases. In the context of fuzzy inference, this concept from functional analysis has an important meaning for fuzzy modeling. If a fuzzy IF-THEN rule base models the way a system changes state, then the input of a fuzzy inference operator is a fuzzy representation of the system's state before state transition, and the output represents the state after state transition. A fixed point of the inference operator then means a fuzzy representation of a state in which the system is

stable. Note that from this discussion, it follows immediately that in this case, $U = V$.

We give one result from [73], concerning the principle FITA; see [73, 74, 75] for further results.

Let a fuzzy IF-THEN rule base of the form (2.9) be given.

**Theorem 6.** *Let* $\text{INT}_{\text{FI}} = [\pi, \dots, \pi, \kappa, \dots, \kappa, \sup, \dots, \sup, \max]$ *such that* $\pi$ *is a t-norm and* $\kappa$ *is a left-continuous t-norm and after defining a fuzzy relation* $S : U \times U \to \langle 0, 1 \rangle$ *by*

$$S(x, y) =_{\text{def}} \max\left(\pi(F_1(x), G_1(y)), \dots, \pi(F_n(x), G_n(y))\right), \qquad (x, y \in U)$$

*for all* $y \in U$ *there exists* $x \in U$ *such that*

$$\sup\left\{\kappa\left(S(z, x), S(x, y)\right) \mid z \in U\right\} = \sup\left\{S(z, y) \mid z \in U\right\}.$$

*Then the fuzzy set* $F$ *defined for* $x \in U$ *by*

$$F(x) = \sup\left\{\kappa\left(S(z, z), S(z, x)\right) \mid z \in U\right\}$$

*is a fixed point of* $\text{FITA}^{\text{RB}}_{\text{INT}_{\text{FI}}}$, *i. e.* $\text{FITA}^{\text{RB}}_{\text{INT}_{\text{FI}}}(F) = F$.

## 2.8 Approximate Reasoning with Context-Dependent Fuzzy Sets

In this section, we present a generalized concept of fuzzy sets and the foundations of fuzzy inference with this type of generalized fuzzy sets.

The idea behind the generalization is the modeling of context dependency of the information represented by the fuzzy set. To this end, fix a non-empty set $W$ of worlds. What kind of context dependency is represented by $W$ (for instance, temporal dependency), is not fixed at this moment?

A context-dependent fuzzy set [86] is a fuzzy set the value of which depends not only on the universe $U$, but also on the worlds $W$.

**Definition 16 (Context-dependent fuzzy set).** $F^W : W \times U \to \langle 0, 1 \rangle$ is said to be a context-dependent fuzzy set on $U$ with respect to $W$.

It is easy to extend the inference methods discussed in Section 2.5 to context-dependent fuzzy sets by carrying out inference within a world $w \in W$ [90]. The issue gets more interesting when quantifying over worlds in inference. This leads to a synthesis between approximate reasoning and classical modal logic.

Let a crisp reachability relation $R \subseteq W \times W$ and a fuzzy reachability relation $S : W \times W \to \langle 0, 1 \rangle$ be given. We define modal operators based on $R, S$ in

the usual way. In the fuzzy case, two additional functions $\pi, \kappa : \langle 0,1 \rangle^2 \to \langle 0,1 \rangle$ are needed for combining the membership degrees of $F^W$ and $S$.

$$
\begin{aligned}
\left( \langle R \rangle F^W \right)(x,w) &=_{\text{def}} \sup \left\{ F^W(x,w') \,\middle|\, w' \in W \wedge wRw' \right\} \\
\left( [R] F^W \right)(x,w) &=_{\text{def}} \inf \left\{ F^W(x,w') \,\middle|\, w' \in W \wedge wRw' \right\} \\
\left( \langle S, \kappa \rangle F^W \right)(x,w) &=_{\text{def}} \sup \left\{ \kappa \left( S(w,w'), F^W(x,w') \right) \,\middle|\, w' \in W \right\} \\
\left( [S, \pi] F^W \right)(x,w) &=_{\text{def}} \inf \left\{ \pi \left( S(w,w'), F^W(x,w') \right) \,\middle|\, w' \in W \right\}
\end{aligned}
$$

Now, it is interesting to study generalized fuzzy modal inference rules, where generalization is understood in the same sense as for the generalized modus ponens (2.3). Investigations of this subject are ongoing [86, 90, 105]. Here, we consider only the approximate reasoning scheme

$$
\begin{array}{|l|}
\hline
[R]\,(F_1^W \Rightarrow G_1^W) \\
[R]\,F^W \\
\hline
\qquad\quad [R]\,G^W \\
\hline
\end{array}
\tag{2.12}
$$

which is a generalization of the classical modal deduction rule

$$
\begin{array}{|l|}
\hline
\Box(A \to B) \\
\Box A \\
\hline
\quad\;\; \Box B \\
\hline
\end{array}
\;.
$$

Exactly as with the generalized modus ponens, we are faced with the problem that $G^W$ is not constructed by simply detaching the premise $F_1^W$ from the implication $F_1^W \Rightarrow G_1^W$. Rather to take fuzziness into account, it has to be constructed by a complex and noncanonical inference method. We consider only an adapted compositional rule of inference. Let $W$ and $R$ be fixed, as discussed above.

**Definition 17 (Context-Dependent Interpretation).**
$I^W = [\pi^W, \kappa^W, \mathcal{Q}^W]$ is said to be a context-dependent interpretation with respect to $W$ iff

1. $\pi^W, \kappa^W : W \times \langle 0,1 \rangle \times \langle 0,1 \rangle \to \langle 0,1 \rangle$,
2. $\mathcal{Q}^W : W \times \mathfrak{P}\langle 0,1 \rangle \to \langle 0,1 \rangle$.

Note that in the concept of context-dependent interpretation defined above, the functions used for constructing the output of the inference scheme also depend on the current world. Based on this notion of interpretation, we define a context-dependent functional operator $\Psi_{I^W}^{F_1^W, G_1^W} : \mathfrak{F}(W \times U) \to \mathfrak{F}(W \times V)$ with respect to $W$ as follows:

**Definition 18 (Modal Compositional Rule of Inference).** Given a context-dependent interpretation $I^W = [\pi^W, \kappa^W, \mathcal{Q}^W]$ with respect to $W$, the modal approximate inference scheme (2.12) is interpreted as follows:

$$\Psi_{IW}^{F_1^W, G_1^W}(F^W)(w, y)$$
$$=_{\text{def}} \mathcal{Q}^W\left(w, \left\{\kappa^W\left(w, F^W(w, x), [R]\,\pi^W(w, F_1^W(w, x), G_1^W(w, y))\right)\,\middle|\,x \in U\right\}\right)$$

Because the inference result $\Psi_{IW}^{F_1^W, G_1^W}(F^W)$ is constructed in a noncanonical way, it has to be established that for the input $[R]\,F_1^W$ the inference method behaves in the expected way, i.e., $\Psi_{IW}^{F_1^W, G_1^W}([R]\,F_1^W) = [R]\,G_1^W$ (correctness of $\Psi_{IW}^{F_1^W, G_1^W}$ with respect to the inference scheme (2.12)).

**Theorem 7.** *If*

1. *for every $w \in W$, $\lambda r\,\lambda s\,\kappa^W(w, r, s)$ and $\lambda r\,\lambda s\,\pi^W(w, r, s)$ are t-norms,*
2. $\mathcal{Q}^W = \sup$, *and*
3. *for every $w \in W$, at least one of the following conditions is satisfied:*
   3.1 $\lambda r\,\lambda s\,\kappa^W(w, r, s) = \lambda r\,\lambda s\,\pi^W(w, r, s) = \min$,
       $\sup\left\{[R]\,F_1^W(w, x)\,\middle|\,x \in U\right\} \geq \sup\left\{[R]\,G_1^W(w, y)\,\middle|\,y \in V\right\}$, *and $R$ is reflexive on $W$,*
   3.2 $\pi^W(w, r, s)$ *is independent of $w$, $\lambda r\,\lambda s\,\kappa^W(w, r, s)$ and $\lambda r\,\lambda s\,\pi^W(w, r, s)$ are continuous, and $\sup\{[R]\,F_1^W(w, x)|x \in U\} = 1$,*
       *or*
   3.3 *there exists an $x \in U$ such that $[R]\,F_1^W(w, x) = F_1^W(w, x) = 1$*

*then $\Psi_{IW}^{F_1^W, G_1^W}([R]\,F_1^W) = [R]\,G_1^W$.*

So far, nothing has been said about the interpretation of the set $W$ of worlds. First, it is obvious that the meaning of a generic fuzzy linguistic term such as "high" is dependent on the context in which it is used. Even when the context is fixed in some way, e. g., high room temperature, there are a number of ways left in which context-dependent differences in the interpretation of the concept can play a role, e. g., which person is asked, which season, what time of day is it, what is the air humidity, and so on. Furthermore, note that the way inference is carried out also depends on the current world $w$ (via the definition of $\pi^W, \kappa^W, \mathcal{Q}^W$). See [86] for a comprehensive discussion.

A second application of context-dependence is temporal approximate reasoning. Here, worlds $w \in W$ represent points in time and the fuzzy modal operators, given appropriate fuzzy or crisp reachability relations $R, S$, allow quantification into the past or into the future. Investigations of this subject are ongoing [30, 31, 32, 33, 102]. This can also be seen as a contribution towards actually incorporating time into approximate reasoning with fuzzy IF-THEN rule bases by providing linguistic variables representing points or intervals in time.

## 2.9  Approximate Reasoning Under Uncertainty

For a long time, different aspects of fuzziness have been considered in approximate reasoning, for example vagueness, uncertainty, probability, and belief. Recently, the distinction and systematic combination of different aspects has become an important topic of investigation. In particular, the combination of the aspects vagueness (in the sense of gradual truth) and uncertainty has been a topic of research. For this combination, so-called type-2 fuzzy sets play an important role. The notion of type-2 fuzzy set was introduced in 1973 by Zadeh [111]. Since then a lot of papers were published, for instance [10, 23, 24, 25, 45, 46, 47, 52, 78, 107]. The book [45] provides a comprehensive survey of the state of art in this field, including a long list of references.

In this particular case, assume that complete lattices $[T, \leq]$ and $[C, \sqsubseteq]$ where $T$ is called set of truth degrees and $C$ is called set of certainty degrees are given.

**Definition 19 (Type-2 fuzzy set).** $F^{T,C}$ is said to be a type-2 fuzzy set on $U$ with respect to the set $T$ of truth degrees and the set $C$ of certainty degrees iff

$$F^{T,C} : U \to C^T.$$

Intuitively, this definition means that by a type-2 fuzzy set, every element $x$ from the universe $U$ gets assigned not only a membership degree (or truth value) from $\langle 0, 1 \rangle$, but a fuzzy set $F^x = F^{T,C}(x)$ on the universe $T$, taking membership degrees from $C$. The membership degree of a truth value $t \in T$ in $F^x$ tells how certain it is for $x$ to have membership degree $t$ in $F^{T,C}$. Hence, classical fuzzy sets are a special case, where one membership degree gets assigned certainty 1 and all other membership degrees get assigned certainty 0.

In most cases, $T = C = \langle 0, 1 \rangle$ is used with the order of real numbers. In this section, retaining the denotations $T$ and $C$ makes it easier to recall the semantic difference between the sets.

It is easy to see that in the case that $C = \langle 0, 1 \rangle$ and the lattice structure of $T$ is neglected, there is a one-to-one correspondence between type-2 fuzzy sets and context-dependent fuzzy sets. Let the set $W$ of worlds be identical with $T$. Then, given a type-2 fuzzy set $F^{T,C}$ on $U$, define a context-dependent fuzzy set $F^W$ for $w \in T$ and $x \in U$ by

$$F^W(w, x) =_{\text{def}} F^{T,C}(x)(w).$$

Given a context-dependent fuzzy set $F^W$, define a fuzzy set $F^x : T \to \langle 0, 1 \rangle$ for every $x \in U$ by $F^x(t) =_{\text{def}} F^W(t, x)$ and define a type-2 fuzzy set $F^{T,C}$ on $U$ for $x \in U$ by

$$F^{T,C}(x) =_{\text{def}} F_x.$$

Obviously, both fuzzy sets are equivalent from a purely mathematical point of view. This relationship as well as more general definitions of type-$n$ fuzzy sets are discussed in [96].

The important difference between type-2 fuzzy sets and context-dependent fuzzy sets lies in the structure. While $W$ is structured by crisp or fuzzy accessibility relations, $T$ and $C$ are structured algebraically as lattices, MV-algebras, or residual monoids.

While the additional operators considered for context-dependent fuzzy sets are modal quantifiers $\langle R \rangle, [R], \langle S, \kappa \rangle, [S, \pi]$, for type-2 fuzzy sets the interesting operations are derived from the algebraic operations defined on $C$ and $T$ and lifted to type-2 fuzzy sets (see the paragraph after the next).

Note that when partially ordering $C^T$ by the canonical partial order induced by $\sqsubseteq$ (independent of $\leq$), $C^T$ is a complete lattice. In [101], a general version of the compositional rule of inference for arbitrary lattices and a specific one for $C^T$ are defined and the correctness with respect to the generalized modus ponens is proved.

Of course in the context of fuzzy inference with type-2 fuzzy sets, operators on type-2 fuzzy sets playing the role of the functions $\pi, \kappa$ in the compositional rule of inference are required. It is interesting to ask whether a function $\pi^T : T \times T \to T$ or a function $\pi^C : C \times C$ can be lifted to a function $\pi^{T,C} : C^T \times C^T \to C^T$. Ideally, an interpretation $[\pi^T, \kappa^T, \mathcal{Q}^T]$ for the generalized compositional rule of inference with respect to $T$ can be lifted to an interpretation $[\pi^{T,C}, \kappa^{T,C}, \mathcal{Q}^{T,C}]$ of the generalized compositional rule of inference with respect to $C^T$ in a way that correctness with respect to the generalized modus ponens is preserved.

To this end, in [101], a principle of external lifting and a principle of internal lifting are defined. Research on this topic is ongoing.

Judging from recent trends in the published literature, the subject of approximate reasoning under uncertainty will be a major topic in the coming years, and a systematic research on type-2 fuzzy sets (and the incorporation into the functional analytic framework used in this project) can be a valuable contribution.

In a different context (incorporation of uncertainty into classical many-valued logics), a systematic study involving the use of type-2 fuzzy sets has just been completed [8, 38, 39, 41, 42]. It can be hoped that valuable contributions to the research in this project can be gained by investigating the relationship with fuzzy inference.

## 2.10   Axiomatic Characterization of Fuzzy Approximation Operators

Let $R$ be an equivalence relation on $U$ and $U|_R$ the partition of $U$ induced by $R$.

In rough set theory, for every $X \subseteq U$, the $R$-upper and $R$-lower approximation of $X$ are defined, respectively, by

$$\overline{R}X =_{\text{def}} \bigcup \{P \mid P \in U|_R \wedge P \cap X \neq \emptyset\}$$
$$\underline{R}X =_{\text{def}} \bigcup \{P \mid P \in U|_R \wedge P \subseteq X\}.$$

In classical modal logic, for every binary relation $R$, a diamond operator and a box operator based on $R$ are defined, respectively, as follows, for $X \subseteq U$:

$$\langle R \rangle\, X =_{\text{def}} \{y \mid \exists x (x\, R\, y \wedge x \in X)\}$$
$$[R]\, X =_{\text{def}} \{y \mid \forall x (x\, R\, y \to x \in X)\}.$$

It is well known that if $R$ is an equivalence relation, then for all $X \subseteq U$, $\overline{R} = \langle R \rangle$ and $\underline{R} = [R]$, and furthermore, the operator $\Phi_R$ defined for $X \subseteq U$ by $\Phi_R(X) =_{\text{def}} \langle R \rangle\, X$ is a closure operator on $U$. For a deeper understanding of the concepts defined above and their relationship, the following questions have to be asked:

1. Under which sufficient conditions for a closure operator $\Phi$ on $U$ does there exist a binary relation $R$ on $U$ such that $\Phi = \Phi_R$?
2. If $R$ is an arbitrary binary relation, how do properties of $R$ translate to $\Phi_R$?

These questions have been answered and systematically investigated in [91].

There are two ways for to fuzzify these concepts. For the fuzzy rough set based case, let a binary fuzzy relation $S : U \times U \to \langle 0, 1 \rangle$ be given. Define a fuzzy diamond operator based on $S$ as follows, for $X \subseteq U$ where $\chi_X$ denotes the characteristic function of $X$:

$$(\langle S \rangle\, X)(y) =_{\text{def}} \sup \{\min(S(x, y), \chi_X(x)) \mid x \in X\}.$$

Obviously, $\langle S \rangle\, X$ induces a fuzzy approximation operator $\Phi_S : \mathfrak{P}U \to \mathfrak{F}U$ in the same way as in the crisp case, and it is possible to ask exactly the same questions as above. These questions have been answered and systematically investigated in [92].

For the rough fuzzy set based case, let a binary relation $R \subseteq U \times U$ be given. Define a fuzzy diamond operator based on $R$ as follows, for $F \in \mathfrak{F}U$:

$$(\langle R \rangle\, F)(y) =_{\text{def}} \sup \{F(x) \mid x \in U \wedge [x, y] \in R\}.$$

Obviously, $\langle R \rangle\, F$ induces a functional operator $\Psi_R : \mathfrak{F}U \to \mathfrak{F}U$ in the same way as in the crisp case, and it is possible to ask exactly the same questions as above. These questions have been answered and systematically investigated in [98]. These investigations add significantly to the understanding of foundations of functional operators for fuzzy inference.

## 2.11   Algebraic Foundations of Information Granulation

The concept of information granulation is of vital importance for applications
of fuzzy inference, as it provides a basis for the construction of linguistic
variables which are used as premises of fuzzy IF-THEN rules.

In his paper "Towards a theory of fuzzy information granulation and its
centrality in human reasoning and fuzzy logic" [115], Zadeh writes,

> There are three basic concepts that underline human cognition: gran-
> ulation, organization and causation. Informally, granulation involves
> decomposition of whole into parts; organization involves integration
> of parts into whole; and causation involves association of causes with
> effects.
>
> Granulation of an object $A$ leads to a collection of granules of $A$,
> with a granule being a clump of points (objects) drawn together by
> indistinguishability, similarity, proximity or functionality.

From a mathematical point of view, the canonical representation of sim-
ilarity, proximity, or functionality with respect to a universe $U$ is a binary
fuzzy relation on $U$ and the canonical representation of a collection of gran-
ules of $A$, with a granule being a clump of points (objects) with respect to a
universe $U$ is a set of fuzzy sets on $U$. This leads to the following fundamental
problems:

1. Given a binary fuzzy relation $R$ on $U$, how can an operator GRAN such
   that GRAN($R$) is a set of fuzzy sets on $U$ that can be called the granu-
   lation associated with $R$ be defined?
2. Given a set $\mathfrak{G}$ of fuzzy sets on $U$, how can an operator REL such that
   REL($\mathfrak{G}$) is a binary fuzzy relation on $U$ which can be called the relation
   associated with $\mathfrak{G}$ be defined?
3. How can it be assured that $GRAN$ and REL have the following recon-
   struction property, for a binary fuzzy relation $R$ on $U$ and a set $\mathfrak{G}$ of fuzzy
   sets on $U$: GRAN(REL($\mathfrak{G}$)) = $\mathfrak{G}$ and REL(GRAN($R$)) = $R$? Among
   other beneficial effects, this property guarantees that no information is
   lost in converting between relations and granulations.
4. In what way do properties of relations translate to granulations and vice
   versa?
5. How can certain classes of relations (for instance equivalence, tolerance,
   and order relations) be characterized by associated classes of granulations
   and vice versa?

These questions have been answered and systematically investigated in
[87, 88, 89, 97]. One particular type of fuzzy granulation is a fuzzy clustering,
in the sense of a fuzzy partition as it could be generated by a fuzzy clustering
algorithm. This special type of granulation and its associated relations have
been investigated in [93].

## 2.12   Algebraic Foundations of Modeling with Words

In [95], the concept of linguistic variable is extended to the concept of linguistic algebra by adding a set of operations on linguistic terms. The concept of a linguistic algebra to be a linguistic model for a given universal algebra is introduced. Methods for constructing linguistic models for universal algebras are described, and the relationship with the subject modeling with words is clarified.

## 2.13   Conclusions and Further Research

Hopefully, it has become clear that the research project reported in this paper provides a significant contribution to the basic theoretical investigations of the concept fuzzy IF-THEN rule base and its handling, as motivated in the introduction, and furthermore to the whole area of fuzzy inference and approximate reasoning, in general.

As the project and its goals are designed for a long term perspective, the state of research in some areas is still intermediary, but it is clear that a substantial and sustaining base of concept definitions and results has been achieved so far.

Further developments are planned mainly in the following areas:

**Complex fuzzy inference mechanisms.**
- Systematic development of the semantical foundations of complex fuzzy inference mechanisms (Section 2.2), using concepts of classical switching theory.
- Incorporation of further inference modules (modifiers, fuzzy arithmetic operators) into the functional analytic perspective.
- Investigation of the expressive power of sets of inference modules.
- Study of functional analytic properties of complex fuzzy inference mechanisms.
- Study of dynamic semantics of inference mechanisms with feedback (fuzzy automata).
- Incorporation of temporal aspects into fuzzy inference.

**Fuzzy IF-THEN rule based inference methods.**
- Systematic study of further inference methods (Section 2.5) from a functional analytic perspective. Further study of similarity-based inference.
- Systematic study of the restriction problem (Section 2.3) for solutions of fuzzy IF-THEN rule bases, using axiomatic approaches. Axiomatic characterizations of solutions of fuzzy IF-THEN rule bases.

**Functional analytic properties of fuzzy inference operators.**
- Systematic study of properties like continuity, linearity, and fixed points for functional operators induced by fuzzy IF-THEN rule bases.

**Approximate reasoning with context-dependent fuzzy sets.**
- Systematic study of generalized fuzzy modal inference rules (Section 2.8). Investigation of applications.
- Further study of temporal approximate reasoning.

**Approximate reasoning under uncertainty.**
- Systematic study of algebraic structures induced on spaces of type-2 fuzzy sets by structures of component algebras by the principle of external lifting and the principle of internal lifting (Section 2.9).
- Formal investigation of approximate reasoning under uncertainty as fuzzy inference with type-2 fuzzy sets.
- Incorporation of results from uncertainty logics.

Research on these and other subjects concerning fuzzy logic as a whole has been going on for a long time at the Department of Computer Science Informatik I, not only in the research project reported here, but in a larger group of researchers. The activities of this research group have yielded, in addition to this project and numerous publications (see list of references), an international conference on computational intelligence taking place regularly in Dortmund [58, 59, 60, 61, 62, 63] as well as specialized workshops on computational intelligence [64].

# References

1. J. F. Baldwin and B. W. Pilsworth. Axiomatic approach to implication for approximate reasoning with fuzzy logic. *Fuzzy Sets and Systems*, 3:193–219, 1980.
2. M. Beyer and S. Lehmke. Ein Software-Werkzeug zur Konstruktion und zum empirischen Test komplex strukturierter Fuzzy-Inferenzmechanismen. In *Fachtagung Computational Intelligence — CI 2000*, Baden-Baden, May 11–12, 2000. VDI/VDE-GMA.
3. A. Di Nola, S. Sessa, W. Pedrycz, and E. Sanchez. *Fuzzy Relation Equations and Their Applications to Knowledge Engineering.* Kluwer, Dordrecht, 1st edition, 1989.
4. P. Diamond and P. Kloeden. Metric spaces of fuzzy sets. *Fuzzy Sets and Systems*, 100(Supplement):63–71, 1999.
5. L. Ding, Z. Shen, and M. Mukaidono. Approximate reasoning based on revision principle. In *NAFIPS '90*, volume 1, pages 94–97, Toronto, 1990.
6. D. Driankov, H. Hellendoorn, and M. Reinfrank. *An Introduction to Fuzzy Control.* Springer, Berlin, 1993.
7. D. Driankov, R. Palm, and H. Hellendoorn. Fuzzy control with fuzzy inputs: The need for new rule semantics. In *FUZZ-IEEE '94 — Third IEEE International Conference on Fuzzy Systems*, pages 13–21, Orlando, FL, June 26–July 2, 1994.
8. D. Dubois, S. Lehmke, and H. Prade. A comparative study of logics of graded uncertainty and logics of graded truth. In S. Gottwald and E. P. Klement, editors, *Linz '97 — Enriched Lattice Structures for Many-Valued and Fuzzy Logics*, pages 10–15, Linz, Austria, February 25–March 1, 1997.

9. D. Dubois and H. Prade. Fuzzy sets in approximate reasoning, Part 1: Inference with possibility distributions. *Fuzzy Sets and Systems (Special Memorial Volume: 25 years of fuzzy sets)*, 40:142–202, 1991.

10. Didier Dubois and Henri Prade. Operations in a fuzzy-valued logic. *Information and Control*, 43(2):224–240, November 1979.

11. U. Fieseler. An axiomatic approach to fuzzified set inclusion and set equality. In *EUFIT '96 — Fourth European Congress on Intelligent Techniques and Soft Computing*, pages 45–49, Aachen, September 2–5, 1996.

12. U. Fieseler. On the interpretation and chaining of fuzzy IF-THEN rule bases using fuzzy equality indicators. In *IFSA '97 — Seventh International Fuzzy Systems Association World Congress*, volume I, pages 78–83, Prague, Czech Republic, June 25–29, 1997.

13. S. Fukami, M. Mizumoto, and K. Tanaka. Some considerations on fuzzy conditional inference. *Fuzzy Sets and Systems*, 4:243–273, 1980.

14. S. Gottwald. On the existence of solutions of systems of fuzzy equations. *Fuzzy Sets and Systems*, 12:301–302, 1984.

15. S. Gottwald. Characterizations of the solvability of fuzzy equations. *Elektron. Informationsverarb. Kybernet.*, 22:67–91, 1986.

16. S. Gottwald. On some theoretical results concerning the construction of fuzzy controllers. In S. F. Bocklisch et al., editors, *Fuzzy Sets Applications, Methodological Approaches, and Results*, volume 30 of *Math. Research*, pages 45–55. Akademie-Verlag, Berlin, 1986.

17. S. Gottwald and W. Pedrycz. Solvability of fuzzy relational equations and manipulation of fuzzy data. *Fuzzy Sets and Systems*, 18:1–21, 1986.

18. S. Gottwald and W. Pedrycz. On the methodology of solving fuzzy relational equations and its impact on fuzzy modelling. In M. M. Gupta and T. Yamakawa, editors, *Fuzzy Logic in Knowledge-Based Systems, Decision and Control*, pages 197–210. North-Holland, Amsterdam, 1988.

19. M. Grabisch, H. Prade, and D. Dubois. Synthesis of real-valued mappings based on gradual rules and interpolative reasoning. In A. L. Ralescu, editor, *Proceedings on the IJCAI-93 Workshop on Fuzzy Logic in Artificial Intelligence*, volume 847 of *LNAI*. Springer, Berlin, August 1994.

20. L. Hildebrand and B. Reusch. Fuzzy color processing. In E. Kerre and M. Nachtegael, editors, *Fuzzy Techniques in Image Processing*. Springer, Berlin, 2000.

21. J. Hiltner, M. Fathi, and B. Reusch. MR-analysis using fuzzy expert knowledge represented by linguistic and model-based descriptions. In *6th International Workshop Fuzzy-Neuro Systems'99 (FNS'99)*, pages 13–22, Leipzig, March 18–19, 1999.

22. J. Hiltner, M. Fathi, and B. Reusch. New model for medical image analysis based on computational intelligence. In *Medical Imaging 2000, SPIE*, San Diego, CA, February 2000.

23. R. I. John. Type 2 fuzzy sets: An appraisal of theory and applications. *International Journal of Uncertainty, Fuzziness and Knowledge Based Systems*, 6(6):563–576, 1998.

24. N. N. Karnik and J. M. Mendel. Operations on type-2 fuzzy sets. *Fuzzy Sets and Systems*, 122:327–348, 2001.

25. N. N. Karnik, J. M. Mendel, and Q. Liang. Type-2 fuzzy logic systems. *IEEE-FS*, 7(6):643, December 1999.

40    Lehmke et al.

26. H. Kiendl. *Fuzzy Control methodenorientiert.* Oldenbourg, München, 1997.

27. H. Kiendl. Non-translation-invariant defuzzification. In *FUZZ-IEEE '97 — Sixth IEEE International Conference on Fuzzy Systems*, pages 737–742, Barcelona, July 1–5, 1997.

28. H. Kiendl. Decision analysis by advanced fuzzy systems. In J. Kacprzyk and L. A. Zadeh, editors, *Computing with Words in Information / Intelligent Systems*, pages 223–242. Physica-Verlag, Heidelberg, 1999.

29. H. Kiendl. Design of advanced fuzzy systems. In L. Yliniemi and Esko Juuso, editors, *TOOLMET '99 Symposium - Tool Environments for Intelligent Systems*, pages 57–76, University of Oulu, Finland, 1999.

30. T. Kiseliova and C. Moraga. On incorporation of time in medical expert systems. In *FSTA 2000 — Fifth International Conference on Fuzzy Set Theory and its Applications*, Liptovsky Mikulas, Slovak Republic, January 30–February 4, 2000.

31. T. Kiseliova and C. Moraga. Using terms to process time in medical expert systems. In *IASTED International Conference Signal Processing and Communications*, pages 487–492, Marabella, Spain, September 19–22, 2000.

32. T. Kiseliova and C. Moraga. Modelling temporal distribution of symptoms and diseases with fuzzy logic. In *9th IFSA World Congress and 20th NAFIPS International Conference*, pages 1637–1642, Vancouver, July 25–28, 2001.

33. T. Kiseliova and H. Wagner. Temporal approximative reasoning in medicine. *Submitted to Fuzzy Sets and Systems*, 2002.

34. F. Klawonn and R. Kruse. Fuzzy control as interpolation on the basis of equality relations. In *Proceedings of the Second IEEE International Conference on Fuzzy Systems*, pages 1125–1130, San Francisco, CA, March 28–April 1, 1993.

35. L. T. Kóczy, editor. *Fuzzy Logic, Texts*, volume 2 of *Fuzzy Systems II (Fuzzy Reasoning and Control), TEMPUS-JEP 07759-94-MODIFY.* Budapest, 1997.

36. R. Kruse, J. Gebhardt, and F. Klawonn. *Foundations of Fuzzy Systems.* Wiley, Chichester, 1994.

37. W. Van Leekwijck and E. E. Kerre. Defuzzification: criteria and classification. *Fuzzy Sets and Systems*, 108:159–178, 1999.

38. S. Lehmke. Weight structures for approximate reasoning with weighted expressions. In *Proceedings of the Twenty-Sixth International Symposium on Multiple-Valued Logic*, pages 178–183, Santiago de Compostela, Spain, May 29–31, 1996.

39. S. Lehmke. On the semantics of uncertainty logics. In *EUFIT '97 — Fifth European Congress on Intelligent Techniques and Soft Computing*, volume 1, pages 137–141, Aachen, September 8–11, 1997.

40. S. Lehmke. Konstruktion von Fuzzy-Inferenzmechanismen aus Inferenzbausteinen unter Anwendung von Methoden aus der Schaltwerktheorie. In *Informatik '99 — 29. Jahrestagung der GI*, Informatik aktuell, pages 322–329, Paderborn, October 5–9, 1999. Springer, Berlin, 1999.

41. S. Lehmke. Degrees of truth and degrees of validity — two orthogonal dimensions of representing fuzziness in logical systems. In V. Novák and I. Perfilieva, editors, *Discovering the World with Fuzzy Logic*, volume 57 of *Studies in Fuzziness and Soft Computing*, pages 192–236. Physica-Verlag, Heidelberg, 2000.

42. S. Lehmke. *Logics which allow degrees of truth and degrees of validity — a way of handling graded truth assessment and graded trust assessment within a single framework*. Dissertation, University of Dortmund, Department of Computer Science, June 2001.
43. S. Lehmke, B. Reusch, K.-H. Temme, and H. Thiele. On interpreting fuzzy if-then rule bases by concepts of functional analysis. Technical Report CI-19/98, Collaborative Research Center 531, University of Dortmund, 1998.
44. J. Martin, J. Hiltner, M. Fathi, J. Statthaus, T. Hackländer, and B. Reusch. System for motion elimination in evaluation of renal function supported by vague knowledge. In *Medical Imaging 2000, SPIE*, San Diego, CA, February 2000.
45. J. M. Mendel. *Uncertain Rule-Based Fuzzy Logic Systems: Introduction and New Directions*. Prentice Hall, Englewood Cliffs, NJ, 2001.
46. M. Mizumoto and K. Tanaka. Some properties of fuzzy sets of type 2. *Information and Control*, 31(4):312–340, August 1976.
47. M. Mizumoto and K. Tanaka. Fuzzy sets under various operations. In *4th International Congress of Cybernetics and Systems*, Amsterdam, The Netherlands, August 21–25, 1978.
48. C. Moraga and K.-H. Temme. *s*-neural networks are fuzzy models. In *8th International Conference on Information Processing and Management of Uncertainty in Knowledge-Based Systems (IPMU 2000)*, Madrid, July 3–7, 2000.
49. C. Moraga, K.-H. Temme, K. Strehl, and R. Stankovic. Decision diagrams applied to Mamdani fuzzy systems. In *International Workshop on Spectral Transforms and Logic Design*. TICSP-Press, Tampere, Finland, 2000.
50. G. C. Mouzouris and J. M. Mendel. Non-singleton fuzzy logic systems. In *FUZZ-IEEE '94 — Third IEEE International Conference on Fuzzy Systems*, pages 456–461, Orlando, FL, June 26–July 2, 1994.
51. T. Murai and H. Thiele. On an interpretation of some real-valued logics by means of probability theory. In *IFSA '99 — Eighth International Fuzzy Systems Association World Congress*, Taipei, Taiwan, August 17–20, 1999.
52. J. Nieminen. On the algebraic structure of fuzzy sets of type-2. *Kybernetica*, 13(4), 1977.
53. R. Palm and D. Driankov. Fuzzy inputs. *Fuzzy Sets and Systems*, 70:315–335, 1995.
54. W. Pedrycz. Fuzzy relational equations with generalized connectives and their applications. *Fuzzy Sets and Systems*, 10(2):185–201, 1983.
55. W. Pedrycz. Approximate solutions of fuzzy relational equations. *Fuzzy Sets and Systems*, 28(2):183–202, 1988.
56. I. G. Perfilieva and A. S. Tonis. Criteria of solvability of fuzzy relational equations systems. In *IFSA '97 — Seventh International Fuzzy Systems Association World Congress*, pages 90–95, Prague, Czech Republic, June 25–29, 1997.
57. H. Prade and D. Dubois. Gradual inference rules in approximate reasoning. *Information Sciences*, 61:103–122, 1992.
58. B. Reusch, editor. *Fuzzy Logic — Theorie und Praxis. 2. Dortmunder Fuzzy-Tage, June 9–10, 1992*. Springer, Berlin, 1993.
59. B. Reusch, editor. *Fuzzy Logic — Theorie und Praxis. 3. Dortmunder Fuzzy-Tage, June 7–9, 1993*. Springer, Berlin, 1993.
60. B. Reusch, editor. *Fuzzy Logic — Theorie und Praxis. 4. Dortmunder Fuzzy-Tage, June 6–8, 1994*. Springer, Berlin, 1994.

61. B. Reusch, editor. *Computational Intelligence — Theory and Applications. International Conference, 5th Fuzzy Days*, Dortmund, April 28–30, 1997. Springer, Berlin, 1997.

62. B. Reusch, editor. *Computational Intelligence — Theory and Applications. International Conference, 6th Fuzzy Days*, volume 1625 of *LNCS*, Dortmund, May 25–28, 1999. Springer, Berlin, 1999.

63. B. Reusch, editor. *Computational Intelligence — Theory and Applications. International Conference, 7th Fuzzy Days*, Dortmund, October 1–3, 2001. Springer, Berlin, 2001.

64. B. Reusch and K.-H. Temme, editors. *Computational Intelligence in Theory and Practice*. Physica-Verlag, Heidelberg, 2001.

65. T. A. Runkler. Extended defuzzification methods and their properties. In *FUZZ-IEEE '96 — Fifth IEEE International Conference on Fuzzy Systems*, volume 1, pages 694–700, New Orleans, September 8–11, 1996.

66. T. A. Runkler and M. Glesner. Defuzzification as crisp decision under fuzzy constraints. new aspects of theory and improved defuzzification algorithms. In *Fuzzy-Systeme '93. Management unsicherer Informationen. Workshop. Proceedings*, pages 156–164, Braunschweig, October 20–22, 1993.

67. T. A. Runkler and M. Glesner. Defuzzification with improved static and dynamic behavior: Extended center of area. In *EUFIT '93 — First European Congress on Fuzzy and Intelligent Technologies*, volume 2, pages 845–851, Aachen, September 7–10, 1993.

68. T. A. Runkler and M. Glesner. A set of axioms for defuzzifications strategies: Towards a theory of rational defuzzification operators. In *Proceedings of the Second IEEE International Conference on Fuzzy Systems*, pages 1161–1166, San Francisco, CA, March 28–April 1, 1993.

69. E. Sanchez. Resolution of eigen fuzzy set equations. *Fuzzy Sets and Systems*, 1(1):69–74, 1978.

70. A. Stoica. Fuzzy processing based on alpha-cut mapping. In *Proceedings: Fifth International Fuzzy Systems Association World Congress '93*, volume I, pages 179–182, Seoul, July 4–9, 1993.

71. K. Strehl, C. Moraga, K.-H. Temme, and R. S. Stankovic. Fuzzy decision diagrams for the representation, analysis and optimization of rule bases. In *The Thirtieth International Symposium on Multiple-Valued Logic*, pages 127–132, Portland, OR, May 23–25, 2000.

72. K.-H. Temme. Fuzzy inference methods as functional operator definitions. In *Proceedings of the International Conference on Soft Computing SOCO '99*, Genova, June 1–4, 1999.

73. K.-H. Temme and M. Fathi. First steps towards a theory for fix-mundis (fixed points for fuzzy if then rule bases). In *Proceedings of the X Spanish Conference on Fuzzy Logic and Technologies (ESTYLF 2000)*, Sevilla, September 20–22, 2000.

74. K.-H. Temme and M. Fathi. Fix-mundis for fuzzy if-then rule bases with standard compositional rule of inference interpretation. In *Proceedings of the ACM Symposium on Applied Computing (SAC 2000)*, Como, Italy, March 19–21, 2000.

75. K.-H. Temme and M. Fathi. Fix-mundis for fuzzy if-then rule bases with t-norm based compositional rule of inference interpretation. In *Proceedings of the International Symposium on Computational Intelligence (ISCI 2000)*, Kosice, Slovakia, August 29–September 1, 2000.

76. K.-H. Temme, C. Moraga, and R. Heider. Generalized neural networks for fuzzy modeling. In *Proceedings of the 1999 EUSFLAT - ESTYLF Joint Conference*, pages 626–632, Palma de Mallorca, September 22–25, 1999.

77. H. Thiele. Investigation of IF-THEN rule bases by methods of mathematical logic. In *FUZZ-IEEE '95 — Fourth IEEE International Conference on Fuzzy Systems*, volume III, pages 1391–1399, Yokohama, March 20–24, 1995.

78. H. Thiele. On logical systems based on fuzzy logical values. In *EUFIT '95 — Third European Congress on Intelligent Techniques and Soft Computing*, volume 1, pages 28–33, Aachen, August 28–31, 1995.

79. H. Thiele. On the calculus of fuzzy IF-THEN rules. In H.-J. Zimmermann, M. Gh. Negoiţă, and D. Dascălu, editors, *Real World Applications of Intelligent Technologies*, pages 91–111. Black Sea University, Editura Academiei Române, Mangalia, Romania, September 1995.

80. H. Thiele. Investigating approximate reasoning and fuzzy control by concepts of functional analysis. In *First International Conference on Conventional and Knowledge-Based Intelligent Electronic Systems*, volume 2, pages 493–500, Adelaide, May 21–23, 1997.

81. H. Thiele. Interpreting linguistic hedges by concepts of functional analysis and mathematical logic. In *EUFIT '98 — Sixth European Congress on Intelligent Techniques and Soft Computing*, pages 114–119, Aachen, September 7–11, 1998.

82. H. Thiele. Investigating the logical structure of fuzzy IF-THEN rule bases using concepts of mathematical logic and of functional analysis. In M. Jamshidi et al., editors, *World Automation Congress 1998 — Advances in Robotics, Manufacturing, Automation, Control, Soft Computing, Multimedia, and Image Processing (WAC '98): Proceedings*, pages ISSCI 132.1–132.6, Anchorage, AK, May 10–14 1998. NASA, 1998.

83. H. Thiele. Towards axiomatic foundations for defuzzification theory. Technical Report CI-31/98, Collaborative Research Center 531, University of Dortmund, 1998.

84. H. Thiele. Towards axiomatic foundations for defuzzification theory. In *Second International Conference on Knowledge-Based Intelligent Electronic Systems — Proceedings*, pages 243–250, Adelaide, April 21–23, 1998.

85. H. Thiele. Funktionalanalytische Methoden zur Interpretation von Fuzzy Control Regelbasen (methods of functional analysis for interpreting fuzzy control rule bases). In *9. Workshop "Fuzzy Control" des GMD-Fachausschusses 5.22*, Forschungsbericht Nr. 0499, Faculty of Electrical Engineering and Information Technology, pages 51–75, University of Dortmund, November 4–5, 1999.

86. H. Thiele. On the concept of qualitative fuzzy set. In *Proceedings of the Twenty-Ninth International Symposium on Multiple-Valued Logic*, pages 282–287, Freiburg, May 20–22, 1999.

87. H. Thiele. On algebraic foundations of information granulation I. The crisp case. In *FUZZ-IEEE 2000 — Ninth IEEE International Conference on Fuzzy Systems*, volume 2, pages 780–785, San Antonio, TX, May 7–12, 2000.

88. H. Thiele. On algebraic foundations of information granulation II. The fuzzy case. In *8th International Conference on Information Processing and Management of Uncertainty in Knowledge-Based Systems (IPMU 2000)*, volume 3, pages 1389–1396, Madrid, July 3–7, 2000.

89. H. Thiele. On algebraic foundations of information granulation III. Investigating the hata-mukaidono approach. In *The Thirtieth International Symposium on Multiple-Valued Logic*, pages 133–138, Portland, OR, May 23–25, 2000.

90. H. Thiele. On approximate reasoning with context-dependent fuzzy sets. In *WAC 2000 — Fourth Biannual World Automation Congress*, Wailea, Maui, HI, June 11–16, 2000.

91. H. Thiele. On axiomatic characterisations of crisp approximation operators. In *7th international Conference on Fuzzy Theory and Technology*, Atlantic City, NJ, February 27–March 3, 2000.

92. H. Thiele. On axiomatic characterizations of fuzzy approximation operators I. The fuzzy rough set based case. In *Second International Conference on Rough Sets and Current Trends in Computing (RSCTC'2000)*, pages 239–247, Banff Park Lodge, Banff, Canada, October 16–19, 2000.

93. H. Thiele. On fuzzy clusterings with variable boundary conditions. In *FSTA 2000 — Fifth International Conference on Fuzzy Set Theory and its Applications*, Liptovsky Mikulas, Slovak Republic, January 30–February 4, 2000.

94. H. Thiele. Generalizing the explicit concept of rough set on the basis of modal logic. In Reusch and Temme [64], pages 93–120.

95. H. Thiele. Linguistic algebras. An algebraic concept for describing modelling with words. In *FUZZ-IEEE 2001 Workshop on Modelling with Words (FUZZ-IEEE 2001 Conference)*, Melbourne, December 2–5, 2001.

96. H. Thiele. A new approach to type-2-fuzzy sets. In *The Second Congress of Logic Applied to Technology(LAPTEC2001)*, volume 1, pages 255–262, São Paulo, November 12-14, 2001.

97. H. Thiele. On algebraic foundations of information granulation. In B. Bouchon-Meunier, J. Gutiérrez-Ríos, L. Magdalena, and R. R. Yager, editors, *Volume of Selected and Updated Papers of the IPMU'2000 Conference held in Madrid*. Springer, Berlin, 2001.

98. H. Thiele. On axiomatic characterisations of fuzzy approximation operators II. The rough fuzzy set based case. In *The 31st International Symposium on Multiple-Valued Logic*, pages 330–335, Warsaw, May 22–24, 2001.

99. H. Thiele. On axiomatic characterisations of fuzzy approximation operators III. The fuzzy diamond and fuzzy box based cases. In *FUZZ-IEEE 2001 — 10th IEEE International Conference on Fuzzy Systems*, Melbourne, December 2–5, 2001.

100. H. Thiele. On the uniqueness of semantic interpretations for fuzzy IF-THEN rule base. In Reusch [63], pages 506–525.

101. H. Thiele. On approximative reasoning with type-2 fuzzy sets. In *9th International Conference on Information Processing and Management of Uncertainty in Knowledge-Based Systems (IPMU 2002)*, Annecy, France, July 1–5, 2002. (Accepted).

102. H. Thiele. Zur Behandlung von Zeitaspekten im approximativen Schließen (on treatment of time aspects in approximate reasoning). Technical report, Collaborative Research Center 531, University of Dortmund, 2002. (To appear).

103. H. Thiele and R. Mesiar. On T-quantifiers and S-quantifiers. In V. Novák and I. Perfilieva, editors, *Discovering the World with Fuzzy Logic*, volume 57 of *Studies in Fuzziness and Soft Computing*, pages 310–326. Physica-Verlag, Heidelberg, 2000.

104. H. Thiele and B. Reusch. On interrelations between flou sets, fuzzy sets, and rough sets. In *19. Workshop "Interdisziplinäre Methoden in der Informatik" — Arbeitstagung des Lehrstuhls Informatik I*, Forschungsbericht Nr. 729, Hotel zur Post, Lüdinghausen, September 21–24, 1999. Department of Computer Science, University of Dortmund.
105. H. Thiele and B. Reusch. On modal approximate reasoning. In *9th IFSA World Congress and 20th NAFIPS International Conference*, pages vol. 4, 1930–1934, Vancouver, July 25–28, 2001.
106. I. B. Türkşen and Y. Tian. Combination of rules or their consequences in fuzzy expert systems. *Fuzzy Sets and Systems*, 58:3–40, 1993.
107. M. Wagenknecht and K. Hartmann. Application of fuzzy sets of type 2 to the solution of fuzzy equation systems. *Fuzzy Sets and Systems*, 25:183–190, 1988.
108. R. R. Yager, S. Ovchinnikov, R. M. Tong, and H. T. Nguyen, editors. *Fuzzy Sets and Applications — Selected Papers by L. A. Zadeh*. Wiley, 1987.
109. L. A. Zadeh. A rationale for fuzzy control. *J. Dynamic Systems, Measurement and Control, Series 6*, 94:3–4, 1972.
110. L. A. Zadeh. Outline of a new approach to the analysis of complex systems and decision processes. *IEEE Transactions on Systems, Man and Cybernetics*, 3(1):28–44, 1973. Reprinted in [108].
111. L. A. Zadeh. The concept of a linguistic variable and its application to approximate reasoning — I. *Information Sciences*, 8:199–249, 1975. Reprinted in [108].
112. L. A. Zadeh. A theory of approximate reasoning. *Machine Intelligence*, 9:149–194, 1979. Reprinted in [108].
113. L. A. Zadeh. The calculus of fuzzy if-then rules. In *Fuzzy Engineering toward Human Friendly Systems — Proceedings of the International Fuzzy Engineering Symposium '91*, volume 1, pages 11–12, Yokohama, November 13–15, 1991.
114. L. A. Zadeh. The calculus of fuzzy if/then rules. In Reusch [58], pages 84–94.
115. L. A. Zadeh. Toward a theory of fuzzy information granulation and its centrality in human reasoning and fuzzy logic. *Fuzzy Sets and Systems*, 90:111–127, 1997.
116. B. Zeungnam and W. Yu. Design of fuzzy logic controllers with inconsistent rule base. *Journal of Intelligent and Fuzzy Systems*, 2:147–159, 1994.
117. Z. Zhong and I.B. Türkşen. An approximate analogical reasoning approach based on similarity measures. *IEEE Transactions on Systems, Man, and Cybernetics*, 18(6):1049–1056, 1988.
118. Z. Zhong and I.B. Türkşen. An approximate reasoning schema based on similarity measures and interval-valued fuzzy sets. *Fuzzy Sets and Systems*, 34:323–346, 1990.

# 3 Data-Based Fuzzy Modeling for Complex Applications

Harro Kiendl, Peter Krause, Daniel Schauten, and Timo Slawinski

University of Dortmund, Faculty of Electrical Engineering and Information
Technology, Control Engineering, 44221 Dortmund, Germany
{kiendl, krause, schauten, slawinski}@esr.e-technik.uni-dortmund.de

**Summary.** As a consequence of the continually increasing complexity of technical
systems and processes, the requirement for high-performing modeling techniques is
increasing. The methods of Computational Intelligence (CI), such as fuzzy model-
ing or artificial neural networks, are specifically designed to deal with imprecise,
incomplete, and partially incorrect information in dynamically changing environ-
ments and large variable spaces. Fuzzy modeling has the additional advantage that,
with certain restrictions, interpretable models are obtained. However, especially
for data-based approaches, there are still difficulties in generating efficiently inter-
pretable rule bases with the required accuracy. Moreover, particularly with regard
to complex applications, efficiency, interpretability, and accuracy are often partly
contradictory modeling objectives. The main focus of this contribution is on de-
termining an adequate compromise in this conflict area. The applicability of the
concepts presented is demonstrated by means of complex real-world applications
in the domains of power management control, classification in quality control, and
prediction in financial service.

## 3.1 Introduction

The use of fuzzy modeling is often motivated by the interpretability of the
resulting model. A fuzzy model consists of qualitative IF-THEN rules that
correspond to the way in which human knowledge is usually presented. When
this knowledge can be directly transformed by experts into corresponding
fuzzy rules, we speak of the knowledge-based modeling approach. However,
for complex applications with many input variables, the knowledge-based
approach may be very time consuming or not feasible. Furthermore, process
knowledge can be uncertain or incomplete such that the knowledge-based ap-
proach fails completely. To overcome these disadvantages, various approaches
have been developed for data-based fuzzy modeling. In contrast to other data-
based modeling approaches, such as artificial neural networks, the generated
fuzzy model is, under certain restrictions, interpretable and allows additional
insight into the process being considered. This is especially important for
complex systems if only limited prior knowledge is available. Moreover, inter-
pretable models are usually more acceptable to the process experts and can
be refined by hand.

Other well-known approaches for data-based fuzzy modeling are based on clustering algorithms (CA), classical optimization methods, decision trees, artificial neural networks (NN) and evolutionary algorithms (EA). Clustering algorithms have been established as methods for search space structuring and preliminary modeling. In a second step, the model quality is improved by optimization, which for example is supplied by an EA or NN technique. In other approaches, the rules, and the composition of the rule base or both are optimized by classical methods or EA. In order to exploit the high modeling quality of NN, in recent research NN structures that can be converted into fuzzy rule bases have been proposed. However, there are still difficulties in efficiently generating small interpretable rule bases with the required accuracy, especially for applications with many input variables.

This leads to the conclusion that applicability of different existing methods for data-based rule generation with respect to complex tasks are mainly limited by the following three factors: the availability of a systematic modeling procedure, the efficiency of the rule-generation process and the accuracy and interpretability of the model obtained. In the following we discuss these factors more in detail.

In order to cope with more and more complex applications, a broad spectrum of sophisticated fuzzy modeling methods has been developed. This introduces the problem of selecting the most favorable method to solve the modeling task at hand. Moreover, most of the methods require a problem-dependent adaptation, which considerably inhibits many non-experts, even when these methods are integrated into commercial software tools, such as MATLAB/Simulink. It would be preferable for the fuzzy modeling to proceed nearly automatically or, at least to have a systematic approach with respect to different problem classes. In Section 3.2, we give a brief introduction to systematic rule generation with the fuzzy rule-oriented statistical analysis (Fuzzy-ROSA) method, which is the predominate modeling approach that is considered in this contribution.

The complexity of a modeling task increases substantially with the number of input variables, i.e., with the size of the search space. Therefore, it is essential that the computational effort for the data-based rule generation is acceptable, even if many linguistic variables and values are considered. However, in the literature, only a few examples with five to ten input variables are found, and more complex cases are rarely published. In addition, if the rule base is optimized as a whole, by simply minimizing the modeling error, it cannot be guaranteed that each individual rule of the rule base obtained is reasonable, i.e., locally relevant. For these reasons, we favor such approaches as the Fuzzy-ROSA method (FRM), which is based on testing individual rules with respect to their relevance. This reduces the problem of finding a good rule base to the problem of finding individual relevant rules. Furthermore, because each rule with high relevance is supposed to express an important aspect of the system, such a rule is meaningful by itself. Therefore,

the rule base obtained is more transparent and comprehensible. A suitable structure of the rules is also important for handling complex applications. In this context, it would be better if the FRM could also handle generalizing (incomplete) rules[1]. These rules need not consider all the linguistic input variables provided, and therefore, the length of the premises may differ. If a single rule is composed of fewer expressions than the total number of input variables, the rule covers several linguistic input situations. In particular, for large search spaces with a large number of input variables, such rule bases may be much smaller than rule bases consisting of complete rules only. However, this leads to a more complex rule generation problem. In Section 3.2, we present several strategies for efficient rule generation of individually tested generalizing rules and discuss the manageable size of the search space.

An important effect on the complexity of the modeling problem is due to difficult dependencies between the input and output variables, e.g., non-linear relations, ambiguities, or stochastic influences. As a result, for a complex application, a high accuracy often means high complexity of the fuzzy model. When considering the interpretability of a complex fuzzy model, the number and structure of the rules as well as the number and shape of the membership function are critical factors. Obviously, high accuracy and good interpretability are partly contradictory modeling objectives. In Section 3.3, we present sophisticated concepts for finding a good balance between interpretability and high accuracy.

In order to demonstrate that the developed concepts can meet the requirements of real-world applications, we applied them to several industrial tasks in the domains of power station management, classification in quality control, and prediction of contract duration (Section 3.4.1).

## 3.2    Efficient Rule Generation in High-dimensional Search Spaces

In this Section, we first present a brief survey of the rule base generation process with the FRM (Section 3.2.1). The aim of this introduction is to provide all the necessary information for the following Sections and to refer to further reading. Then we describe two sophisticated rule search concepts for the FRM: the tree-oriented complete search (Section 3.2.2) and the hybrid evolutionary search concept (Section 3.2.3). Although both were originally developed for the FRM, they may also be applied to any rule generation process based on individually tested rules. Using these advanced search concepts, high-dimensional systems are manageable, with up to 150 input variables. An alternative approach to coping with high-dimensionality is a preliminary complexity reduction, based on feature selection. In Section 3.2.4, we present an

---

[1] A complete rule/input situation is determined by assigning a linguistic value to each input variable.

evolutionary method for this task, which also provides granularization, i.e., the shape of the membership functions of the input variables.

### 3.2.1  Systematic Rule Generation with the Fuzzy-ROSA Method

The basic idea of the FRM is to apply a relevance test to individual fuzzy rules in order to assess their ability to describe a relevant aspect of the system under consideration [10, 31, 39, 57]. This reduces the problem of finding good rule bases to the problem of finding relevant rules. As a result, the FRM can handle complex applications but does not, in general, achieve the maximal possible accuracy. Moreover, since each rule with high relevance is supposed to express an important aspect of the process under consideration, such rules are meaningful by themselves. This, in turn leads to more transparent and comprehensible rule bases.

The rule base generation process of the FRM is divided into five main steps. There are alternative strategies available for each step so the method can be adapted to differing application requirements (e.g., for modeling, classification, approximation or prediction) and problem sizes (e.g., number of variables, linguistic values, and data sets). In order to support non-experts applying the FRM, we propose a systematic procedure[2] for each of these steps. Results obtained with the systematic approach and more general views with details can be found in [20, 49].

**Project Definition:** An essential prerequisite of high quality and efficiency in data-based methods is appropriate feature generation and feature selection. The quality of the resulting model depends on the choice of input variables (features) used. Furthermore, the complexity of the corresponding search space can be reduced only if relevant and nonredundant features are used for rule generation. Such a feature generation and selection process can be based on statistical methods (e.g., correlation analysis), on NN methods, or on fuzzy-based methods [44]. Prior to rule generation, the membership functions for the linguistic values of the input and output variables in the system being considered must be defined. This can be done as a knowledge-based process by an expert, if a corresponding domain of knowledge is available. Otherwise it can be done either heuristically or by a data-based process. Common heuristic approaches suggest the choice of an equidistant partition or careful provision of evenly distributed data with regard to the membership functions. In data-based approaches, cluster algorithms are usually used to define membership functions [5, 36]. Furthermore in Section 3.2.4, we present an evolutionary approach to automatic feature selection and definition of the membership functions specifically designed for supporting users of the FRM. In order to enforce interpretability of the generated rules, triangular or trapezoidal

---

[2] The systematic approach is designed for the WINROSA 2.0 software tool [32, 34]). Thereby, the default settings are used, except when otherwise described.

membership functions are predominantly used in the FRM in the form of a fuzzy information system.

The size of the search space can be restricted by allowing a maximum combination depth, i.e., the maximum number of linguistic expressions in the premise. Usually we start with a preliminary model with a low combination depth to become familiar with the problem being considered. If satisfactory settings for the project definition are found, a search with a higher maximum combination depth can be started to improve the results. For modeling time series or for considering time-varying variables of a dynamic process, a maximum time depth can be chosen in the FRM. Then the modeling is based not only on the linguistic values of the actual data set but also on the linguistic values of the data sets corresponding to preceding time steps. Within the FRM, time depth is treated as an additional linguistic variable, e.g., temperature two time steps ago. Therefore, time depth is not explicitly considered in the following. Moreover, we found that it is sometimes more profitable to consider the time dependency by aggregated features, as, for example, the trend or the derivative of an input variable.

**Systematic Project Definition:** *If no prior knowledge is available, we consider all available variables. For each continuous variable, we use an equidistant partition with usually about five trapezoidal, 50 percent overlapping membership functions in the form of a fuzzy information system. In contrast, discrete variables are mapped into numbers if necessary, and for each number we basically define a corresponding singleton. The maximum combination depth is chosen to be two.*

**Rule Search:** The basic task of the rule search is to set up potential rules (hypotheses) and apply a statistical relevance test to the hypotheses. Only hypotheses that pass this test are accepted as relevant fuzzy rules. As a result of the test, each rule is given a rating index, which can be interpreted as a certainty factor. The rule base is generated by collecting relevant rules iteratively. To meet different modeling goals, three different rule test and rating strategies have been developed [9, 10, 11, 12, 28, 38, 39, 49, 57]. In Section 4.3.2, an advanced strategy to cope with ambiguous data is presented. This approach is based on universal rules and an incremental relevance test [14].

Depending on the size of the search space, a complete search, an evolutionary search, or a combination of both is applied. The complete search, which is described in Section 3.2.2, guarantees that all relevant rules are found. We have found that a complete search can usually be applied successfully up to a maximum combination depth of four. For higher combination depths, we usually apply the hybrid evolutionary search concept (Section 3.2.3).

**Systematic Rule Generation:** *The modeling result strongly depends on the rule test and rating strategy that have been chosen. Therefore we propose applying the following deterministic scheme for each test and rating strategy (with a confidence level $\alpha = 0.05$): (1) perform a complete tree-oriented search (TOCS) with maximum combination depth of two and a threshold for the evaluation index of $J_{min} = 0$, (2) compute the modeling error and the data sets of learning data not covered, (3) reduce the rule base by increasing the threshold by $\Delta J = 0.1$ and (4) repeat (2) until the rule base is empty. Finally, choose the rule test and rating strategy and the threshold $J_{min}$ by considering the modeling error, the number of rules and the number of data sets not covered.*

**Restriction:** The FRM allows the introduction of the following restrictions in order to prohibit the generation of undesired or infeasible rules. To preserve maximum flexibility, these restrictions are optional.

*Combination Restriction:* The default prohibits the combinations of linguistic expressions that refer to the same linguistic variables, because a premise of the type IF temperature is high AND temperature is low is usually not desired. A deactivation of this restriction, in the case of a fuzzy information system, is only useful for the purpose of analysis. In that case, rules might be generated that combine two neighboring membership functions in the premise. This means that decisive data sets are in the fuzzy range of both membership functions.

*Minimal Data Support:* The trustworthiness of a rule strongly depends on the number of supporting data sets. In the FRM, the data support of a given premise is defined by the sum over the degrees of membership for each data set [49, 57]. This restriction allows us to define a minimal data support. The default value of the minimal data support is chosen as one.

*Complement Restriction:* In the FRM, each linguistic expression can be used in a negated form to formulate premises of the type IF temperature is not high. This restriction is activated as the default because few of the commonly available fuzzy tools can process negated statements. In the following, negated statements in the premise are not considered.

*Only Positive Rules:* As described in [17], conventional fuzzy systems can process only positive rules that express recommendations. This is a serious disadvantage of conventional fuzzy systems. In many applications, both positive experience, in the form of recommendations, as well as negative experience, in the form of warnings and prohibitions to avoid undesirable operating situations, are important. Negative rules can be processed – together with positive rules – by the two-way fuzzy system based on hyperinference introduced by Kiendl [15, 16, 17]. An essential advantage of the FRM is that, as a result of the rule test, it extracts both positive and negative rules from the data [31, 57]. If a fuzzy tool

that cannot process negative rules is used, this restriction should be activated. Furthermore, if the hybrid evolutionary search concept is applied, positive rules should be searched separately, because otherwise the population is usually dominated by negative rules.

It must be pointed out that, especially in high-dimensional search spaces, the combination/complement restriction and minimal data support can lead to a drastic reduction in the number of rules that need to be tested (Section 3.2.2).

**Systematic Restrictions Settings:** *By default, the combination, complement and only positive rules restrictions are activated. The minimal data support is chosen to be one.*

**Rule Reduction:** The rule generation process considers only the quality (relevance) of the individual rules and not that of the complete rule base. Consequently, there is still potential for optimization. In order to create small transparent rule bases or to improve the model quality, subsequent rule reduction steps can be applied. There are two main types of reduction strategies [27, 28]. Online reduction aims at preventing redundant rules. It is applied if a new rule is found and added to the rule base. Rules are considered as redundant in the rule base if the latter already contains a more general rule with a higher rating index and the same conclusion. The methods for offline reduction are developed to meet different requirements. On the one hand, it is of interest to reduce the number of rules in order to create small and transparent rule bases. On the other hand, it is possible to improve the modeling error. The following offline reduction strategies are available:

*Situation-Based Conflict Reduction:* Situation-based conflict reduction deletes rules that do not belong in any complete linguistic input situation, leaving the best rules (rules with the highest rating indices). This reduction strategy guarantees that the resulting rule base covers as many linguistic input situations after applying the reduction as it did before. Therefore, if the original rule base covers all possible linguistic input situations, the reduced rule base will also cover all possible linguistic input situations. Situation-based conflict reduction requires computing time that is proportional to the number of complete input situations. Therefore, situation-based conflict reduction can only be applied to smaller problems.

*Structure-Based Conflict Reduction:* To handle more complex problems, we have developed structure-based conflict reduction. This strategy removes rules that do not belong to the best rules (rules with the highest rating indices) in any generalizing input situation (structure) with a maximum combination depth. It also maintains the completeness of a rule base. However, this approach may remove fewer rules than situation-based conflict reduction.

*Data-Based Conflict Reduction:* In contrast to situation-based conflict reduction, this method is not based on analyzing input situations, instead it is based on analyzing data points supporting multiple rules. Rules that are mainly based on data points that also support better rules are removed. Therefore, the rule test is applied again to each rule of the rule base, beginning with the best. The difference to the relevance test of the rule search is that the data points can only support a restricted number of rules and then are removed from the learning data. The disadvantage of this method is that it cannot be guaranteed that the number of input situations covered is preserved. However, the great advantage is that this rule reduction method always has a practical computing time, i.e., a small fraction of the computing time of the rule search. Additionally, the amount of reduction is usually higher than in the case of situation-based conflict reduction.

*Optimizing Conflict Reduction:* This strategy aims at increasing the quality of the model by removing contradictory rules. Starting with the given rule base, certain combinations of rules are tested with respect to their input/output behavior in order to find the rule base with the best model quality. For this complex optimization problem, we have developed a special genetic algorithm (GA), which is described in Section 3.3.1. The main advantage of this approach, compared to other approaches in the literature, is that the GA only needs to optimize a rule base of relevant rules and not the set of all possible rules.

**Systematic Reduction:** *The optimizing conflict reduction is optionally performed in a subsequent step.*

**Analysis and Optimization:** The final step of the rule generation process is analysis and optimization. The analysis of the rule base allows us to rate the modeling process as well as the obtained model quality and provides feedback that can be used to modify the problem formulation. Furthermore, in Section 3.3, we present concepts for direct feedback of the modeling quality in the rule generation process. Finally, the input/output behavior of the resulting fuzzy system can be optimized by tuning the remaining free parameters, such as the output membership functions [23].

**Systematic Analysis:** *For the computation of the modeling error, we propose using problem dependent settings of the fuzzy model. For approximation tasks, we usually choose the center of gravity defuzzification and, if no rule is activated, the mean value of the output variable is recommended. For classification tasks, we usually choose the mean of maximum defuzzification and, if no rule is activated, a supplementary class 'not classified' is recommended.*

## 3.2.2   Tree-Oriented Complete Search (TOCS)

The complete search method described in this Section was especially designed for efficient rule generation in high-dimensional search spaces [49, 56, 57]. As logical conjunctions of output variables are not supported by the FRM, the complexity arises predominantly from the number of input variables and their linguistic values. Therefore, in the following, the influence of the output variables is neglected and only the premises are considered.

On the one hand, the number of feasible premises increases exponentially with the number of input variables. While the number of available data sets is strictly limited in most applications. In this case, the number of data sets per premise decreases drastically if the combination depth is increased. By exceeding a certain combination depth, a large number of the premises are no longer supported by any data point, and when combined with any conclusion, will not pass the rule tests [49, 57]. If further linguistic expressions are added to such a premise in order to generate a more special premise, the derived premise is also not supported by the data. Thus, it need not be considered in the generation process.

In the tree-oriented complete search, premises are set up and tested with respect to all linguistic output values. The TOCS concept is based on linguistic expressions $e_{ij} = X_i = s_{ij}$, where $X_i$ denotes the linguistic variable and $s_{ij}$ the linguistic value. For the tree-oriented search, the linguistic expressions are numbered serially: $\tilde{e}_1 = (X_1 = s_{11}), \tilde{e}_2 = (X_1 = s_{12}), \ldots, \tilde{e}_E = (X_V = s_{VS_V})$, where the total number of linguistic expressions is denoted by $E = \sum_{i=1}^{V} S_i$ and $S_i$ is the number of linguistic values of the $i$th variable. Starting from the first linguistic expression $\tilde{e}_1$, all more special premises are generated, as shown in Figure 3.1.

Consider now the $n$th linguistic expression $\tilde{e}_n$. For the combination depth $c = 1$, the premise is given by $P = \tilde{e}_n$. Based on this premise, rules are set up by combining it with all linguistic output values as conclusions. The chosen test and rating strategy is applied, and each rule that is found to be relevant is added to the rule base.

A new linguistic expression $\tilde{e}_{n+1}$ is added to the premise, until the maximum combination depth $c'_{max}$ is reached. In order to avoid infeasible premises, the maximum combination depth is set as $c'_{max} = \min(c_{max}, E - n + 1)$. If the TOCS is applied in this way, all possible rules are set up and tested. By activating combination restriction and minimal data support (Section 3.2.1), the TOCS can be accelerated considerably as follows:

- A rule is unreasonable in the sense of combination restriction if two linguistic expressions refer to the same linguistic variable. If this occurs at a certain point, the premise remains restricted, even if it is specialized. Consequently, this branch need not be considered further.
- Rules with a high combination depth are more likely not to be supported by any data sets. Therefore, if a rule is not supported, no additional

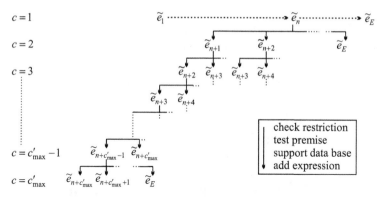

**Fig. 3.1.** The tree-oriented complete search. Starting from a given premise, the necessary specialization (generation) steps are indicated in the box. First, the restrictions are checked. If the premise is restricted, the branch need not be considered further. Then, the premise is tested. A result of the test is the supporting data sets, which are taken as a data base in the next specialization step. Finally, a new term is added to the premise.

specialized rule with additional linguistic expressions based on that rule can be found to be relevant. This means that if such a point is reached in a branch of the TOCS, the branch need not be followed further. The computational savings are estimated in [49, 56, 57].

- Following one branch, a decreasing number of supporting data sets need to be considered. Therefore, the rule test and rating [49, 57] can be implemented very efficiently by considering only the supporting data sets.

In this way, the computational effort can be reduced drastically, especially for sparsely populated search spaces. Consequently, the manageable search space increases at least by a factor of 100 for the examples shown in Figure 3.2. This provides the basis for very complex examples as described in [49, 57].

### 3.2.3 Hybrid Evolutionary Search Concepts (HESC)

Although the performance of the complete search could be increased drastically by the tree-oriented search concept, presented in the previous Section, a complete search still fails for very complex search spaces. In this case, an alternative approach is to use probabilistic searches as evolutionary algorithms.

In the literature, we find three main application areas of evolutionary algorithms in the field of fuzzy modeling: optimization of membership functions, optimization (generation) of rules, and simultaneous optimization of both. In the case of rule base optimization (generation), most evolutionary algorithms use a fixed rule base structure of complete rules, where one individual in the evolving population represents a whole rule base (Pittsburgh style). The Pittsburgh style is often not practical for more than three or four

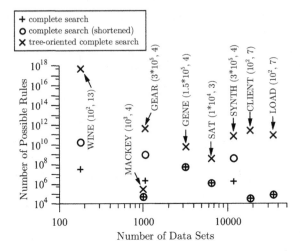

**Fig. 3.2.** For the benchmark problems and applications considered in [49], the manageable complexity of the search space is indicated for the tree-oriented search concept and the prior existing complete searches of the FRM. Measures for the complexity are the number of data sets (abscissa) and the number of possible rules (ordinate). Manageable means that all relevant rules are found in a given search time. This search time in seconds and the chosen maximum combination depth $c_{max}$ are denoted in brackets for each example.

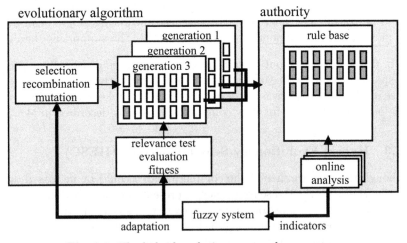

**Fig. 3.3.** The hybrid evolutionary search concept.

variables, and more complex applications are rarely published. However, as shown in Section 3.3.1, a subsequent evolutionary optimization of rule bases consisting of individually tested rules can lead to very good results.

In our HESC [49, 53], which is depicted in Figure 3.3, each individual represents a single fuzzy rule (Michigan style). While the evolutionary algorithm is searching for the best rule, many good rules are generated. The basic concept is to collect these good (relevant) rules during the course of evolution. In comparison to other approaches that also take single rules as individuals, our concept is based on one population and a single evolution process. Since the FRM makes use of generalizing (incomplete) rules, it was necessary to develop a sophisticated genetic representation and extended genetic operators [37, 49, 54].

As described above, the fitness evaluation of individuals is based solely on the relevance of individual rules. Thus, the overall quality of the rule base collected at the current generation does not influence the evolutionary search process directly. In some situations, this might lead to a rule base consisting of a few rules of high relevance, but poor overall performance. Consequently, additional measures have been introduced to avoid premature stagnation and to keep the algorithm exploring the search space. This has been achieved by dynamically adapting some of the strategy variables depending on the performance of the current rule base collected during the course of evolution. The implementation is based on a hybrid fuzzy evolutionary approach, described in [49, 52, 53], where a fuzzy system is used to adapt the strategy parameters of the evolutionary algorithm.

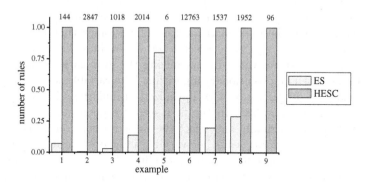

**Fig. 3.4.** The performance of the HESC in comparison to the previous evolutionary search (ES), measured by the number of generated rules (in a given number of generations). The number of rules is normalized by the maximum number of rules found (depicted above the bars for each example).

In Figure 3.4, the improvement of performance obtained with this hybrid evolutionary search concept is compared with the previous evolutionary search for the FRM described in [28, 29]. For more details, such as the examples considered, see [49].

### 3.2.4   Feature Selection and Automatic Dimension Partitioning

The approaches to coping with high-dimensional search spaces presented so far are based on improvement of the rule search performance. An alternative approach is to reduce the complexity of the search space. As described in [43, 45, 46], the success of the data-based modeling approach depends decisively on the suitable selection of the input variables that are actually considered from a great number of influence variables. To reduce the number of input variables, and subsequently the complexity of the fuzzy model, preprocessing of the data is desirable to obtain a set of relevant and nonredundant variables, which is both as small as possible and as big as necessary.

A well-known method for this task is correlation analysis. However, it reveals only linear dependencies between two variables. In [43], a fuzzy method for detecting both linear and non-linear dependencies is presented, tailored to data-based modeling with the Fuzzy-ROSA method [34, 57] by taking fuzzy partition granularity into account. Input variables are only kept if they affect the output variable and are not correlated with any other kept input variable. However, this method evaluates the relevance of each variable individually and does not consider the relevance of a set of variables, which would be more adequate. To overcome this disadvantage, a new approach for selecting relevant sets of variables by evaluating the relevance of sets of influence variables depending on the chosen granularity has been developed [46, 61]. This approach allows consideration and optimization of the granularities and therefore also copes with ambiguities, as described in Section 4.5.

For this we proceed stepwise. In each step, we determine for each as yet unselected variable the maximum increase of relevance that results if this variable is added to the previous selected set of variables. Therefore we maximize the increase of relevance by optimizing the granularity of the considered variables with an evolutionary algorithm. The variable that supplies the greatest increase in relevance with optimized granularities is added to the previously selected variables. In the first step, we determine the relevance of each variable individually and select the most relevant one. The increase of the relevance by adding a further variable indicates the improvement to be expected by the subsequent fuzzy modeling. This iterative evolutionary feature-selection process terminates if no further significant improvement of the relevance is found by adding another variable. Consequently, the compact and relevant set of influencing variables with associated optimized granularities that is finally obtained does not contain superfluous, i.e., redundant, variables. The efficiency of the method is demonstrated in [45], using it in various well-known benchmark problems.

## 3.3   Interpretable Fuzzy Models with High Accuracy

The concepts presented so far are mainly based on individually tested rules. However, the modeling quality and the interaction of the rules are not directly

considered in the rule generation process. This may lead to a lower accuracy and to large and redundant rule bases. To overcome these disadvantages, in this Section we present three concepts for feedback on the modeling quality in the rule generation process of the individually tested rules. The general purpose is the generation of small interpretable rule bases with high accuracy even for high-dimensional problems. The three concepts were originally developed for the Fuzzy-ROSA method. However, under certain conditions, the concepts can also be used for other approaches, based on the testing of individual fuzzy rules.

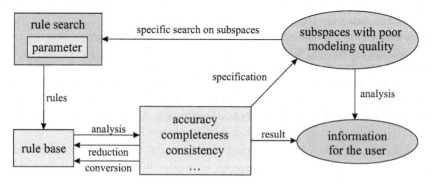

**Fig. 3.5.** Concepts for the feedback of modeling quality.

On the left of Figure 3.5, the FRM is displayed in its initial form. Depending on certain parameters, the rule search sets up and tests individual rules. All relevant rules are collected in the rule base. The overall performance of the rule base, generated so far, can be analyzed and assessed with respect to several indicators for the modeling quality. The following indicators are widely used:

*Accuracy:* An appropriate measure for the accuracy of the rule base is the modeling error. Here we use the mean absolute/square error (MAE/MSE) for approximation tasks and the relative classification error (RCE) for classification tasks.

*Completeness:* Often a complete coverage, i.e., an activation of at least one rule in each input situation, is desired. The coverage rate (the number of input situations covered divided by the total number of input situations) is a suitable measure for the completeness of a rule base.

*Consistency:* Especially in data-based approaches, the obtained rule bases can produce multimodal output membership functions. Sometimes these ambiguities are caused by inconsistencies and therefore are hard to interpret. Approaches to quantifying of the consistency of a rule base are presented as examples in [41].

In addition to providing information for the user, these indicators can be used for optimization. An example is the optimizing conflict reduction presented in Section 3.3.1. In this example, a genetic algorithm is used to simultaneously reduce the modeling error and the number of rules. Optionally, the coverage with respect to the input situations supported by the learning data can be considered. However, it is quite evident that so far these indicators only provide a numerical value for the accuracy, completeness, or consistency, respectively and supply no direct information as to how to improve the modeling quality.

The finally reached modeling quality can be improved, if a method would be available that allows to specify the subspace with poor modeling quality during the process of modeling: Such a method would allow us to generate rules exclusively in the subspace with poor modeling quality. Due to this restriction on the search space, such a specific generation is generally more efficient, and in addition smaller rule bases are obtained. In order to profit from these advantages, we have developed and implemented a specific search on subspaces with poor modeling quality (Section 3.3.2). An essential prerequisite was the design of efficient algorithms and suitable data structures for the specification of the subspace with poor modeling quality.

Another approach for improving the accuracy was developed especially for approximation tasks. By means of a transformation, the Mamdani rules are converted into Takagi-Sugeno-Kang (TSK) rules. As described in Section 3.3.3, the function described in the conclusion of the TSK rules is identified with the modeling error. In contrast to other TSK approaches, the initial Mamdani rules, which are less complex, can be used for interpreting the rule base.

The results obtained by applying these three concepts to several benchmark problems are discussed and compared in Section 3.3.4. It shows that, with the new concepts, a data-based generation of small interpretable rule bases with high accuracy is possible even for high-dimensional problems.

### 3.3.1   Optimizing Conflict Reduction (OCR)

The aim of optimizing conflict reduction [28, 33] is to reduce the number of rules by consideration of the modeling or classification error. In contrast to other GA-based approaches, in this reduction step of the FRM only the set of all relevant rules and not the set of all possible rules has to be considered. However, to find the optimal subset of relevant rules is still a complex optimization problem, and because of the high complexity and the large number of local optima, a genetic algorithm has been designed for optimization. Furthermore, a parallel version has been implemented to handle very high dimensional problems [25]. In the following, we describe the fitness function and the genetic operators in detail.

**Fitness Function:** For the definition of the fitness function, three aspects are considered:

1. Small modeling and classification errors are desired.
2. The number of rules should be reduced as much as possible.
3. It is often desirable that the number of covered data sets is preserved.

Since the first aspects are partly competitive, the user has to determine a weighting factor $w \in [0,1]$ in order to define a cost function. This weighting factor is used to evaluate the fitness function:[3]

$$
f(\mathbf{a}_i) =
\begin{cases}
w \cdot \dfrac{\epsilon(\mathbf{a}_i)^2}{\epsilon(\mathbf{a}_0)^2} + \dfrac{(1-w) \cdot R(\mathbf{a}_i)}{R(\mathbf{a}_0)} & \text{if } \epsilon(\mathbf{a}_0) \neq 0 \\[2ex]
w \cdot \epsilon(\mathbf{a}_i)^2 + \dfrac{(1-w) \cdot R(\mathbf{a}_i)}{R(\mathbf{a}_0)} & \text{if } \epsilon(\mathbf{a}_0) = 0
\end{cases}
\tag{3.1}
$$

Here $\mathbf{a}_0$ is the initial individual representing the initial rule base and $\mathbf{a}_i$ is the actual individual under investigation that represents a possible, reduced rule base. Further on $\epsilon(\mathbf{a}_0)$ is the modeling error of the initial individual and $\epsilon(\mathbf{a}_i)$ is the modeling error of the actual individual. $R(\mathbf{a}_0)$ is the number of rules of the initial individual and $R(\mathbf{a}_i)$ is the number of rules of the actual individual. Depending on the application, the function $\epsilon$ can vary as described in [51].

Small values for the weighting factor $w$ predominantly lead to a reduction of the number of rules, whereas high values of $w$ predominantly lead to a reduction of the modeling or classification error. The optimum for $w = 0$ is the empty rule base, and the optimum for $w = 1$ is a rule base with the smallest modeling or classification error, independent on the number of rules.

In contrast to the similar approach in [7], the number of rules here plays an important role in the evaluation of the fitness values. For small initial rule bases (less than 50 rules), the number of rules is less important. However, for large rule bases, there is a substantial difference between 100 rules or 2000 rules. In such a case, a slightly higher modeling error is accepted in favor of further rule reduction.

To consider the third aspect, the fitness $f(\mathbf{a}_i)$ is multiplied by a penalty factor $p(\mathbf{a}_i)$ in the form of:

$$
p(\mathbf{a}_i) = \frac{D - D_0}{D - D_i} ,
\tag{3.2}
$$

where $D$ is the number of data points, $D_0$ is the number of data points that are not covered by the initial rule base, and $D_i$ is the number of data points that are not covered by the actual rule base.

---

[3] Good individuals have low fitness values.

**Genetic Operators:** One individual in a population consists of $r$ bits, with $r$ being the number of rules of the initial rule base. Each bit represents the state of a rule. A rule can have the state activated or deactivated. The genetic operators are standard operators [4]. The recombination is realized by a uniform crossover operator with fitness proportional selection. The $n$-point-crossover operator leads to worse results as the algorithm converges too quickly. For the mutation operator, a number $z \in [0, 1]$ is selected randomly for each bit. For $z \leq p_m$, the bit is inverted. As selection operator, a tournament selection with two descendants is chosen. The actual best individual of the evolution process is kept outside of the populations. The initialization of the starting population has been found to be of strategic importance [33]. The standard initialization[4] has proved to be unfavorable for complex rule bases. Instead the number $z$ is chosen by the user so that only a few rules are activated.

### 3.3.2  Specific Search on Subspaces (SSS) with Poor Modeling Quality

The specific search on subspaces was divided into two main steps. First, efficient data structures and algorithms were developed for the specification of subspaces with poor modeling quality [49, 51]. Second, a rule search was designed for setting up rules covering the specified subspace. In our approach, presented in [49, 51], an iterative search concept is applied for the specific search (see Figure 3.6).

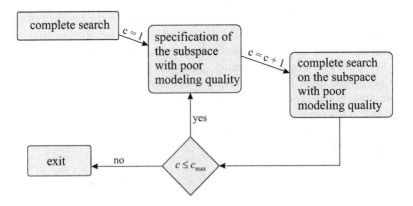

**Fig. 3.6.** Specific search on subspaces.

---

[4] For each individual, a number $z \in [0; 1]$ is randomly selected. For each bit, $b_j$, a second number $z_b \in [0; 1]$ is randomly selected. If $z_b \leq z$ holds, the bit is activated.

Starting with an initial rule set with a combination depth $c = 1$, the subspaces of poor modeling quality are specified as described in [49, 51]. Therefore, a modeling objective must be chosen, e.g., a high accuracy, completeness of consistency (compare to Figure 3.5). Then the complete search on subspaces is applied to generate exclusively rules, which partly cover the subspaces with poor modeling quality. These two iteration steps are repeated until the chosen maximum combination depth $c_{max}$ is reached. In Section 3.3.4, we demonstrate that, with the specific search on subspaces with poor accuracy, a compact rule base with high accuracy can be obtained. Due to space limitations, we do not present here the results of the specific search on the subspaces not covered. As shown in [49], this approach is useful for efficiently generating a complete rule base.

### 3.3.3  Transformation in Takagi-Sugeno-Kang Fuzzy Systems

The FRM aims to generate rule bases consisting of Mamdani fuzzy rules. The advantage of such rules is that they can be easily interpreted and comprehended. This advantage is often lost if a TSK fuzzy model is generated directly. The reasons can be that for each of the rules individually membership functions are provided and that they are strongly overlapping. Another reason may be that the conclusions become hard to interpret because of the high number of parameters that come with high dimension. On the other hand, the advantage of TSK fuzzy models is that they generally allow a higher accuracy to be achieved. Therefore in [21, 22, 24, 51], the FRM is extended for the generation of TSK rules as follows: Starting with a generated Mamdani rule base, this rule base is converted into a TSK rule base. It is essential that the conversion is designed to increase the model accuracy while preserving the interpretability of the rules.

In contrast to a Mamdani rule (3.6), the conclusion of a TSK rule consists of a function $f(\mathbf{x})$ depending on the values of the input variables $\{x_1, \ldots, x_V\}$:

IF $P$ THEN $f(\mathbf{x})$ . $\hspace{6cm}$ (3.3)

Most commonly, a linear combination of the input values is used as a conclusion:

$$f(\mathbf{x}) = a_0 + \sum_{i=1}^{V} a_i \cdot x_i \ . \hspace{4cm} (3.4)$$

The value of the output of the fuzzy model is calculated by:

$$y_D = \frac{\sum_{j=1}^{R} p_j \cdot f_j(\mathbf{x})}{\sum_{j=1}^{R} p_j} \ . \hspace{4cm} (3.5)$$

**Motivation:** In order to clarify the method proposed here, we consider a single input single output (SISO) system with the input variable $X$ and the output variable $Y$. The data points describing the system are distributed as shown in Figure 3.7. For this system, the relevant rule:

$$\text{IF } X = medium \text{ THEN } Y = medium \tag{3.6}$$

is found.

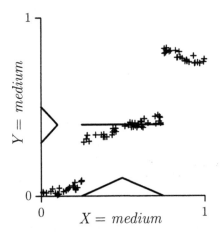

**Fig. 3.7.** Example of an SISO system. The solid line shows the output value of the fuzzy model using a Mamdani fuzzy model.

**Fig. 3.8.** Example of an SISO system. The solid line shows the output value of the fuzzy model using a TSK fuzzy model with rule (3.7).

If only rule (3.6) is active, the resulting output value of the model is for $X = medium$ constant as shown in Figure 3.7 by the solid line. As the visible trend in the data cannot be modeled adequately by the linguistic value of the Mamdani rule, the resulting modeling error becomes relatively high. If instead the rule:

$$\text{IF } X = medium \text{ THEN } 0.25 + 0.125 \cdot x \tag{3.7}$$

with a function as conclusion is used, the modeling error is reduced considerably, as depicted in Figure 3.8. Here, the solid line shows the new output values, which better reflect the true behavior of the system. Thus, to make use of the advantage of the TSK rule, we transform the Mamdani rule by replacing the linguistic expression with a function. In general, an arbitrary function with some free parameters can be used to replace the linguistic expression of the conclusion. In our approach, we provide linear/multilinear combinations as well as polynomials [21, 22].

Furthermore, it is necessary to choose a suitable set of data points for the identification process in order to identify the free parameters. As described in [22, 51], we use those data points $\mathbf{x}$ that fulfill, i.e., $\mu_P(\mathbf{x}) > 0$, the premise $P$ of the rule being considered.

### 3.3.4 Results and Comparison

Finally, we present a comparison of the different concepts for the feedback on the modeling quality in the rule-generation process [51]. For the comparison, four well-known benchmark problems were chosen.[5] As described in [57], the WINE and SAT problems are examples of a small and a complex classification problem, respectively. The MACKEY and KIN problems are a small approximation problem and a complex approximation problem, respectively [57].

The modeling results are assessed by the obtained accuracy with regard to the modeling error on validation data, and the interpretability, i.e., the complexity of the final rule base (number of rules). As references, we use the best results previously obtained with the FRM as well as the results obtained by the systematic approach $FRM_{sys}$ presented in [24, 49]. The systematic approach was originally developed to support non-experts by applying the FRM. It is based on some heuristic procedures for the project definition and a deterministic scheme for the adjustment of the strategy parameters of the FRM (Sect 3.2.1). Following the heuristic procedures the rule generation is performed with a maximum combination depth of $c_{max} = 2$. It has been found that, in this way, satisfactory results can be obtained in an acceptable time.

**Table 3.1.** Modeling results (error MAE/RCE on validation data) for the different benchmark problems. The number of rules is indicated in brackets. For the best result, $FRM_{best}$, the corresponding literature is cited.

|  | $FRM_{sys}$ | $FRM_{sys}^{OCR}$ | $FRM_{SSS}$ | $FRM_{SSS}^{OCR}$ | $FRM_{best}$ |
|---|---|---|---|---|---|
| WINE | 11.2 (105) | 10.7 (15) | 11.3 (83) | 9.5 (17) | 6.2 (131) [49] |
| SAT | 18.2 (2683) | 18.2 (1044) | 17.9 (574) | 16.0 (99) | 12.7 (204) [33] |
| MACKEY | 0.07 (59) | 0.05 (20) | 0.07 (46) | 0.05 (21) | 0.01 (92) [21] |
| KIN | 0.22 (1530) | 0.12 (457) | 0.24 (308) | 0.17 (127) | 0.16 (309) [49] |

As shown in Table 3.1, the systematic approach $FRM_{sys}$ can lead to a large rule base, especially if it is applied to the more complex problems SAT

---

[5] The corresponding data sets are available at the UCI Repository of Machine Learning Databases and Data for Evaluating Learning in Valid Experiments:
http://www.ics.uci.edu/~mlearn/MLRepository.html
http://www.cs.utoronto.ca/~delve/data/datasets.html

and KIN. In order to reduce the number of rules, the optimizing conflict reduction presented in Section 3.3.1 is performed in a subsequent step. For the easier problems WINE and MACKEY, the OCR leads to a significant decrease of the number of rules as well as to a slight improvement in accuracy. With 15 and 20 rules, with a maximum combination depth $c_{max} = 2$, the rule bases are easily interpreted. In contrast, the rule bases of the more complex problems SAT and KIN are still large after applying the OCR. Considering the accuracy, the results are different. On the one hand, the modeling error is drastically reduced by the OCR for the KIN problem, yet on the other hand, there is no improvement obtained for the SAT problem. To summarize, with the $FRM_{sys}^{OCR}$ approach, a satisfying compromise between accuracy and interpretability is found for WINE and MACKEY. For KIN, high accuracy is obtained, but with nearly 500 rules, the rule base is still complex and consequently hard to interpret. Finally, the rule base for the SAT problem is neither interpretable nor does it yield high accuracy.

We now consider the specific search on the subspaces with poor accuracy presented in Section 3.3.2. In order to allow a fair comparison, the iterative search concept depicted in Figure 3.6 is also restricted to a maximum combination depth $c_{max} = 2$. The results for this approach $FRM_{SSS}$ are shown in Table 3.1. It has to be pointed out that, for the SAT and KIN problems, the number of rules is considerably smaller than in the $FRM_{sys}$ and even smaller than in the $FRM_{sys}^{OCR}$ approach. Nevertheless, the accuracy is still comparable to the $FRM_{sys}$ approach. These results can be significantly improved by also applying the OCR. In contrast to the $FRM_{sys}^{OCR}$ approach described above, the $FRM_{SSS}^{OCR}$ approach leads to much smaller rule bases for SAT and KIN, with only about 100 rules. From our point of view, a rule base of this size is still interpretable. It should be noted that, in principle, it would be possible to produce the same small rule base with the $FRM_{sys}^{OCR}$ approach. This is because the initial rule base obtained by the $FRM_{sys}$ approach is a superset of the $FRM_{SSS}$ rule base. However, it seems that the high number of rules in the $FRM_{sys}$ rule base makes it too difficult for the OCR to find a small subset with acceptable accuracy. This explanation is in agreement with the results obtained for the easier problems WINE and MACKEY. Due to the smaller initial rule bases for the OCR, the final rule bases of the $FRM_{sys}^{OCR}$ and the $FRM_{SSS}^{OCR}$ approach are comparable.

All together, the results show that, with feedback on the modeling quality, it is possible to greatly improve the interpretability without impairing the accuracy. In particular, for the more complex applications, the accuracy can be improved simultaneously with interpretability. Depending on the complexity of the considered problem, different approaches are recommended. For easier problems with a small initial rule base ($\ll 1000$), it is sufficient to only apply the OCR. For more complex problems, it is promising to first generate a smaller initial rule base with the specific search on subspaces with poor modeling quality (SSS), then in a subsequent step, apply the OCR. A comparison

with the best results $FRM_{best}$ obtained with the Fuzzy-ROSA method lead to the conclusion that a further improvement in the accuracy is possible at the expense of interpretability.

So far, only fuzzy models of the Mamdani type have been considered. As described in Section 3.3.3, for approximation tasks, a conversion in TSK fuzzy models can lead to a much higher accuracy without loss of interpretability. The modeling results for the conversion of the rule bases described above are shown in Table 3.2. Thereby, a multilinear function is used for MACKEY and a linear function is used for KIN. It should be noted that the OCR is applied to the converted fuzzy models $FRM_{sys}$ and $FRM_{SSS}$ in Table 3.2 respectively.[6] Therefore, the number of rules of the OCR approaches $FRM_{sys}^{OCR}$ and $FRM_{SSS}^{OCR}$ in Table 3.2 are different from the approaches in Table 3.1.

**Table 3.2.** Results (error MAE on validation data) after the conversion in TSK fuzzy models. The number of rules after the OCR is indicated in brackets.

| | $FRM_{sys}$ | $FRM_{sys}^{OCR}$ | $FRM_{SSS}$ | $FRM_{SSS}^{OCR}$ |
|---|---|---|---|---|
| MACKEY | 0.009 | 0.006  (19) | 0.011 | 0.007 (16) |
| KIN | 0.19 | 0.12 (359) | 0.15 | 0.10 (62) |

To summarize, the conversion in TSK fuzzy models leads in all approaches considered in Table 3.2 to the highest accuracy yet obtained with the FRM. More importantly, the size of the rule base is almost the smallest. Similar to the results discussed above for the easier MACKEY problem with the $FRM_{sys}^{OCR}$ and the $FRM_{SSS}^{OCR}$ approach, comparable rule bases are obtained. For the more complex problem KIN, the $FRM_{SSS}^{OCR}$ approach also leads to the best results.

## 3.4   Applications

In this Section, we show that the FRM allows us to meet the requirements of real-world applications. Due to space limitations, only three applications in the domains of power station management (Section 3.4.1), classification in quality control (Section 3.4.2), and prediction of contract duration (Section 3.4.3) are described here. They are chosen as examples of fuzzy modeling in high-dimensional search spaces and learning from noisy and contradictory data. Additional successful applications of the FRM, which are not presented here, are:

- semiconductor rare fault detection (Section 8.1.4),

---

[6] The OCR is applied in this way, because a direct conversion of the $FRM_{sys}^{OCR}$ and $FRM_{SSS}^{OCR}$ of Table 3.1 does not achieve the same accuracy.

- stock exchange market analysis for the Greek market index [59],
- access to the creditworthiness of enterprises [6],
- L-dopa dosing in Parkinson's disease pharmacotherapy [8],
- design of a fuzzy controller for robot motion [60],
- automatic generation of fuzzy systems for combustion processes [26],
- data-based generation of a fuzzy controller for an alkoxylation process [30, 42],
- controller adaptation in robotics [35, 48, 55],
- dynamic simulation of a solar power plant [13],
- improvement of a recycling process [28], and
- modeling of an industrial batch process [49].

### 3.4.1    Application to Power Station Management

One main task of an electrical power supply company is to guarantee a safe, adequate, and cost-efficient electrical power supply for all online consumers at all times. This task is supported by the power station management, which calculates the set point values for all remote-controlled generation units under real-time conditions. The power station management also has to meet many contradictory requirements simultaneously. It should work in a fast and dynamically efficient manner to meet specific criteria for optimization to reduce costs with regard to unit-specific limits and to conserve the resources of the units in order to increase their lifetime. Therefore, a hierarchical control environment, known as the automatic generation control module (AGC), typically contains two main modules: load frequency control and economic dispatch.

The optimization module for economic dispatch calculates the set point values for the several power stations by considering the predicted required load, so that all the power required is generated as cost-efficiently as possible. Unavoidable differences between the predicted load and the actual demand for power occur due to the stochastic behavior of the consumers. To prevent a compensating for these load differences per unrequested power exchange over the interconnected grid, the module for load frequency control is quickly activated to modify the set point values of the elected units.

The prerequisite for this control task is sufficient free resources, the power reserve, which has to be kept free by the optimization module. A special difficulty of the process-control task is that the generation systems in question consist of different types of generation units, which differ greatly in their specific production costs and dynamic properties. For example, coal-fired generation units are characterized by low specific production costs compared with gas-fired units. Otherwise, gas-fired units can respond faster to changes of the set point values. A further difficulty lies in adapting the time for powering up or down the generation units to affect the auxiliary power reserve.

Overall, this leads to a very complex optimization problem with at least three main input variables (strategy parameters): the adaptation to the short-time load prediction, the power reserve, and the auxiliary power reserve. In

previous works, in cooperation with an electrical power supply company, we have developed both an advanced optimization and an advanced load-frequency control strategy by means of classical control methods [1, 2, 3, 18]. In our simulation, it was demonstrated that these advanced strategies lead to a considerably more cost-efficient energy production. Consequently, the company placed an order to implement these strategies in a new automatic generation-control module (AGC) [40].

An analysis of this new automatic generation-control module leads to the conclusion that, in principle, a further improvement by means of CI methods is possible. However, it should be noted that, due to the deregulation of the German energy market and the reorganization of the company, our results cannot be tested in practice. Nevertheless, it was found that this real-world task is very useful for a prototype testing and evaluation of data-based modeling techniques. For example, substantial parts of the Chap. 4 arise from this problem.

A promising approach to improving modeling techniques is a dynamic adaptation of the chosen values of the strategy parameters mentioned above, which are currently fixed. In order to learn from good and poor strategy parameter settings [48], we performed a data-based rule generation, as described in [47]. However, the application of these situation-dependent decision rules fails. Further investigations have shown that this failure is mainly caused by differences in the degree of suboptimality of the real-time optimization algorithm that was used. This observation prompted a fundamental analysis and classical redesign of the optimization algorithm so that it is no longer suboptimal but optimal [19].

An alternative approach [47] was based on a direct data-based modeling of decisions from process experts, who were able to adapt the strategy parameters by hand. However, it was found that the modeling is strongly affected by ambiguities and/or contradictories contained in the data, which were caused by different decisions of the different experts. To overcome these problems, we developed new strategies for measuring and coping with ambiguities (Chap. 4).

In the existing system for short-term load prediction, a primary forecast is generated by statistical methods of regression analysis based on historical data consisting of temperature, day of the week, time of day, and the corresponding load. The forecast is adapted to the actual load using a simple exponential smoothing technique. A model of the existing adaptation method was implemented under Matlab/Simulink in a simulation tool for the development and evaluation of the AGC and is used as a reference.

As the prediction error of the existing system shows a structure (Figure 3.9, left), it was a challenging question, whether the accuracy of the prediction may be improved by a data-based modeling based on the past recorded data.

In [9, 10], a fuzzy system was generated using the Fuzzy-ROSA method [28, 42, 49] to predict the difference of the load in 15 minutes from the current load. A central problem is the selection of relevant input variables. Here, expert knowledge was used, and different sets of input variables were investigated. The best obtained fuzzy system, which has a parallel structure, uses four input variables (*time of day*, *type of day*, *season*, and *dawn*). With this system, the mean value of the error between the actual load and the predicted load was reduced to 40.9 MW on the validation data compared to 46.3 MW achieved with the reference system.

To automate the selection of relevant input variables, the method of complexity reduction [43] was applied. As a result, four relevant nonredundant input variables (*current load*, *time of day*, *type of day*, and *season*) were extracted from 17 possibly relevant input variables. Based on these selected input variables, the model accuracy could be slightly improved to a mean absolute error of 40.5 MW.

**Fig. 3.9.** Forecast error of the reference system (left) compared to the fuzzy load prediction module generated by the Fuzzy-ROSA method (right).

For a further improvement, a new approach for feature selection and dimension partitioning (Section 3.2.4) was applied [46]. As a result, a set of three relevant input variables was obtained: *time of day*, *type of day*, and *dawn*. In taking these variables into account, a TSK fuzzy load predictor was designed using the Fuzzy-ROSA method. This leads to a considerable reduction of the prediction error to 39.8 MW. In Figure 3.9, the forecasting error of the reference system and the fuzzy model for the complete year is shown in a contour plot. The error is coded in a greyscale shading given the time of the day (abscissa) and the day of the year (ordinate). Obviously, the reference

model shows a considerable influence of day type and time (horizontal and vertical structures). Moreover, a significant influence of daylight (dawn) can be observed. The fuzzy model generated by the Fuzzy-ROSA method nearly eliminates these influences and consequently reduces the prediction error.

### 3.4.2   Classification of Automatic Gearboxes in Quality Control

In quality control, different parameters are inspected during a test to detect material faults, production faults, assembly faults, or unusual sounds during operation. In our application, automatic gearboxes produced by an automobile manufacturer are tested acoustically by human specialists. With 149 acoustic input characteristics, the design of a fuzzy classifier is a very complex problem (possible rules with maximum combination depth of six). As described in [42, 43, 44, 49, 50, 58], data-based generation of a fuzzy classifier is possible using the FRM based on 1060 data sets (1000 O.K. and 60 not O.K.). Five equidistant membership functions were chosen heuristically for each characteristic. After a 20-hour hybrid evolutionary search and subsequent rule reduction, the resulting rule base (with less than 100 rules and a maximum combination depth of six) achieved a 100 percent correct classification on the learning data. In order to test the system's capability of generalizing classification results, learning was repeated on 90 percent of the data set. The remaining 10 percent were taken as test data. The classification error was 8 percent (5 percent for O.K. and 30 percent for not O.K.).

### 3.4.3   Prediction of the Duration of Insurance Contracts

Insurance companies are interested in contracts with long duration to avoid high administration costs. Therefore, statistical methods are used to analyze the dependency of contract duration on client profiles. Contract duration is the time between the start of the contract and the date of cancellation. Here, the profile of a client is characterized by the following seven socio-demographic characteristics: social status, profession, sex, age, type of first contract, place of residence, and state of residence. Data-based rule generation of a fuzzy system to predict contract duration is a transparent approach, as the premise of a fuzzy rule can be interpreted as a profile of a client. The premise is a fuzzy conjunction of the socio-demographic characteristics. Initial results are presented in [50, 58]. In a 13-minute tree-oriented rules search, a rule base of 6755 rules with a specified minimal rating index was generated and used directly without any subsequent reduction for prediction [49]. The comparisons[7] in Figure 3.10 between real and predicted contract duration

---

[7] If the prediction error is larger than four years, the result is called wrong; if it is between two and four years, it is called acceptable; and if it is between four months and two years, it is called near. An exact prediction means that the error is less than four months.

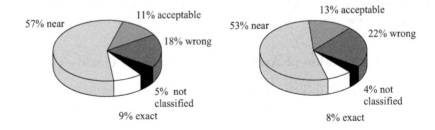

**Fig. 3.10.** Quality of the prediction on learning data sets (left chart) and validation data sets (right chart).

lead to the conclusion that in principle it is possible to learn fuzzy rules for typical client profiles. We are at present designing a fuzzy classifier.

## 3.5   Concluding Remarks

The visionary superordinate target in the domain of fuzzy modeling may be stated as follows: *A systematic data-based generation of interpretable models with a high accuracy in an acceptable time for complex applications.* Although these modeling objectives are partly contradictory, we show in this Chapter the way in which to obtain a good balance by using individually tested rules. As discussed in the first part of this contribution, our approach distinguishes itself by extremely efficient rule-generation methods and locally reasonable rules. Furthermore, a systematic procedure for its application is available. However, this approach may lead to large rule bases and does not achieve the intended accuracy. To overcome these disadvantages, we present in Section 3.3 concepts for feedback of the modeling quality in the rule generation process. It was found that, with these concepts, a data-based generation of small and interpretable rule bases with a high accuracy is possible, even for high-dimensional real-world problems.

## References

1. J. Albrecht. *Vorausschauende optimale Steuer- und Regelstrategien zur Verbesserung der Kraftwerksführung.* Fortschritt-Berichte VDI, Reihe 8, Nr. 616. VDI Verlag, Düsseldorf, 1997.
2. J. Albrecht, K. Albers, and P. Stelzner. Sequentielle vorausschauende Vorsteuerung eines Kraftwerksparks durch ein Führungssystem. *at-Automatisierungstechnik,* 44(8):381–390, 1996.
3. J. Albrecht, H. Kiendl, A. Michalske, K. Albers, and P. Stelzner. *Verfahren zum Regeln der Leistung eines Kraftwerksparks,* 1995. Patent, DE 195 10 342.

4. Th. Bäck, U. Hammel, and H.-P. Schwefel. Evolutionary computation: Comments on the history and current state. *IEEE Transactions on Evolutionary Computation*, 1:1–15, 1997.
5. J.C. Bezdek. *Pattern Recognition with Fuzzy Objective Function Algorithms*. Plenum Press, New York, 1981.
6. S. Blochwitz and J Eigermann. Creation and application of fuzzy rules to access the creditworthiness of enterprises. In *Proceedings of the Fourth European Congress on Intelligent Techniques and Soft Computing (EUFIT '99)*, volume CD-ROM. Verlag Mainz, Aachen, 1999.
7. O. Cordon, A. Gonzalez, F. Herrera, and R. Perez. Encouraging cooperation in the genetic iterative rule learning approach for qualitative modeling. In *Computing with Words in Information/Intelligent Systems*, pages 95–117. Physica-Verlag, Heidelberg, 1999.
8. K. Honczarenko, A. Jardzioch, A. Piegat, and J Honczarenko. Application of soft computing method to Parkinson's disease therapy. In *Proceedings of the Fourth European Congress on Intelligent Techniques and Soft Computing (EUFIT '99)*, volume CD-ROM. Verlag Mainz, Aachen, 1999.
9. H. Jessen. *Test- und Bewertungsverfahren zur regelbasierten Modellierung und Anwendung in der Lastprognose*. Fortschritt-Berichte VDI, Reihe 8, Nr. 836. VDI Verlag, Düsseldorf, 2000.
10. H. Jessen. Test and rating strategies for automatic fuzzy rule generation and application to load prediction. In M. Mohammadian, editor, *New Frontiers in Computational Intelligence and its Applications*, pages 11–21. IOS Press, Amsterdam, 2000.
11. H. Jessen and T Slawinski. Mittelwertbasierter Regeltest und Bewertung für das Fuzzy-ROSA-Verfahren und Anwendung zur Lastprognose. In *Tagungsband des 8. Workshops Fuzzy Control*, number 0298, pages 67–81. VDI/VDE GMA-FA 5.22, Faculty of Electrical Engineering and Information Technology, University of Dortmund, 1998.
12. H. Jessen and T. Slawinski. Test and rating strategies for data-based rule generation. Technical Report CI-39/98, Collaborative Research Center 531, University of Dortmund, 1998.
13. E. Juuso, D. Schauten, T. Slawinski, and H. Kiendl. Combination of linguistic equations and the fuzzy-ROSA method in dynamic simulation of a solar collector field. In *Proceedings of TOOLMET2000 Symposium – Tool Environments and Development Methods for Intelligent Systems*, pages 63–77. Oulun Yliopistopaino, Oulu, Finland, 2000.
14. H. Kiendl. Implizite Modellierung, inkrementeller Relevanzindex und Rauigkeitsmaß: neue Strategieelemente für die datenbasierte Modellierung. In *Tagungsband des 10. Workshop Fuzzy Control des GMA-FA 5.22*, pages 1–14. VDI/VDE GMA-FA 5.22, Research Reports Forschungszentrum Karlsruhe (FZKA 6509), Karlsruhe, 2000.
15. H. Kiendl. Verfahren zur Erzeugung von Stellgrößen am Ausgang eines Fuzzy-Reglers und Fuzzy-Regler hierfür. Technical report, 1994. Patent, DE 43 08 083.
16. H. Kiendl. System of controlling or monitoring processes or industrial plants employing a dual-line fuzzy unit. Technical report, 1998. Patent, U.S. 5,826,251.
17. H. Kiendl. Decision analysis by advanced fuzzy systems. In J. Kacprzyk L. Zadeh, editor, *Computing with Words in Information/Intelligent Systems*, pages 223–242. Physica-Verlag, Heidelberg, 1999.

74     Kiendl et al.

18. H. Kiendl and J. Albrecht. *Verfahren zur sequentiellen Vorsteuerung eines Prozesses*, 1997. Patent, DE 195 10 343.

19. H. Kiendl and W. Hansen. Quadratische Optimierung unter Nebenbedingungen und Anwendung in der Kraftwerksführung. Technical Report CI-119/02, Collaborative Research Center 531, University of Dortmund, 2002.

20. R. Klinkenberg and T. Slawinski. Wissensmanagement in der Computational Intelligence: Systematisierung der Beschreibung von Problemen, Methoden und Methodeneinsätzen. In *Proceedings 11. Workshop Fuzzy Control des GMA-FA 5.22, Dortmund*, pages 78–99. VDI/VDE GMA-FA 5.22, Research Reports Forschungszentrum Karlsruhe (FZKA 6660), Karlsruhe, 2001.

21. P. Krause. Generierung von Takagi-Sugeno-Fuzzy-Systemen aus relevanten Fuzzy-Regeln. In *Proceedings 10. Workshop Fuzzy Control des GMA-FA 5.22, Dortmund*, pages 84–97. VDI/VDE GMA-FA 5.22, Research Reports Forschungszentrum Karlsruhe (FZKA 6509), Karlsruhe, 2000.

22. P. Krause. *Datenbasierte Generierung von transparenten und genauen Fuzzy-Modellen für mehrdeutige Daten und komplexe Systeme*. Fortschritt-Berichte VDI, Reihe 10, Nr. 691. VDI Verlag, Düsseldorf, 2001.

23. P. Krause, A. Krone, and T. Slawinski. Fuzzy System Identification by Generating and Evolutionary Optimizing Fuzzy Rule Bases Consisting of Relevant Fuzzy Rules. Technical Report CI-84/00, Collaborative Research Center 531, University of Dortmund, 2000.

24. P. Krause and T. Slawinski. Das Fuzzy-ROSA-Verfahren: Von der regelorientierten statistischen Analyse zur datenbasierten Generierung von interpretierbaren Takagi-Sugeno-Systemen. *at-Automatisierungstechnik*, 9(49):391–399, 2001.

25. P Krause, D. Wiesmann, and T. Slawinski. Parallel evolutionary algorithms for optimizing data-based generated fuzzy systems. Technical Report CI-101/00, Collaborative Research Center 531, University of Dortmund, 2000.

26. T. Kretschmer, T. Born, C Gierend, and M. Born. Automatische Generierung von Fuzzy-Systemen in Verbrennungsprozessen. In *Tagungsband 5tes Nationales DataAnalysis Symposium*, pages 45–50. MIT GmbH, Aachen, 1999.

27. A. Krone. Advanced rule reduction concepts for optimizing efficiency of knowledge extraction. In *Proceedings of the Fourth European Congress on Intelligent Techniques and Soft Computing (EUFIT '96)*, volume 2, pages 919–923. Verlag Mainz, Aachen, 1996.

28. A. Krone. *Datenbasierte Generierung von relevanten Fuzzy-Regeln zur Modellierung von Prozesszusammenhängen und Bedienstrategien*. Fortschritt-Berichte VDI, Reihe 10, Nr. 615. VDI Verlag, Düsseldorf, 1999.

29. A. Krone, T. Bäck, and P. Teuber. Evolutionäres Suchkonzept zum Aufstellen signifikanter Fuzzy-Regeln. *at-Automatisierungstechnik*, 44(8):405–411, 1996.

30. A. Krone, C. Frenck, and O. Russak. Design of a fuzzy controller for an alkoxylation process using the ROSA method for automatic rule generation. In *Proceedings of the Third European Congress on Intelligent Techniques and Soft Computing (EUFIT '95)*, volume 2, pages 760–764. Verlag Mainz, Aachen, 1995.

31. A. Krone and H. Kiendl. Automatic generation of positive and negative rules for two-way fuzzy controllers. In *Proceedings of the Second European Congress on Intelligent Techniques and Soft Computing (EUFIT '94)*, volume 1, pages 438–447. Verlag Mainz, Aachen, 1994.

32. A. Krone and H. Kiendl. *WINROSA von Daten zu Regeln.* MIT-Managment Intelligenter Technologien GmbH, Aachen, 1st edition, 1997. Manual.
33. A. Krone, P. Krause, and T. Slawinski. A new rule reduction method for finding interpretable and small rule bases in high dimensional search spaces. In *Proceedings of the Ninth IEEE International Conference on Fuzzy Systems, (FUZZ-IEEE '00), San Antonio, TX,* volume 2, pages 696–699. IEEE Press, Piscataway, NJ, 2000.
34. A. Krone, P. Krause, T. Slawinski, and R. Knicker. WINROSA 2.0 and DORA for Windows 6.2. In *Symposium on System Identification (SYSID 2000), Santa Barbara, CA,* pages 521–526. Elsevier Science, Amsterdam, 2000.
35. A. Krone and U. Schwane. Generating fuzzy rules from contradictory data of different control strategies and control performances. In *Proceedings of the Fifth IEEE International Conference on Fuzzy Systems (FUZZ-IEEE '96), New Orleans,* volume 1, pages 492–497. IEEE Press, Piscataway, NJ, 1996.
36. A. Krone and T. Slawinski. Data-based extraction of unidimensional fuzzy sets for fuzzy rule generation. In *Proceedings of the Seventh IEEE International Conference on Fuzzy Systems (FUZZ-IEEE '98), Anchorage, AK,* volume 2, pages 1032–1037. IEEE Press, Piscataway, NJ, 1998.
37. A. Krone and T. Slawinski. A distance measure for the mutation of fuzzy-rules. Technical Report CI-83/00, Collaborative Research Center 531, University of Dortmund, 2000.
38. A. Krone and H. Taeger. Relevance test for fuzzy rules. Technical Report CI-40/98, Collaborative Research Center 531, University of Dortmund, 1998.
39. A. Krone and H. Taeger. Data-based fuzzy rule test for fuzzy modelling. *Fuzzy Sets and Systems,* 126:343–358, 2001.
40. K. Linke. Kraftwerksführungssystem der Verbundebene – Struktur, Entwicklungsstand, neue Konzepte. In *FGE-Tagung 1996,* pages 618–624. Energiewirtschaft und Technik Verlagsgesellschaft mbH, Essen, Energiewirtschaftliche Tagesfragen, Heft 10/1996, 1996.
41. W. Pedrycz. *Fuzzy Control and Fuzzy Systems.* Research Studies LTD, Taunton, UK, 1993.
42. B. M. Pfeiffer, J. Jäkel, A. Kroll, C. Kuhn, H. B. Kuntze, B. Lehmann, T. Slawinski, and V. Tews. Tutorial: Erfolgreiche Anwendungen Fuzzy Control. In *Proceedings 11. Workshop Fuzzy Control des GMA-FA 5.22, Dortmund,* pages 1–27. VDI/VDE GMA-FA 5.22, Research Reports Forschungszentrum Karlsruhe (FZKA 6660), Karlsruhe, 2001.
43. J. Praczyk. *Entwurf von Fuzzy-Gütemaßen zur Prozeßbewertung.* Fortschritt-Berichte VDI, Reihe 8, Nr. 796. VDI Verlag, Düsseldorf, 1999.
44. J. Praczyk, H. Kiendl, and T. Slawinski. Finding relevant process characteristics with a method for data-based complexity reduction. In B. Reusch, editor, *Computational Intelligence: Theory and Applications (Proceedings of 6th Fuzzy-Days),* pages 548–555. Springer, Berlin, 1999.
45. D. Schauten, B. Nicolaus, and H. Kiendl. Evolutionäres Verfahren zur Selektion relevanter Merkmalssätze für die datenbasierte Fuzzy-Modellierung. In *Proceedings 11. Workshop Fuzzy Control des GMA-FA 5.22, Dortmund,* pages 133–147. VDI/VDE GMA-FA 5.22, Research Reports Forschungszentrum Karlsruhe (FZKA 6660), Karlsruhe, 2001.
46. D. Schauten, B. Nicolaus, and H. Kiendl. An evolutionary concept for selecting relevant sets of input variables for data-based fuzzy modeling. In *Proceedings of*

the European Symposium on Intelligent Technologies, Hybrid Systems and their Implementation on Smart Adaptive Systems (EUNITE '01), volume CD-ROM. Verlag Mainz, Aachen, 2001.

47. D. Schauten, T. Slawinski, and H. Kiendl. Datenbasierte Generierung von situationsbezogenen Entscheidungsregeln für die Kraftwerksführung. Technical Report CI-110/01, Collaborative Research Center 531, University of Dortmund, 2001.

48. U. Schwane. Datenbasierte Generierung von Adaptionsregeln und Anwendung zur Erhöhung der Bahngenauigkeit eines Industrieroboters. Fortschritt-Berichte VDI, Reihe 8, Nr. 748. VDI Verlag, Düsseldorf, 1999.

49. T. Slawinski. Analyse und effiziente Generierung von relevanten Fuzzy Regeln in hochdimensionalen Suchräumen. Fortschritt-Berichte VDI, Reihe 10, Nr. 686. VDI Verlag, Düsseldorf, 2001.

50. T. Slawinski, H. Jessen, J. Praczyk, P. Krause, A. Krone, and H. Kiendl. Einsatz der datenbasierten Fuzzy-Modellierungen für komplexe Anwendungen. In Computational Intelligence im industriellen Einsatz, Tagung Baden-Baden, VDI-Berichte, Nr. 1526, pages 119–124. VDI/VDE GMA und GI, VDI-Verlag, Düsseldorf, 2000.

51. T. Slawinski, P. Krause, and H. Kiendl. Using individually tested rules for the data-based generation of interpretable rule bases with high accuracy. In J. Casillas, O. Cordón, F. Herrera, and L. Magdalena, editors, Trade-off Between Acuracy and Interpretability in Fuzzy Rule-Based Modeling, page (Accepted). Physica-Verlag, Heidelberg, 2002.

52. T. Slawinski, P. Krause, A. Krone, U. Hammel, and D. Wiesmann. Fuzzy Adaption der evolutionären Regelsuche im Fuzzy-ROSA-Verfahren. In Tagungsband des 9. Workshops Fuzzy Control des GMA-FA 5.22, Dortmund, number 0499, pages 14–39. VDI/VDE GMA-FA 5.22, Faculty of Electrical Engineering and Information Technology, University of Dortmund, 1999.

53. T. Slawinski, A. Krone, U. Hammel, D. Wiesmann, and P. Krause. A hybrid evolutionary search concept for data-based generation of relevant fuzzy rules in high dimensional spaces. In Proceedings of IEEE International Conference on Fuzzy Systems (FUZZ-IEEE '99) Seoul, volume 3, pages 1432–1437. IEEE Press, Piscataway, NJ, 1999.

54. T. Slawinski, A. Krone, U. Hammel, D. Wiesmann, and M. Lindenblatt. A hybrid evolutionary search concept for data-based generation of relevant fuzzy rules in high dimensional spaces (extended version). Technical Report CI-50/98, Collaborative Research Center 531, University of Dortmund, 1998.

55. T. Slawinski, A. Krone, and H. Kiendl. Automatisierung durch datenbasierte Fuzzy-Modellierung von Prozessbedienern. In Computational Intelligence: neuronale Netze, evolutionäre Algorithmen, Fuzzy-Control im industriellen Einsatz, Tagung Berlin, 1998, VDI-Berichte, Nr. 1381, pages 203–219. VDI/VDE GMA, VDI-Verlag, Düsseldorf, 1998.

56. T. Slawinski, A. Krone, and P. Krause. Efficient design of a complete rule search in sparsely populated search spaces. Technical Report CI-77/99, Collaborative Research Center 531, University of Dortmund, 1999.

57. T. Slawinski, A. Krone, P. Krause, and H. Kiendl. The fuzzy-ROSA method: A statistically motivated fuzzy approach for data-based generation of small interpretable rule bases in high-dimensional search spaces. In M. Last, A. Kan-

del, and H. Bunke, editors, *Data Mining and Computational Intelligence*, pages 141–166. Physica-Verlag, Heidelberg, 2001.

58. T. Slawinski, J. Praczyk, U. Schwane, A. Krone, and H. Kiendl. Data-based generation of fuzzy rules for classification, prediction and control with the fuzzy-ROSA method. In *European Control Congress (ECC '99), Karlsruhe*, volume CD-ROM. Karlsruhe, 1999.

59. N. Thomaidis, G. Dounias, and Tselentis. Stock exchange market analysis using data engine and WINROSA. In *Proceedings of the Fourth European Congress on Intelligent Techniques and Soft Computing (EUFIT '99)*, volume CD-ROM. Verlag Mainz, Aachen, 1999.

60. G. Vachkov. Fuzzy modeling examples by WINROSA. In *Proceedings of the 2nd DataAnalysis Symposium*. MIT GmbH, Aachen, 1998. Artikel Nr. 25299.

61. N. Xion. *Designing Compact and Comprehensible Fuzzy Controllers Using Genetic Algorithms (Entwurf kompakter und interpretierbarer Fuzzy Controller mittels Genetischer Algorithmen)*. Berichte aus der Automatisierungstechnik. Shaker Verlag, Aachen, 2001.

# 4 Fuzzy Modeling to Cope with Ambiguities

Harro Kiendl, Peter Krause, Daniel Schauten, and Timo Slawinski

University of Dortmund, Faculty of Electrical Engineering and Information
Technology, Control Engineering, 44221 Dortmund, Germany
{kiendl, krause, schauten, slawinski}@esr.e-technik.uni-dortmund.de

**Summary.** For the design and management of complex technical systems, data-based fuzzy modeling methods such as the fuzzy rule-oriented statistical analysis (Fuzzy-ROSA) method, have often proved to be useful. However, they may encounter problems if the recorded data on which the modeling is based contains ambiguities or contradictions. We present various approaches that aim at overcoming these problems. They allow the rating and reduction of the roughness of the data and the resulting model behavior. They serve to reduce ambiguities using negative rules and advanced defuzzification strategies. We also present a method for an ambiguity-based optimization of membership functions, both the selection and construction of the sets of relevant and nonredundant input variables. Finally, the implicit modeling approach allows decisions on how the ambiguities in the data can be treated after the actual modeling has been performed. The potential of these approaches is demonstrated partly by illustrative examples and partly by real-world applications.

## 4.1 Introduction

Data-based modeling of processes may encounter problems if the recorded process data show ambiguities and/or contradictions. As the sources of these ambiguities may be quite diverse, and since the modeling issues may differ greatly, we require a broad repertoire of strategies to overcome these problems in data-based fuzzy modeling. Here, we describe the most recent progress in methods used to handle ambiguities.

To illustrate the different modeling issues, we consider a process with one input variable $x$ and one output variable $y$. Let the data, which are depicted in Figure 4.1 (left), be recorded from the process as learning data $(x_1, y_1), ..., (x_m, y_m)$ for a data-based modeling method, where each pair $(x_i, y_i)$ describes the corresponding values of the input and output variable. In this case, visual inspection of the data suggests that there are no ambiguities in the data, so we consider a model behavior as depicted in Figure 4.1 (right) to be adequate. In contrast to this, the learning data depicted in Figure 4.2 (a) show ambiguities/contradictions. Depending on the assumption about the source of the ambiguities/contradictions, we will consider different models as appropriate.

**Fig. 4.1.** Learning data without ambiguities (left), and behavior of a corresponding adequate model (right).

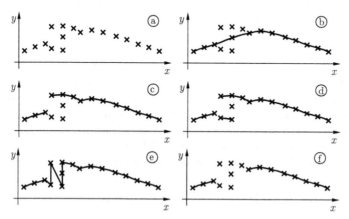

**Fig. 4.2.** Unclear learning data (**a**), and behavior of three different models (**b**, **c**, **d**, **e**, **f**) that correspond to different interpretations of the lack of clarity and to different modeling issues.

We may assume that the data points are affected by disturbances, such as measurement errors. In this case a model's behavior, according to Figure 4.2 (b), is adequate. We may assume that the modeled process itself, a human process operator, for instance, behaves ambiguously. In this case, only one of the different operator actions in the same input situation may be correct and the others wrong. Then a model's behavior, according to Figure 4.2 (c), may be adequate. Alternatively, different operator actions performed in the same input situation may be correct, so a model, according to Figure 4.2 (d), is adequate. Finally, the underlying process may behave clearly, but may show strong variations of the output value for very small variations of the input value. In this case, a model behavior as depicted in Figure 4.2 (e) may be adequate.

Furthermore, the required model properties essentially depend on the intended use of the model: we may intend to use the model for a process observation in order to predict the values of the output variables for new input situations. In this case, we want the correct model output values, whether they show a causal relationship with respect to the input values or not. Alternatively, we may use the model to influence the process output values by suitable settings of its input values. In this case, we need a model that

describes statistically significant relationships between the output and input values of the process. It may happen that such relationships cannot be established for all possible input situations. Then we have to accept a model, such as shown in Figure 4.2 (f), which does not supply an output value for any input value of interest.

This chapter is organized as follows: In Section 4.2, we present a method for rating and reducing ambiguities, which can be applied both to the learning data and to the resulting model, whether or not fuzzy modeling or any other modeling is performed. In Sections 4.3, 4.4, and 4.5, we present different strategies to cope with ambiguities in fuzzy modeling. These strategies address the use of negative rules, the defuzzification, and the predesign, that is the choice of membership functions and selection of relevant input variables.

In the last section, Section 4.6, the implicit modeling approach is described. This approach differs fundamentally from the usual data-based modeling approaches in so far as possible ambiguities in the data are preserved in the resulting model and resolved after the so-called implicit model has been generated. This approach makes use of defuzzification, which need not be fuzzy, i.e., it can be any other type of decision making.

## 4.2   Rating and Reduction of Ambiguities

### 4.2.1   Motivation

Visual inspection of the learning data of a process with only one input and one output variable reveals whether there are outliers which should be ignored and whether the relationship between input and output values is good-natured (smooth) or more bizarre (rough). From this, indications of the difficulty of the modeling and the choice of an adequate model structure can be derived. In addition, visual inspection of the behavior of an obtained model in view of the learning data, reveals whether the smoothness or roughness of the model behavior corresponds to the assumptions about the nature of the process to be modeled. This allows us to judge whether the model is satisfactory for its intended purpose. If, for instance, we design a controller by modeling the behavior of a human operator, it is undesirable for small variations of the input values of the controller to cause great changes of its output value. In addition, if we design a fuzzy system for rating the performance of a technical system by modeling the rating behavior of human experts, it may be undesirable for a monotonic variation of an input value to lead to a nonmonotonic variation of the output value (theoretical aspects of monotonicity are discussed in Section 2.3). This would be disadvantageous if the rating system were to be used as a criterion for optimization purposes.

### 4.2.2   Roughness Measure

The above considerations show that we need a measure that allows us to rate the roughness of the learning data and of the resulting model for the general

$n$-dimensional case. For this, we consider the set of $m$ learning data points $z_j = (x_j, y_j)$, $j = 1, 2, \ldots, m$, where each $x_j$ is an $n$-dimensional vector, which comprises the values of $n$ input variables of the process, and $y_j$ is the corresponding output value. We determine for each point $z_j$ a set of neighbors $S_j$ consisting of the $n + 1$ nearest neighbors $z_{j,1}, z_{j,2}, \ldots, z_{j,n}, z_{j,n+1}$ in the input space (Figure 4.3). From each $S_j$, we derive the $n+1$ subsets $S_{j,i} \subset S_j$, each consisting of $n$ different neighbors. Each subset $S_{j,i}$ defines a hyperplane $H_{j,i}$, which intersects all points $z_j$ that correspond to the points of $S_{j,i}$. We determine all normal vectors $h_{j,i}$ that are perpendicular to the hyperplanes $H_{j,i}$ by solving a system of linear equations. To allow for an ill-conditioned system of equations affecting the reliability of the obtained $h_{j,i}$, we attach to each vector $h_{j,i}$ a suitable weighting factor. Based on the weighted vectors $h_{j,i}$, the average normal vector $\overline{h}_j$ and the average angle $\Delta \varphi$ of all angles between $h_{j,i}$ and $\overline{h}_j$ with $1 \leq i \leq n + 1$ are determined. The value $\Delta \varphi$ is taken as a measure for rating the local roughness [10]. Because the range of input values to be considered as local may depend on the application, we consider the condition that the $n + 1$ nearest neighbors of $x_j$ lie outside the sphere $|x_j - x| \leq r$, where $r$ is a design parameter.

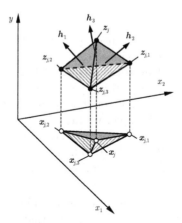

**Fig. 4.3.** Definition of the roughness measure.

This roughness measure allows analysis of the learning data and the model behavior in the $n$-dimensional case. Moreover, we use it for smoothing the learning data: we modify the output values $y_j$ of all data points $z_j$, where the local roughness exceeds a certain threshold so that the local roughness is reduced slightly. This smoothing procedure is repeated until the roughness has reached a sufficiently small value.

### 4.2.3    Applications

One application involving the roughness measure was to model the decisions of two process experts concerning the settings of three strategy parameters for the control of a complex technical process that depended on the values of 20 process variables. With the roughness measure, we detected that the decisions of expert 2 were much rougher compared with those of expert 1. Consequently, the decisions of the latter were considered to be more reliable and were used for data-based modeling using the Fuzzy-ROSA method (Chapter 3). The resulting model showed small modeling errors and less roughness than the learning data. This was essential for the acceptance of the model [37].

Another task called for the optimization of the settings of approximately 20 controller parameters of a position-controlled valve. The established practice was for experts to do this by hand because the rating of the overall system performance required consideration of many different factors. To automate the optimization, we designed a fuzzy rating system by modeling the rating behavior of the experts using the Fuzzy-ROSA method. The fuzzy model obtained fitted the learning data well but showed more roughness than was acceptable.

This result highlights an essential problem that comes with modeling based on the generation of individual rules. This approach reduces the complexity of the design problem drastically [21, 24, 27, 28, 39, 40]. However it may require subsequent model tuning to obtain the desired model properties, such as good accuracy and small roughness, which result from the cooperation of several rules. For example, we may tune the membership functions involved and/or the certainty factors of the rules. The accuracy has been successfully achieved [26, 29], and the roughness measure that was developed to supply a suitable optimization criterion.

However, in the above example of the hydraulic valve, we took a more direct approach: we developed a fuzzy structure that guarantees certain monotonic properties of the input/output behavior and yet provided sufficient flexibility to be fitted to the learning data. In that fashion, we obtained a satisfactory rating system [1, 18, 32, 38].

We consider the systematic investigation of this procedure to be promising, especially as regards how to set up a model structure that guarantees certain qualitative desired input/output properties. These properties may reflect qualitative knowledge of the behavior of the process to be modeled that is available or may be required for the model while providing sufficient flexibility to fit the model to the data.

## 4.3    Reduction of Ambiguities by Negative Rules

### 4.3.1    Motivation

A conventional Mamdani fuzzy system allows processing of only positive rules in the form of

$$\text{IF } < premise > \text{ THEN } < y = L_i > . \tag{4.1}$$

Such a rule recommends the linguistic output value $L_i$, which is modeled by a membership function $\mu_i(y)$, if the rule premise is met. Consequently, a more precise notation of the conclusion of the rule (4.1) would be $< y = L_i \text{ is recommended} >$.

The more that rules are activated, the more extensive is the support of the resulting overall output membership function $\mu(y)$, which in general represents the recommendations of all the activated rules.

The two-way fuzzy system [11, 14, 16, 17] can process both positive rules (4.1) and negative rules in the form of

$$\text{IF } < premise > \text{ THEN } < y = L_i \text{ is forbidden} > . \tag{4.2}$$

The superposition of the recommendations of all activated positive rules supplies the membership function $\mu^+(y)$, and the superposition of the warnings/vetoes of all activated negative rules supplies the membership function $\mu^-(y)$. Considering all recommendations and all warnings, a hyperinference strategy such as

$$\mu(y) = \mu^+(y) \wedge \neg \mu^-(y) \tag{4.3}$$

produces the resulting membership function from which the hyperdefuzzification calculates the crisp output value. The contribution of the negative rules often results in the support of $\mu(y)$ being considerably smaller than that of $\mu^+(y)$. This reduces the ambiguity with respect to the determination of the crisp value.

The Fuzzy-ROSA method generates both positive and negative rules. However, the existing strategies for the acceptance and rating of rules, although they have proven to be successful in many applications [41], are not satisfactory in all situations, especially with regard to the negative rules. This motivates the test and rating strategies proposed below.

### 4.3.2    Existing Test and Rating Strategies

The Fuzzy-ROSA method (Chapter 3) aims at generating IF-THEN rules from the learning data in the form of $m$ data points $z_j = (x_j, y_j)$. Having chosen certain linguistic output values $L_i$ (modeled by membership functions $\mu_i(y)$), the central idea is to specify suitable hypotheses (potential rules) in

the form of (4.1) and to investigate which of these hypotheses pass a certain test, considering the data. Each hypothesis for which this applies is adopted as a relevant rule $R_j$, and receives a rating index $\sigma_j$ that expresses how well the test is passed. All adopted rules $R_j$ are collected to form the rule base of the fuzzy model and are processed with the rating index $\sigma_j$ serving as the certainty factor.

The existing test and rating strategies of the Fuzzy-ROSA method [9, 28, 40] (Section 3.2.1) are based on the following parameters derived from the learning data (Figure 4.4). These are the relative frequency $\hat{p}(y = L_i)$ of the conclusion of the hypothesis to be fulfilled, and the confidence interval $I(y = L_i)$, which estimates the unknown probability $p(y = L_i)$ of the conclusion of the hypothesis being fulfilled. These parameters and the resulting rating indices are well-motivated both for the fuzzy and the crisp case; however, for the latter case they are more transparent. Additionally, the parameters used are the relative frequency $\hat{p}(y = L_i|P)$ of the fulfillment of the conclusion, given that the premise $P$ of the hypothesis is met, and a confidence interval $I(y = L_i|P)$ that estimates the unknown probability $p(y = L_i|P)$ of the fulfillment of the conclusion, given that the premise $P$ is met.

A first test strategy [27, 30] evaluates whether the two confidence intervals $I(y = L_i)$ and $I(y = L_i|P)$ are overlapping. If this is not the case, the unknown probabilities $p(y = L_i)$ and $p(y = L_i|P)$ are considered to differ significantly so that the tested hypothesis is adopted as a relevant rule. The case $\hat{p}(y = L_i|P) > \hat{p}(y = L_i)$ is considered as a positive rule which recommends $y = L_i$ as a well-motivated action or decision if the premise of the rule is met (Figure 4.4, left). The case $\hat{p}(y = L_i|P) < \hat{p}(y = L_i)$ is considered a negative rule that warns against the action or decision $y = L_i$ if the rule premise is met (Figure 4.4, right). For the rating of the relevance of the rule, the suitably normalized distance of the confidence intervals is used (relevance index). This test strategy aims at finding statistically relevant causal dependencies between input and output values, and finds in a natural way both positive and negative rules. However, it is not free of drawbacks: if the confidence intervals overlap and $\hat{p}(y = L_i)$ is large (say 0.8), then the hypothesis is not adopted as a relevant rule, even though it has a high hit rate (i.e., the conclusion is often met). If the confidence intervals are not overlapping and $\hat{p}(y = L_i)$ is large (say 0.95), and $\hat{p}(y = L_i|P)$ is only slightly smaller (say 0.8), then the hypothesis is adopted as a negative rule, although the conclusion is mostly fulfilled if the premise is met.

A second family of test and rating strategies [8, 9] is based on the hit rate. If $\hat{p}(y = L_i|P) > 0.5$ is valid (or if the left corner point of the confidence interval $I(y = L_i|P)$ is greater than 0.5), then the hypothesis is adopted as a positive rule. If $\hat{p}(y = L_i|P) < 0.5$ (or if the right corner point of $I(y = L_i|P)$ is smaller than 0.5), then the hypothesis is adopted as a negative rule. The suitably normalized difference between 0.5 and $\hat{p}(y = L_i|P)$ (or between 0.5 and the nearest corner point of the interval $I(y = L_i|P)$ ) is taken as the

**Fig. 4.4.** Data-based generation of positive and negative rules (left and right, respectively).

rating index (hit rate or confident hit rate). This strategy aims at finding successful rules whether or not the success is due to causal relationships. However, the criterion "< 0.5" is questionable for the acceptance of negative rules: assume, for example, that three linguistic output values $L_1, L_2$, and $L_3$ are provided and that the output values $y_i$ of the data are distributed equally over $L_i$. In this case, for each conclusion $\hat{p}(y = L_i) = 0.33$ is valid. Consequently, the criterion "> 0.5" for the acceptance of positive rules makes sense because the threshold 0.5 is considerably larger than 0.33. However, the criterion "< 0.5" is questionable for the acceptance of negative rules because this will lead to many weakly supported negative rules.

A third test-and-rating strategy [6, 7, 8] compares the mean value of the output value of all learning data (unconstrained mean value) with the corresponding mean value that refers to only those learning data that meet the premise (constrained mean value). If the difference between these mean values is statistically significant, then the hypothesis is adopted as a relevant positive rule that recommends the constrained mean values as output values. This strategy for highly ambiguous learning data can be handled: it resolves the ambiguities by supplying a compromise output value. However, this is not adequate in all applications. Moreover, negative rules are not generated.

### 4.3.3  Universal Rules and Incremental Test and Rating Strategy

The above analysis motivates the following approach [10]: We introduce for each of the $r$ linguistic output values provided, $L_i$ rules in the form of

$$\text{IF} \ < 1 = 1 > \ \text{THEN} \ < y = L_i > . \tag{4.4}$$

We call these universal rules because their premises are always met. In addition, for the modeling of the $L_i$ we provide a system of membership functions $\mu_i(y)$ so that the degree of membership for any value $y$ sums to unity (Figure 4.5). To each of the rules (4.4) we attach the rating index

$$\rho_i = \lambda \frac{\frac{F}{F_i}\hat{p}(y = L_i)}{\sum_{i=1}^{r} \frac{F}{F_i}\hat{p}(y = L_i)}, \tag{4.5}$$

which is used as the certainty factor for processing of the rules. Here, $F_i$ is the area below the graph of $\mu_i(y)$, $F$ is the sum of all $F_i$, and $\lambda$ is a design

parameter which allows modification of the influence of the universal rules. To understand the effect of the universal rules, we assume that these are

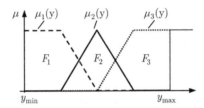

**Fig. 4.5.** Membership functions $\mu_i(y)$ which model the linguistic values $L_i$.

the only rules provided and that $\lambda = 1$ holds. In this case, the weighted superposition of all corresponding membership functions $\mu_i(y)$ produces the output membership function

$$\mu_{universal}(y) = \bigvee_{i=1}^{r} \rho_i \mu_i(y). \tag{4.6}$$

Choosing the sum for the operator $\vee$ and the COG defuzzification, we obtain the unconstrained mean value of the output values of the learning data as the output value.

In addition to the above-mentioned rules, we now consider the usual hypotheses of the form (4.1) with input value dependent premises. These are rated according to the difference $\sigma_i$ between the relative frequencies $\hat{p}(y = L_i|P)$ and $\hat{p}(y = L_i)$ (or between the nearest end point of the confidence interval $I(y = L_i|P)$ and $I(y = L_i)$). Depending on whether or not $\sigma_i > 0$ or $\sigma_i < 0$ is valid, the hypothesis is adopted as a positive or a negative rule. As $\sigma_i$ determines to what extent the premise of rule (4.1) is fulfilled, it modifies the influence of the universal rule (4.3), and we denote $\sigma_i$ as an incremental rating index. Obviously, this index prefers neither negative nor positive rules.

The generated positive and negative rules (4.1) and (4.3) are processed by a two-way fuzzy system. By adjusting the parameter $\lambda$, we can influence the model behavior as to the desired relationships of hit rate or causal relationship between input and output values.

### 4.3.4   Example of Application

Figure 4.6 shows an ambiguous distribution of learning data $z_j = \{x_j, y_j\}$ in the $x$–$y$–space. For 70 percent of the data $y_j \approx 0$ is valid, while for the remaining 30 percent of data $y_j \approx -0.5 + x_j$ holds. The mean value $\bar{y}$ of all values $y_i$ is given by $\bar{y} = 0$. The above strategy supplies both positive and negative rules, both in the form of universal and standard rules. The figure

shows that the resulting model allows adjustment to the desired modeling objective, after the model has been established, by tuning the value of $\lambda$.

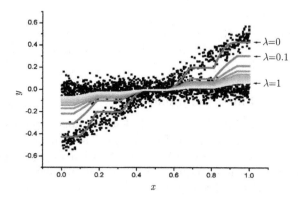

**Fig. 4.6.** Learning data (partially hidden) and behavior of the model, obtained by data-based generation of universal rules and of standard rules with the incremental test and rating strategy, for the values $\lambda = 0, 0.1, 0.2, \ldots, 1$.

## 4.4  Resolution of Ambiguities by Defuzzification

### 4.4.1  Motivation

The output membership function $\mu(y)$ of a conventional Mamdani fuzzy system is ambiguous as it usually recommends different possible output values (see Section 2.2). The task of the defuzzification unit is to resolve such ambiguities and to decide the best motivated output value.

### 4.4.2  COG, MAX and TOR Defuzzification

The most frequently used defuzzification strategies are the maximum (MAX) and the center of gravity (COG) defuzzification. However, in some applications, if the rule premises do not consider all relevant influencing factors, we have to resolve the ambiguities of the output membership function according to the principle that equidirectional recommendations of different rules must be superimposed so as to amplify one another [14, 15, 17, 18]. That is, if the linguistic values *positive small* and *positive large* are recommended by two rules, an output value may be desired that corresponds to the linguistic value *positive very large*.[1] The two standard defuzzification strategies above

---

[1] We realized the need for this type of superposition while designing a fuzzy system for interpreting music (compare [20]).

do not allow realization of this. However, the recently introduced defuzzification strategy

$$y_{TOR} = c \int_{y_{min}}^{y_{max}} \mu(y)y\,dy \qquad (4.7)$$

supplies the property [13, 17]: note, that for $c = 1$ the resulting value $y_{TOR}$ may be interpreted as the torque with respect to the neutral point $y = 0$ induced by the area below the graph of $\mu(y)$. Therefore, we speak of torque (TOR) defuzzification.

### 4.4.3   Inference Filter Defuzzification

Another recent defuzzification approach does not follow the usual interpretation of the output membership function $\mu(y)$ as being the attractiveness function (in this case only the MAX defuzzification would be justified). Instead of this, the idea is introduced that the real attractiveness function is evaluated from $\mu(y)$ by a prescription

$$\hat{\mu}(y) = \mathcal{F}\{\mu(y)\}, \qquad (4.8)$$

called inference filter. It is based on the idea, that the attractiveness of an output value $y_i$ depends not only on $\mu(y_i)$, but also on the values $\mu(y_j)$ for the neighboring points $y_j \neq y_i$ [12, 13, 14, 17]. This leads to the fuzzy system shown in Figure 4.7.

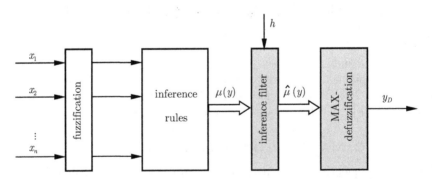

**Fig. 4.7.** Fuzzy system with inference filter and subsequent MAX defuzzification. The thin arrows designate real values and crisp membership degrees, respectively. The thick arrows designate membership functions.

### 4.4.4   Convolution-Based Filtering

Choosing the convolution integral

$$\hat{\mu}(y) = \int_{y_{min}}^{y_{max}} \mu(y')h(y - y')\,dy' \qquad (4.9)$$

as an inference filter, where the functional values of the so-called distance function $h(y - y')$ decrease with increasing distance $|y - y'|$ in a quadratic manner, linear manner, or as a $\delta$-function, the fuzzy system behaves as a conventional fuzzy system with COG, mean of area (MOA) or MAX defuzzification, respectively. Providing a distance function $h_\gamma(y-y')$ depending on a parameter $\gamma$, we can reproduce the above classical defuzzification strategies. Moreover, we obtain a continuum of additional options [14, 17]: by tuning the value of $\gamma$ we obtain a performance that is never worse than the above classical defuzzification strategies, but is often considerably better [34]. In the case of output membership functions in the form of singletons the integral (4.9) is replaced by a sum [14].

### 4.4.5 Pattern-Based Filtering

The above defuzzification strategies do not correspond in all situations to the way human experts resolve ambiguities of the output membership function.

Figure 4.8 shows four different output membership functions in the form of singletons. In cases (a) and (b), the output value $y_H$ of a human expert's decision will correspond to $y_{COG}$. However, in cases (c) and (d) an expert may decide differently. In this case, let the expert assume that the extension of the singleton distribution is not considerably smaller than extension $|y_{max} - y_{min}|$ of the output value range or even compared with the desired, or required, accuracy of the output value. Under this assumption, with regard to the scales that are involved, the expert will ignore the weakly activated isolated singletons and form the output $y_H$ by applying COG defuzzification to the remaining singletons. Obviously, such decision making does not correspond, in all situations, to the COG and MAX defuzzification, respectively. Moreover, the above convolution-based filtering, using any fixed chosen distance function, does not supply the value $y_H$ for all situations. The reason is that the convolution-based filtering considers individual activations and distances, whereas the expert's decision is based on patterns of activations and distances.

These considerations motivate the following pattern-based filtering approach [22, 25]. Let $\mu(y)$ consist of $m$ singletons

$$\mu_i(y) = \begin{cases} \mu_i & \text{,if } y = y_i \\ 0 & \text{,otherwise} \end{cases} \tag{4.10}$$

with $y_1 < y_2 < \cdots < y_m$. We consider all $r = m(m + 1)/2$ membership functions

$$\mu_{j,k}(y) = \bigvee_{i=j}^{k} \mu_i(y), 1 \leq j \leq k \leq n \tag{4.11}$$

consisting of adjacent singletons (4.10). From each of these membership functions (4.11) we derive a singleton $\hat{\mu}_{j,k}(y)$. Its position is given by applying

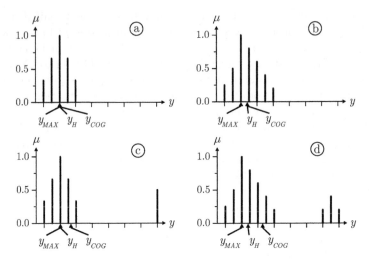

**Fig. 4.8.** Different output membership functions in the form of singletons, and corresponding output values $y_{COG}, y_{MAX}$, and $y_H$ resulting from COG and MAX defuzzification, respectively, and from a human expert.

COG defuzzification to $\mu_{j,k}(y)$. Its activation $\alpha_{j,k}$ is given by

$$\alpha_{j,k} = M_{j,k} \cdot e^{-q \cdot \sigma_{j,k}} \tag{4.12}$$

with

$$M_{j,k} = \sum_{i=j}^{k} \mu_i \quad \text{and} \quad \sigma_{j,k} = \sqrt{\frac{\sum_{i=j}^{k} \mu_i (C - y_i)^2}{M_{j,k}}}. \tag{4.13}$$

where $C$ results from applying COG defuzzification to $\mu_{j,k}(y)$. Equations (4.12) and (4.13) indicate that a human expert considers a group of adjacent singletons to be the more important the more that mass of activation is contained in the group, and the less the mass is distributed over the output value range. Finally, we compose the resulting filtered membership function by the singletons $\hat{\mu}_{j,k}(y)$ in the form of

$$\hat{\mu}(y) = \bigvee_{1 \le j \le k \le n} \hat{\mu}_{j,k}(y). \tag{4.14}$$

In this filtering approach, $q$ is a design parameter which allows adjustment of the model according to the above mentioned assumption concerning the scales involved. If we cannot make such a well-motivated corresponding assumption, we may use $q$ as a tuning parameter. In so doing, the settings $q = 0$ and $q = large$, respectively, correspond to applying COG and MAX defuzzification, respectively, to the original membership function $\mu(y)$. Consequently,

the model quality obtainable here is at least as good as with the above classical defuzzification methods. For values of $q$ with $0 < q < large$, we obtain output values that correspond to the pattern-based human decisions. In the above example, any fixed value of $q$ in the range $0.1 \leq q \leq 0.3$ supplies, for each of the four cases considered, the desired output value $y_H$ (Figure 4.9). Moreover, in the modeling of complex real-world processes, the pattern-based filtering improves the model accuracy noticeably, compared with COG and MAX defuzzification [25].

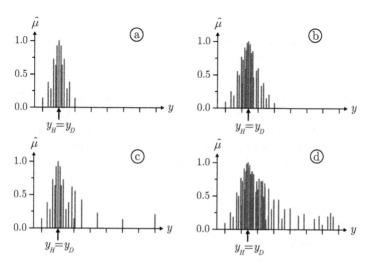

**Fig. 4.9.**   Real attractiveness functions $\hat{\mu}(y)$ obtained by pattern-based filtering of the membership functions $\mu(y)$ as depicted in Figure 4.8. The resulting output values $y_D$, given by the position of the maximum of each function $\hat{\mu}(y)$, correspond in all cases **a**, **b**, **c**, and **d** to the values $y_H$ of Figure 4.8.

### 4.4.6   Example of Application

In an industrial application, a model of a complex chemical process for the production of polyether is required. To achieve the global aim, the process is split into subprocesses. Then, for each subprocess, a model is constructed independently. One such subprocess is the heat exchanger [33]. The input variables for the model are the temperature $\vartheta_R$ of the reactor, the pressure $p_R$ of the reactor, and the manipulated variable $u$ for the regulation of the heat exchanger. The output variable is the temperature $\vartheta_H$ of the cooling fluid in the heat exchanger. The fuzzy model of the heat exchanger was generated with the Fuzzy-ROSA method.

Figure 4.10 shows the resulting error values of the model when applying the advanced inference filter method for different values of $q$. For $q \rightarrow 0$, the

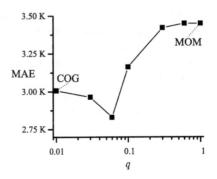

**Fig. 4.10.** Mean absolute temperature error in K of a fuzzy system using the pattern-based filtering with different values of the factor $q$.

same result is obtained as when using the center of gravity defuzzification method. Conversely, if $q$ exceeds a certain value, the same result is obtained as when using the mean of maximum defuzzification method. However, by setting $q = 0.6$, the advanced inference filter method supplies a better model quality than either of the classical methods.

## 4.5   Ambiguity-Based Predesign

### 4.5.1   Motivation

The application of the Fuzzy-ROSA method (Chapter 3) requires a preceding selection of the input variables and membership functions (Section 3.2.4). We will show that these tasks, here addressed as predesign, can be profitably based on the consideration of ambiguities.

### 4.5.2   Specification of Membership Functions

Consider the learning data shown in Figure 4.11 and the two crisp partitions of the one-dimensional input space into intervals. Each interval corresponds to a linguistic value that can be addressed by the premise of a rule in the form of (4.1). The more the output values of all data points that meet the rule premise coincide, the better such a rule is supported by the data. The more ambiguity there is in these output values, the more the problems discussed above emerge. Figure 4.11 shows that the amount of ambiguity greatly depends on the chosen partition. Consequently, it makes sense to reduce the ambiguity as much as possible by specifying suitable membership functions.

For this we propose the following approach [35]. For simplicity, we first consider that we provide crisp partitions (consisting of intervals) for each of the $n$ input variables. These partitions induce a crisp partition of the whole

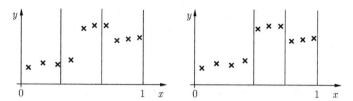

**Fig. 4.11.**   Learning data (crosses) and different partitions of the input space. Obviously, the number of ambiguities depends greatly on the chosen partition.

input space into axes parallel boxes $B(k)$. For each box, we define a partial function

$$y(B(k), \boldsymbol{x}) = \begin{cases} y(B(k)) \text{ , if } \boldsymbol{x} \in B(k) \\ 0 \qquad\quad \text{, otherwise} \end{cases} \tag{4.15}$$

where the definition of the box-specific constant value $y(B(k))$ depends on the purpose of the desired model. To solve an approximation problem, we may take the mean value of the output values $y_i$ of the data with $\boldsymbol{x} \in B(k)$. To solve a classification problem, we may take the most frequent output value (referring to a chosen partition of the output value range).

Based on these partial functions, we compose the partition-dependent function

$$y(\text{partition}, \boldsymbol{x}) = \sum_{\text{all } k} y(B(k), \boldsymbol{x}) \tag{4.16}$$

and rate by a suitable index $Q$ how well this function meets the modeling goal. In an approximation problem, $Q$ may rate the errors $|y_i - y(\text{partition}, \boldsymbol{x}_j)|$ referring to all learning data. In the case of a classification problem, $Q$ may rate the overall classification error. In any case, $Q$ depends on the partition chosen. This allows optimization of the partition by an evolutionary algorithm where $Q$ serves as the fitness function. Here, we may preselect the number of intervals provided individually for each variable, or preselect the total number of intervals and let the algorithm determine the best individual numbers.

For approximation problems, we extend this approach to the fuzzy case as follows: For each input variable, we provide a system of overlapping trapezoidal membership functions (the membership degrees always add up to unity in the interest of providing certain monotony properties [14]). The kernels of the membership functions, referring to all input variables, induce in the input space a family of boxes $B'(k)$ parallel to the axes, separated from each other by the overlapping zones. We define partial functions

$$y(B'(k), \boldsymbol{x}) = \mu(B'(k), \boldsymbol{x}) \cdot y(B'(k)). \tag{4.17}$$

Here, the box-specific constant value $y(B'(k))$ is the mean value of output values $y_i$ for all data $(\boldsymbol{x}_j, y_j)$ with $\boldsymbol{x}_j \in B'(k)$. Furthermore, $\mu(B'(k), \boldsymbol{x})$ is the

degree of membership of $x$ with respect to the box $B'(k)$. For this we take the product of the coordinate-wise membership degrees. Finally, we apply (4.16) as before.

### 4.5.3   Selection of Relevant Sets of Input Variables

In applications, there is an increasing interest in the data-based modeling of processes, where many, 100 or more, potentially relevant input variables are involved. The Fuzzy-ROSA method has recently been further developed to cope with still larger problem sizes [39]. However, mostly it is more efficient to base the modeling on a smaller set of relevant and nonredundant input variables. For the selection of such a set, correlation analysis may serve, however, it detects only linear dependencies between the output and the input variables. In contrast to this, the approach [33] detects nonlinear dependencies and has the additional advantage of considering the given partitions of the variables involved. However, it considers the relevance of individual variables, not of sets of variables, which would be more appropriate. The approach [42] does this, but does not consider partitions.

Progress in this respect supplies an extension of the above ambiguity-based approach for optimizing partitions. Once we have optimized the partitions referring to the considered input variables $x_1, x_2, \ldots, x_n$, the resulting index $Q$ characterizes the fitness of the set of the variables considered for the modeling purpose. Now we consider all reduced input spaces resulting from omitting one or more of the input variables provided originally. If we project the input vectors $x_j$ of all learning data into a reduced input space, we can optimize the partitions of the input variables of this space, using the above value of the index $Q$ for rating the relevance of the set of these variables. For small numbers $n$ of input variables provided originally, we compare all possible subsets of variables and decide on a favorable compromise between the aspects "fitness" and "small number of variables". For larger numbers $n$, we apply a greedy algorithm, starting with the fittest individual variable and then each time adding the variable that supplies the greatest increase of fitness.

The efficiency of this approach can be demonstrated with a well-known benchmark problem, i.e., the wine data set [4]. This resulted from a chemical analysis of three types of wine, and reflected the quantities of 13 wine constituents. The set of 178 data points was randomly divided into two samples of 89 data points for cross validation. We obtained the subset {*flavonoids, color intensity, malic acid*} of relevant and nonredundant input variables. Using the automatically generated membership functions, the Fuzzy-ROSA method supplied the two data sets taken as learning data classifiers with 42 and 25 rules, respectively. The resulting model quality can be averaged to 0.6 percent classification error on the learning data, and to 2.6 percent classification error on the validation data. Thereby, the reduction from 13 potentially relevant

attributes to three relevant nonredundant attributes implies an enormous reduction in the size of the rule search space. The model quality achieved here is the best so far obtained using the Fuzzy-ROSA method, and is similar to other published results [39].

### 4.5.4 Construction of Relevant Sets of Input Variables

In the data-based modeling of real-world problems, it often happens that the variables $w$, which have been formed as a certain function $w = f(x_1, \ldots, x_n)$ of the original input variables, are more relevant in the above sense than each individual $x_j$ [36]. This suggests the approach of constructing relevant input variables $w$ by choosing a flexible function

$$w = f(a_1, a_2, \ldots, a_q; \boldsymbol{x}) \tag{4.18}$$

and optimizing the values of the tuning parameters provided $a_1, \ldots, a_q$ by an evolutionary algorithm taking the above index $Q$ as the fitness function. That way we obtain a population consisting of many relevant variables $w$ from which we select a relevant set as described above.

This approach is of special interest for modeling dynamic processes. If, for instance, a process operator looks at a measuring instrument indicating the actual values $x(t)$ of a process variable, his or her decision may depend not only on the actual value $x(t)$, but also on past values $x(t - \tau), x(t - 2\tau), \ldots$. Providing all these values as components of $\boldsymbol{x}$ in (4.18), we may find a function $f$ which supplies a highly relevant variable $w$. This is illustrated in Figure 4.12 for a synthetic example.

## 4.6 Implicit Modeling

### 4.6.1 Motivation

The inference filter defuzzification provides a tuning parameter that allows us to influence the way ambiguities are resolved *after* the model is established. This may be used for a subsequent model optimization and save a complete redesign, if the model behavior obtained is not satisfactory at the first attempt. However, in the case of ambiguous data, the data-based modeling may lead to a rule base which does not reflect the ambiguities. The Fuzzy-ROSA method (Chapter 3), for instance, may fail to find sufficient relevant rules if the output values $y_j$ of the learning data for closely neighboring input vectors $\boldsymbol{x}_j$ are distributed over a large value range. The following implicit modeling method aims at overcoming this problem [10, 19, 23, 25].

### 4.6.2 The Implicit Modeling Method

The basic idea is to split the modeling into two steps: The first step is to set up a so-called implicit model that describes the distribution of the data

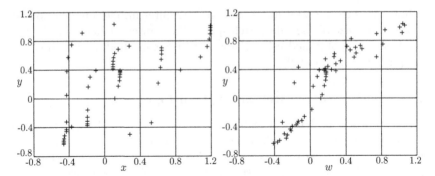

**Fig. 4.12.** Data $(x_j(t), y_j(t))$ recorded from a synthetically dynamic process with input and output variables $x$ and $y$, respectively, and state variable $z$. Here, $y = f(z)$ with a nonlinear function $f$ holds, and $z$ is the output of a linear dynamic subsystem with input $x$ and transfer function $(1 + Ts)^{-1}$. A graph of the values of $y(t)$ over the values of $x(t)$ shows considerable ambiguities (left). However, a graph of the same values $y_j$ over the corresponding values of the variable $w = 0.1(x(t) + 2x(t - \tau) + 4x(t - 2\tau) + 2x(t - 3\tau) + x(t - 4\tau)$ with optimized value $\tau$ shows no ambiguities (right).

$z_j = (x_j, y_j)$ in the input/output space. The second step is to use the implicit model together with a selectable decision strategy, which resolves the ambiguities and thus supplies an appropriate output value (Figure 4.13). This approach has the advantages that the generation of the implicit model is not affected by ambiguities in the learning data, and that the decision as to how to consider the ambiguities has to be made not before or in the modeling process, but afterwards. Consequently, tuning a model by modifying the decision on resolving the ambiguities does not require a repeated modeling.

The input vector $z^T = (x^T\, y)$ of the implicit model consists of the input vector $x$ of the process to be modeled and of its corresponding output variable $y$. The implicit model is designed so that it describes the distribution of the learning data in the $z$-space according to principle

$$\mu(z) = \begin{cases} large \text{ , if there is a learning point } z_j \text{ close to } z \\ small \text{ , otherwise} \end{cases} \quad (4.19)$$

Note that the data distribution in the $z$-space is never ambiguous, even if the input/output behavior of the process to be modeled is ambiguous. Consequently, such ambiguities do not affect the generation of the implicit model.

The use of the implicit model for generating an output value $\hat{y}$ for any given input vector $x$ is depicted in Figure 4.13. Therefore, a $y_j$ generator produces discrete values $y_1, y_2, \ldots, y_k$ that are fed, together with the input vector $x$, as data points $z_i$ into the implicit model. For a constant input vector $x$, the resulting output values $\mu(y_i)$ represent an output membership function in the form of singletons. This function makes possible ambiguities

visible. They are still present in the implicit model. To generate the output value $\hat{y}(x)$, the output membership function is fed into a selectable decision unit, which may be one of the described defuzzification strategies.

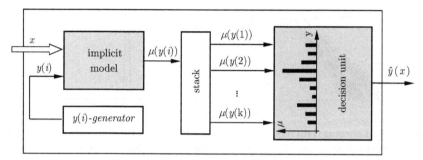

**Fig. 4.13.** Implicit modeling method based on an implicit model (which preserves possible ambiguities of the learning data) and on a decision unit (which resolves the ambiguities by a selectable decision strategy). The thin arrows designate real values and crisp membership degrees, respectively. The thick arrow designates a real-valued vector.

### 4.6.3  Construction of the Implicit Model

One possible method for constructing the desired implicit model is point-based. For each data point $z_j$, an implicit partial function $\nu_j(z)$ is generated according to the principle

$$\nu_j(z) = \begin{cases} large & \text{, if } z \text{ is close to } z_j \\ small & \text{, otherwise} \end{cases}. \tag{4.20}$$

By superimposing all implicit partial functions according to

$$\mu(z) = \sum_{j=1}^{m} \nu_j(z), \tag{4.21}$$

we produce the desired implicit model (Figure 4.14, left). Instead of the operator $\sum$ any fuzzy operator $\bigvee$ may be provided. This approach corresponds to the Parzen-windows method for determining the probability density based on data points [5].

For the design of the partial functions, we use the approach

$$\nu_j(z) = \exp(-(z_j - z)^T Q_j(z_j - z)) \tag{4.22}$$

with $Q_j$ being a positive definite matrix of the form

$$Q_j = \frac{1}{a_j} h_j h_j^T + \frac{1}{b_j}\left(I - h_j h_j^T\right) \text{ with } |h_j| = 1 \tag{4.23}$$

**Fig. 4.14.** Learning data (crosses) and contour lines of a function $\mu(x, y)$ describing the distribution of these data (left). Output membership function $\mu(y)$ resulting from feeding a fixed value of $x$ (dotted line left) into $\mu(x, y)$ (right).

where $I$ is the unit matrix. This matrix $Q_j$ induces an ellipsoid $x^T Q x = 1$, which is rotationally symmetrical with respect to the axis having the direction of the vector $h_j$ and where the ratio of the length of the axis parallel to $h_j$ and of the length of all other axes perpendicular to $h_j$ is given by $a_j/b_j$.

We consider the above approach for the design of the partial functions as a good compromise. On the one hand, it allows modeling local data distributions, which are predominantly allocated in any hyperplane and along any straight line, respectively. On the other hand, we have to determine the values of only $n + 2$ parameters to obtain a partial functional that fits the local data distribution well. For this fitting, we consider well-known features, such as errors and number of involved data points [23, 25]. Additionally, we characterize the cases where a number of closely neighboring data points supply more or less the same information compared with a single data point. (More measurements of the same item may make the result more reliable, however, more proofs of the same theorem do not make it more reliable.) Consequently, we provide a loneliness factor, such as

$$p_j = \frac{1}{\sum\limits_{i=1}^{m} e^{-d|x_j - x_i|}} \tag{4.24}$$

to each data point $z_j$, which considers its degree of isolation within the cloud of all learning data [22, 25]. This factor may be used as weighting factor for all subsequent modeling steps for the implicit modeling, and moreover, for any other data-based modeling, such as the Fuzzy-ROSA method.

Finally, we note that each partial function may be considered as a rule in the sense of the Fuzzy-ROSA method. Consequently, the different rating indices of the Fuzzy-ROSA method may be used as optimization criteria for fitting the partial functions (4.22) to the data distribution around $z_j$ [10].

In the case of many learning data points, the point-based approaches for the construction of the implicit model may supply too many partial functions. To reduce the corresponding effort, we preprocess the learning data in order to detect clusters of neighboring data points. We use the cluster centers as individual points for the design of the partial functions and fit the functions as before [25].

### 4.6.4  Choice of the Decision Unit

To resolve the ambiguities contained in the membership function $\mu(y)$ generated by using the implicit model (Figure 4.14, right), any of the above defuzzification strategies may be applied. Here, the strategies based on the inference filter are especially useful because they provide a parameter that can be adjusted according to the different assumptions on the source of the ambiguities and modeling aims, or used for model tuning.

The above incremental rating index suggests the following relevance-based resolution of ambiguities in the implicit modeling approach. For this, we consider a partition of the given output value range consisting of intervals $I(i)$ with $y(i) \in I(i), i = 1, 2, \dots, k$. For each of these intervals, the relative frequency $\hat{\mu}(i)$ for the validity of $y(i) \in I(i)$ considering *all* learning data is calculated (unconstrained output value distribution) and the modified values

$$\tilde{\mu}(y(i)) = \mu(y(i)) - \hat{\mu}(i) \tag{4.25}$$

are fed into the decision unit. This option is profitable if we want a model that describes causal dependencies between output and input variables.

Figure 4.15 illustrates the options that can be used for resolving the ambiguities contained in the implicit model by the choice of decision unit.

### 4.6.5  Comparisons

In order to compare the implicit modeling to other data-based modeling methods, we have chosen two benchmark problems. The first is the chaotic Mackey-Glass time series. The task is to predict the next value of the time series dependent on three past values and the current value. The second is the Boston housing problem. Here, the value of houses in the Boston, Massachusetts area should be predicted on the basis of 13 area characteristics. The implicit modeling has been used to generate a model for both tasks. Figure 4.16 shows the results of the implicit modeling and other modeling approaches. The approaches with which the implicit modeling is compared for the Mackey-Glass time series include Takagi-Sugeno modeling [2, 3, 31] and Mamdani modeling [25, 27]. The methods chosen for the Boston-housing problem are neural networks, fuzzy modeling, $k$-nearest neighbors, Gaussian processes for regression trained with a maximum a posteriori approach implemented with conjugate gradient optimization, and multivariate adaptive regression splines.

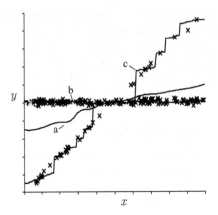

**Fig. 4.15.** Input/output behavior of the model obtained by applying the implicit modeling method to the learning data as depicted in Figure 4.14. The different curves a, b, and c correspond to the different options for the choice of the decision unit: center-of-gravity defuzzification, maximum height defuzzification, and relevance-based decision unit, respectively.

**Fig. 4.16.** Comparison of different data-based modeling methods with the implicit modeling for the Mackey-Glass time series problem and the Boston housing problem (× results from literature, ▼ implicit modeling) .

The figure shows the mean error values that rate the quality of the approaches. It can be seen that the implicit modeling achieves very good results. In the case of the Boston housing problem, it leads to the best of the models considered here. As the data of the Boston housing problem are very ambiguous, other modeling approaches seem to have more difficulties than the implicit modeling in coping with them.

# References

1. W. Blumendeller, V. Leutner, U. Schwane, and H. Kiendl. Regleroptimierung mit evolutionären Algorithmen. In *2. Internationales Fluidtechnisches Kolloquium*, pages 437–452. Dresden, 2000.
2. C.-L. Chen, S.-H. Hsu, C.-T. Hsieh, and W.-K. Lin. Generating crisp-type fuzzy models from operating data. In *Proceedings of the Seventh IEEE International Conference on Fuzzy Systems (FUZZ-IEEE '98), Anchorage, AK*, volume 1, pages 686–691. IEEE Press, Piscataway, NJ, 1998.
3. K.B. Cho and B.H. Wang. Radial basis function based adaptive fuzzy systems and their applications to system identification and prediction. *Fuzzy Sets and Systems*, 83:325–339, 1996.
4. A. L. Corcoran and S. Sandip. Using real-valued genetic algorithms to evolve rule sets for classifications. In *Proceedings of the First IEEE Conference on Evolutionary Computation (ICEC '94), Orlando*, volume 1, pages 120–124. IEEE Press, Piscataway, NJ, 1994.
5. R. O. Duda and P. E. Hart. *Pattern Classification and Scene Analysis*. Wiley, New York, 1973.
6. H. Jessen. A mean-value based test and rating strategy for automatic fuzzy rule generation and application to load prediction. In *International Conference on Computational Intelligence for Modelling Control, Wien*. IOS Press, Amsterdam, 1999.
7. H. Jessen. *Test- und Bewertungsverfahren zur regelbasierten Modellierung und Anwendung in der Lastprognose*. Fortschritt-Berichte VDI, Reihe 8, Nr. 836. VDI Verlag, Düsseldorf, 2000.
8. H. Jessen. Test and rating strategies for automatic fuzzy rule generation and application to load prediction. In M. Mohammadian, editor, *New Frontiers in Computational Intelligence and its Applications*, pages 11–21. IOS Press, Amsterdam, 2000.
9. H. Jessen and T Slawinski. Mittelwertbasierter Regeltest und Bewertung für das Fuzzy-ROSA-Verfahren und Anwendung zur Lastprognose. In *Tagungsband des 8. Workshops Fuzzy Control*, number 0298, pages 67–81. VDI/VDE GMA-FA 5.22, Faculty of Electrical Engineering and Information Technology, University of Dortmund, 1998.
10. H. Kiendl. Implizite Modellierung, inkrementeller Relevanzindex und Rauigkeitsmaß: neue Strategieelemente für die datenbasierte Modellierung. In *Tagungsband des 10. Workshop Fuzzy Control des GMA-FA 5.22*, pages 1–14. VDI/VDE GMA-FA 5.22, Research Reports Forschungszentrum Karlsruhe (FZKA 6509), Karlsruhe, 2000.
11. H. Kiendl. Verfahren zur Erzeugung von Stellgrößen am Ausgang eines Fuzzy-Reglers und Fuzzy-Regler hierfür. Technical report, 1994. Patent, DE 43 08 083.
12. H. Kiendl. Verfahren zur Defuzzifizierung für signalverarbeitende Fuzzy-Baueinheiten und Filtereinrichtungen hierfür. Technical report, 1995. Patent, DE 44 16 465.
13. H. Kiendl. Verfahren zur Defuzzifizierung für signalverarbeitende Fuzzy-Baueinheiten. Technical report, 1996. Patent, DE 19 640 635.
14. H. Kiendl. *Fuzzy Control Methodenorientiert*. Oldenbourg, Munich, 1997.
15. H. Kiendl. Non-translation-invariant defuzzification. In *Proceedings of the Sixth IEEE International Conference on Fuzzy Systems (FUZZ-IEEE '97), Barcelona*, pages 737–742. IEEE Press, Piscataway, NJ, 1997.

16. H. Kiendl. System of controlling or monitoring processes or industrial plants employing a dual-line fuzzy unit. Technical report, 1998. Patent, U.S. 5,826,251.
17. H. Kiendl. Decision analysis by advanced fuzzy systems. In J. Kacprzyk L. Zadeh, editor, *Computing with Words in Information/Intelligent Systems*, pages 223–242. Physica-Verlag, Heidelberg, 1999.
18. H. Kiendl. Design of advanced fuzzy systems. In Leena Yliniemi and Esko Juuso, editors, *Proceedings of TOOLMET'99 Symposium-Tool Enviroments for Intelligent Systems*, pages 57–76. University of Oulu, Oulun, Finland, 1999.
19. H. Kiendl. Verfahren zur datenbasierten Konstruktion eines Moduls mit den Daten entsprechendem Übertragungsverhalten und Modul hierfür. Technical report, 2000. Patent, DE 100 135 509.
20. H. Kiendl, T. Kiseliova, and Y. Rambinintsoa. Fuzzy interpretation of music. Technical Report CI-120/02, Collaborative Research Center 531, University of Dortmund, 2002.
21. H. Kiendl and M. Krabs. Ein Verfahren zur Generierung regelbasierter Modelle für dynamische Systeme. *at-Automatisierungstechnik*, 37(11):423–430, 1989.
22. H. Kiendl and P. Krause. Advanced inference filter defuzzification. In B. Reusch, editor, *Computational Intelligence: Theory and Applications (Proceedings of 7th Fuzzy-Days), Dortmund*, pages 249–260. Springer, Berlin, 2001.
23. H. Kiendl and P. Krause. Implicit modeling to cope with ambigous data. In *Proceedings of the IASTED International Conference Modeling, Identification and Control (MIC 2001), Innsbruck*, volume 1, pages 291–296. ACTA Press, Anaheim, 2001.
24. M. Krabs. *Das ROSA-Verfahren zur Modellierung dynamischer Systeme durch Regeln mit statistischer Relevanzbewertung*. Fortschritt-Berichte VDI, Reihe 8, Nr. 404. VDI Verlag, Düsseldorf, 1994.
25. P. Krause. *Datenbasierte Generierung von transparenten und genauen Fuzzy-Modellen für mehrdeutige Daten und komplexe Systeme*. Fortschritt-Berichte VDI, Reihe 10, Nr. 691. VDI Verlag, Düsseldorf, 2001.
26. P. Krause, A. Krone, and T. Slawinski. Fuzzy System Identification by Generating and Evolutionary Optimizing Fuzzy Rule Bases Consisting of Relevant Fuzzy Rules. Technical Report CI-84/00, Collaborative Research Center 531, University of Dortmund, 2000.
27. A. Krone. *Datenbasierte Generierung von relevanten Fuzzy-Regeln zur Modellierung von Prozesszusammenhängen und Bedienstrategien*. Fortschritt-Berichte VDI, Reihe 10, Nr. 615. VDI Verlag, Düsseldorf, 1999.
28. A. Krone and H. Kiendl. Automatic generation of positive and negative rules for two-way fuzzy controllers. In *Proceedings of the Second European Congress on Intelligent Techniques and Soft Computing (EUFIT '94)*, volume 1, pages 438–447. Verlag Mainz, Aachen, 1994.
29. A. Krone, P. Krause, and T. Slawinski. A new rule reduction method for finding interpretable and small rule bases in high dimensional search spaces. In *Proceedings of the Ninth IEEE International Conference on Fuzzy Systems, (FUZZ-IEEE '00), San Antonio, TX*, volume 2, pages 696–699. IEEE Press, Piscataway, NJ, 2000.
30. A. Krone and H. Taeger. Data-based fuzzy rule test for fuzzy modelling. *Fuzzy Sets and Systems*, 126:343–358, 2001.
31. D. Nauck and R. Kruse. A neuro-fuzzy approach to obtain interpretable fuzzy systems for function approximation. In *Proceedings of the Seventh IEEE Inter-*

*national Conference on Fuzzy Systems (FUZZ-IEEE '98), Anchorage*, volume 2, pages 1106–1111. IEEE Press, Piscataway, NJ, 1998.

32. B. Nicolaus, H. Kiendl, W. Blumendeller, and U. Schwane. Evolutionary optimization of an industrial hydraulic valve with the help of a fuzzy performance-index. In *Proceedings of the Tenth IEEE International Conference on Fuzzy Systems (FUZZ-IEEE '01), Melbourne*, volume CD-ROM. IEEE Press, Piscataway, NJ, 2001.

33. J. Praczyk. *Entwurf von Fuzzy-Gütemaßen zur Prozeßbewertung*. Fortschritt-Berichte VDI, Reihe 8, Nr. 796. VDI Verlag, Düsseldorf, 1999.

34. G. Reil and H. Jessen. Fuzzy contour modelling of roll bent components using inference filter. In *Proceedings of the Third European Congress on Intelligent Techniques and Soft Computing (EUFIT '95)*, volume 2, pages 771–774. Verlag Mainz, Aachen, 1995.

35. D. Schauten, B. Nicolaus, and H. Kiendl. Evolutionäres Verfahren zur Selektion relevanter Merkmalssätze für die datenbasierte Fuzzy-Modellierung. In *Proceedings 11. Workshop Fuzzy Control des GMA-FA 5.22, Dortmund*, pages 133–147. VDI/VDE GMA-FA 5.22, Research Reports Forschungszentrum Karlsruhe (FZKA 6660), Karlsruhe, 2001.

36. D. Schauten, B. Nicolaus, and H. Kiendl. An evolutionary concept for selecting relevant sets of input variables for data-based fuzzy modeling. In *Proceedings of the European Symposium on Intelligent Technologies, Hybrid Systems and their Implementation on Smart Adaptive Systems (EUNITE '01)*, volume CD-ROM. Verlag Mainz, Aachen, 2001.

37. D. Schauten, T. Slawinski, and H. Kiendl. Datenbasierte Generierung von situationsbezogenen Entscheidungsregeln für die Kraftwerksführung. Technical Report CI-110/01, Collaborative Research Center 531, University of Dortmund, 2001.

38. U. Schwane, B. Nicolaus, H. Kiendl, W. Blumendeller, and V. Leutner. Fuzzy-Gütemaß zur evolutionären Optimierung eines industriellen hydraulischen Regelventils. In *Fachtagung Computational Intelligence, VDI-Berichte 1526*, pages 229–234. VDI/VDE GMA und GI, Baden-Baden, 2000.

39. T. Slawinski. *Analyse und effiziente Generierung von relevanten Fuzzy Regeln in hochdimensionalen Suchräumen*. Fortschritt-Berichte VDI, Reihe 10, Nr. 686. VDI Verlag, Düsseldorf, 2001.

40. T. Slawinski, A. Krone, P. Krause, and H. Kiendl. The fuzzy-ROSA method: A statistically motivated fuzzy approach for data-based generation of small interpretable rule bases in high-dimensional search spaces. In M. Last, A. Kandel, and H. Bunke, editors, *Data Mining and Computational Intelligence*, pages 141–166. Physica-Verlag, Heidelberg, 2001.

41. T. Slawinski, J. Praczyk, U. Schwane, A. Krone, and H. Kiendl. Data-based generation of fuzzy rules for classification, prediction and control with the fuzzy-ROSA method. In *European Control Congress (ECC '99), Karlsruhe*, volume CD-ROM. Karlsruhe, 1999.

42. N. Xion. *Designing Compact and Comprehensible Fuzzy Controllers Using Genetic Algorithms (Entwurf kompakter und interpretierbarer Fuzzy Controller mittels Genetischer Algorithmen)*. Berichte aus der Automatisierungstechnik. Shaker Verlag, Aachen, 2001.

# Part II

# Evolutionary Algorithms

# 5 Theory of Evolutionary Algorithms and Genetic Programming

Stefan Droste[1], Thomas Jansen[1], Günter Rudolph[2], Hans–Paul Schwefel[2], Karsten Tinnefeld[1], and Ingo Wegener[1]

[1] University of Dortmund, Department of Computer Science, Informatik II, 44221 Dortmund
{droste, jansen, tinnefeld, wegener}@Ls2.cs.uni-dortmund.de
[2] University of Dortmund, Department of Computer Science, Informatik XI, 44221 Dortmund
{rudolph, schwefel}@Ls11.cs.uni-dortmund.de

**Summary.** Randomized search heuristics are an alternative to specialized and problem-specific algorithms. They are applied to NP-hard problems with the hope of being efficient in typical cases. They are an alternative if no problem-specific algorithm is available. And they are the only choice in black-box optimization where the function to be optimized is not known. Evolutionary algorithms (EA) are a special class of randomized algorithms with many successful applications. However, the theory of evolutionary algorithms is in its infancy. Here many new contributions to constructing such a theory are presented and discussed.

## 5.1 Introduction

An optimization problem consists of a typically infinite set $F$ of functions $f\colon S(f) \to \mathbb{R}$ (in multiobjective optimization $f\colon S(f) \to \mathbb{R}^n$) that share some semantical or syntactical properties. A typical example is the traveling salesperson problem (TSP) where, based on a distance matrix, each permutation $\pi$ is interpreted as a tour, and the cost function $f(\pi)$ measures the cost of $\pi$. Another more syntactical example is the maximization of pseudo-Boolean polynomials $p\colon \{0, 1\}^n \to \mathbb{R}$ of degree at most 2. We have to distinguish the problem from its instances that are described by a single function $f \in F$.

A classical algorithm for an optimization problem is a general procedure to work on all $f \in F$. It may use the special properties of the considered instance $f$, may calculate with the parameters of $f$ and produces a solution $x \in S(f)$. The optimization is called exact if the solution is optimal, it is called an $(1 + \varepsilon)$ approximation if the value of the solution differs at most by a factor of $1 + \varepsilon$ from the value of an optimal solution, and it is called a heuristic if we only hope to produce in many or typical cases a rather good solution. The algorithm is called efficient if the runtime with respect to the input length (and sometimes with respect to the inverse approximation ratio $\varepsilon^{-1}$) is polynomial.

A randomized search heuristic also works on all $f \in F$, but it typically does not perform complex calculations with the parameters of $f$. It interprets

$S(f)$ as search space. The hope is that $f$ and $S(f)$ have some structure to guide the search. In particular, all randomized search heuristics are based on the assumption that it is more likely that very good search points will be found close to good search points than close to bad search points. Nevertheless, the fact that one has to walk through bad regions or jump over them is not ignored.

More formally, a random search heuristic can pick the first search point $a_1$ with respect to some probability distribution. Then it obtains or computes $f(a_1)$. The probability distribution for the choice of $a_t$ may depend on $a_1, f(a_1), \ldots, a_{t-1}, f(a_{t-1})$. A stopping criterion determines when to stop and which of the considered search points is presented as solution. Most search heuristics "forget" many of the considered search points.

Search heuristics use different search operators. Evolutionary algorithms produce new search points with operators motivated by biological evolution. Based on the fitness (the $f$-value) of the individuals (search points) in the current population (which are not forgotten), some individuals are chosen as parents, i. e., to produce children (new search points) based on crossover (recombination of some search points) and mutation (random noise). Another fitness-based selection scheme decides which individuals will survive and become members of the next generation.

This implies that the paradigm of EA describes a large class of randomized search heuristics. Details of specific EA will be presented in later sections. The experimental knowledge about EA is immense (see also Chapter 5.9). For a long time, theory of EA was focused on the analysis of what happens in a short period of time (quality gain, progress rate, and schema theorem), on meta-ideas (building block hypothesis), and on the analysis under assumptions, the consequences of which cannot be controlled (infinite populations).

However, an EA applied to an optimization problem is an optimization algorithm, or an optimization heuristic, which can and should be analyzed like other randomized algorithms. It has been our aim to bridge the gap between results in theoretical computer science on the design and analysis of efficient algorithms and the theory of EA. People in theoretical computer science treating EA as the black sheep of the family of algorithms should be convinced that EA are legal members of the family of algorithms. This means that results on EA should also be published in theoretical conferences and journals. Survey articles of this type are [22, 58, 77, 78], articles with new results are mentioned in the following sections. Moreover, people in the community of EA have needed convincing that this new type of theory of EA is a valuable and perhaps necessary theory [80].

It is perhaps easy to formulate such ideas and concepts, but it is much harder to obtain results that fulfill the aims of the concepts. Some of our contributions will be presented in the rest of this chapter.

Wolpert and Macready [83] have presented their so-called no free lunch (NFL) theorem, which has initiated long and controversial debates in the

EA community. Our contributions to this NFL debate, which clarify some of the issues of the NFL theorem, are presented in Section 5.2. In Section 5.3, we start our analysis of EA on various types of problems. Section 5.3 contains results on probably the simplest EA, the (1+1)-EA, which can also be interpreted as a randomized hill-climber. The expected optimization time for many problems is estimated and methods, such as potential functions or the coupon collector's theorem, are applied on EA for the first time. Section 5.4 is a generalization of the common single-objective optimization scenario to multiobjective optimization. The (1+1)-EA works with population size 1 and mutation as the only search operator. Section 5.5 contains the first proofs that populations without crossover can outperform the (1+1)-EA significantly and that crossover can outperform mutation-based EA significantly, i. e., the expected optimization time decreases from exponential to polynomial. Section 5.6 is focused on a special aspect of these search operators, namely takeover times. Obviously, the choice of the free parameters of EA is crucial. The choice of the mutation probability, dynamic schedules to vary parameters and adaptation schemes is investigated in Section 5.7. Real-world applications of EA are disturbed by the fact that the situation is changing. The robustness of EA with respect to this effect is also considered in Section 5.7. It is obvious that there are no strict results on how to design the best EA. However, what the theoretically founded guidelines might look like is presented in Section 5.8. Finally, the black-box scenario, where the considered problem instance is unknown, offers the possibility to prove explicit lower bounds on the black-box complexity of problems which are presented in Section 5.9 as appropriate lower bounds for randomized search heuristics.

## 5.2   A Contribution to the NFL Debate

As briefly sketched in the previous section, randomized search heuristics are not specialized to the problem at hand. They are intended to work on a broad field of problems by incorporating operators that should lead to a good behavior for many reasonable problems: for EA mutation, crossover, and selection operators are believed by many researchers to guarantee a high robustness over a wide range of problems. This was motivated by the belief that natural evolution is a superior optimization principle, leading to the conclusion that EA are over the range of all optimization problems a superior search heuristic [26].

Hence, the NFL theorem by Wolpert and Macready [83], proving that over all objective functions $f : A \to B$ between two finite sets $A$ and $B$, all optimization algorithms sample on average the same number of different search points until they find an optimum of $f$, initiated many debates. Although the theorem is easy from a mathematical point of view, its statement was contradicting everyday experience.

We gave arguments for this only apparent contradiction by emphasizing the important role of the scenario of the NFL theorem [15, 19]: for the NFL theorem, it is assumed that all functions $f: A \to B$ are instances of the problem to be optimized and, furthermore, have the same likelihood to be chosen. Despite of the validity of the NFL theorem this is just a theoretical scenario. In practice all functions $f$ to be optimized with the help of a computer have to be efficiently representable in soft- or hardware. Hence, time and space needs of evaluating $f$ have to be rather small and $f$ has to be compactly representable, i. e., its Kolmogorov complexity cannot be too large. By simple examples we can show that even for sets $A$ and $B$ of moderate size only a tiny fraction of all functions between $A$ and $B$ can occur in practice. Using sets $A$ and $B$ so small that their complete enumeration is manageable, we exactly computed the number of search points sampled until an optimum was found for different search heuristics in complexity-restricted scenarios and proved that differences occur [15].

This result could be interpreted as a signal of hope for researchers looking for a superior search heuristic for at least the set of all realistic functions. But we also presented a more general form of the NFL theorem [19], which is valid not only in the unrealistic NFL scenario discussed above. The main observation underlying this generalized NFL theorem is the fact that in the NFL scenario no algorithm can learn anything (apart from the points already sampled) about the optimal search points. But this is also true if the set $F \subseteq \{f: A \to B\}$ is closed under permutations, i. e., if for every permutation $\pi$ on $A$ and every $f \in F$, also $f \circ \pi \in F$. In this case there is no possibility of excluding functions of $F$ as the actual one or even to assume another distribution than the uniform one, apart from excluding functions that do not fit the search points already sampled. Hence, all search algorithms sample the same number of points averaged over $F$.

Furthermore, we were able to prove the so-called almost no free lunch (ANFL) theorem, showing that for every randomized search heuristic $H$ and function $f: \{0,1\}^n \to B$, there are $|B|^{2^{n/3}-1}$ many functions $f': \{0,1\}^n \to B'$ that differ from $f$ on at most $2^{n/3}$ points, such that $H$ finds an optimum of $f'$ within $2^{n/3}$ steps only with probability at most $2^{-n/3}$ [19]. Its proof is based on the fact that during $2^{n/3}$ steps $H$ cannot sample all regions of the search space with high probability. More precisely, if we split $\{0,1\}^n$ into $2^{2n/3}$ sets of size $2^{n/3}$ each, following the pigeonhole principle there has to be one set $S$ containing at least one sampled point, with probability at most $2^{-n/3}$. Therefore, the function $f$ can be modified on all points of this set to another function $f'$ and $H$ will behave the same on $f$ and $f'$, with a probability of at least $1 - 2^{-n/3}$. By modifying $f$ to have its new optimal point in $S$, $B$ may even have to be enlarged by one value to $B'$, but $H$ will find the optimum of $f'$ in some exponential time only with an exponentially small probability.

If we split $\{0,1\}^n$ into the sets of size $2^{n/3}$ by a simple scheme (fix the first $2n/3$ bits, for instance) and modify $f$ on $S$ according to a simple func-

tion, $f'$ will only be by an additional factor of $O(n)$ more complex than $f$. Hence, the ANFL theorem additionally states that for all $f$, even for those being efficiently representable, there are exponentially many functions $f'$ being similarly complex as $f$, which cannot be efficiently optimized by $H$.

One could object that all the arguments against EA are not valid for natural objective functions, in the sense that all functions implying this bad behavior do not stem from a function occurring in well-known problems and have a rather complicated definition. Proving that this is not true, we were able to find a function Count : $\{0,1\}^n \to \mathbb{N}$ whose optimum will be found by EA with reasonable properties in some exponential time with only an exponentially small probability [17]. It is based on one instance of the well-known NP-hard MAXSAT problem, where for a number of clauses over Boolean variables, a variable setting should be found that fulfills the maximum number of clauses. Transforming this problem on a for most MAXSAT heuristics hard instance by Papadimitriou [50] into a pseudo-Boolean function, we obtain Count, a polynomial of degree 3. As for almost all search points, all indications for this symmetric function suggest that it is better to decrease the number of ones, reasonable EA and also simulated annealing will tend to a local optimum, a bit-string with only one $\mathbf{1}$, although the optimum is $\mathbf{1}^n$. As the probability of producing search points where the fitness values of the Hamming neighbors increase with the number of ones is exponentially small, EA need exponential runtime almost surely.

All in all, we have presented results explaining the consequences of the NFL theorem more precisely. The ANFL theorem examines the situation for complexity-restricted scenarios. As all such classes, regardless of the EA used, contain functions with exponential runtime, it makes sense to restrict our analyses in the following sections to function classes defined by syntactical or semantical properties.

## 5.3   On the Analysis of Simple Evolutionary Algorithms

Even if we restrict ourselves to syntactically or semantically defined function classes, EA used in practice are often very complicated and specialized to the problem at hand. In order to derive helpful theoretical results on EA, a possible direction of research is to consider simplified EA that are easier to analyze than more complicated variants and still have interesting features that are common to most EA. One extreme example is the (1+1)-EA, the perhaps simplest EA possible. It uses a population consisting of just one individual, in each generation it creates only one child using standard bitwise mutations and it uses a plus selection as known from evolution strategies. We give a formal description of the algorithm in a form that can be used for the maximization of pseudo-Boolean functions $f\colon \{0,1\}^n \to \mathbb{R}$.

## Algorithm 1 (The (1+1)-EA)

1. *Choose* $x \in \{0,1\}^n$ *uniformly at random.*
2. *Create* $y$ *by flipping independently each bit in* $x$ *with probability* $1/n$.
3. *If* $f(y) \geq f(x)$, *then* $x := y$.
4. *Continue at 2.*

The algorithm is considered without a stopping criterion and one is interested in the first point of time $X_f$ when $x$ is $f$-optimal, i. e.,

$$f(x) = \max \left\{ f(x') \mid x' \in \{0,1\}^n \right\}.$$

The average number $\mathrm{E}\left(X_f\right)$ of steps this takes is called expected optimization time. Also other parameters like the variance and the probability distribution of $X_f$ are of interest. However, most often the focus is on $\mathrm{E}\left(X_f\right)$.

This simple evolutionary algorithm is often subject to intense research [25, 31, 49, 56]. Typically, one investigates the expected optimization time on artificial example functions that often support theoretical analysis and are meant to exhibit important, general aspects of fitness landscapes. Probably, the best known example is OneMax, where the function value is OneMax$(x) = ||x||$, the number of ones in $x$. In [18] we have presented a number of fundamental results on the expected optimization time of the (1+1)-EA. The general upper bound $n^n$ on the expected optimization time is proven. Moreover, two example functions are presented that need, on average, asymptotically that many steps. One of these functions can be described as a polynomial of degree two. However, this polynomial is optimized by the (1+1)-EA within $O(n \log n)$ steps with probability $1/2 - o(1)$. Thus, applying multistarts this polynomial becomes easy to optimize. Another polynomial of degree two with the same properties and a detailed analysis of the effects of multistarts is presented by Wegener and Witt [81].

It is a common experience from experiments with evolutionary algorithms that they often get stuck in local optima. This leads to the speculation that functions without local optima, i. e., unimodal functions, are somehow easier to optimize. Horn, Goldberg, and Deb [31] introduce so-called long path functions where the path to the unique global optimum has exponential length. Rudolph [56] proves that the (1+1)-EA has expected optimization time $O\left(n^3\right)$ on this long path and also presents a generalized definition of long paths. He conjectures that on such long paths the (1+1)-EA has exponential expected optimization time. This conjecture is confirmed by Droste, Jansen, and Wegener [12], who prove that the expected optimization time of the (1+1)-EA is $\Theta\left(n^{3/2} \cdot 2^{\sqrt{n}}\right)$ on this unimodal function.

Another unimodal function that is of special interest is called LO (short for Leading Ones). The function value LO$(x)$ equals the length of the longest prefix of $x$ consisting of ones only. For this function, one can prove that the optimization time of the (1+1)-EA is sharply concentrated: Droste, Jansen, and Wegener [18] prove that there exist two constants $c_1 < c_2$, such that,

with probability $1 - 2^{-\Omega(n)}$, the (1+1)-EA optimizes LO within $t$ steps where $c_1 \cdot n^2 \leq t \leq c_2 \cdot n^2$ holds.

Perhaps the most interesting and most important result from [18] concerns the expected optimization time of the (1+1)-EA on the class of linear functions [13, 14]. A function $f \colon \{0,1\}^n \to \mathbb{R}$ is called linear iff it can be written as $f(x) = w_0 + \sum\limits_{1 \leq i \leq n} w_i \cdot x_i$, with $w_i \in \mathbb{R}$. The key idea for a proof is to measure the progress of the algorithm by a kind of potential function that reflects general properties of linear functions. Then, by considering a random process that optimizes a linear function provably slower than the (1+1)-EA, one can prove that the expected optimization time of the (1+1)-EA is $O(n \log n)$ for any linear function. Moreover, if $w_i \neq 0$ holds for all $1 \leq i \leq n$, then the expected optimization time is $\Theta(n \log n)$. This lower bound can be derived with the coupon collector's theorem [48].

The (1+1)-EA changes each bit in a mutation step with probability $1/n$. Of course, other mutation probabilities are possible. A mutation probability of $1/n$ implies that, on average, in each step one bit flips. If the mutation probability is clearly smaller, i.e., $p = o(1/n)$, we expect that in too many steps nothing at all happens. If on the other hand, the mutation probability is clearly larger, i.e., $p = \omega(1/n)$, we expect that in each step too many bits that already have the correct-value flip and the child $y$, i.e., the result of the mutation, does not replace its parent $x$. In fact, both speculations can be proven correct and it can be shown that mutation probabilities that are either $\omega(1/n)$ or $o(1/n)$ lead to larger expected optimization times, thus mutation probabilities $\Theta(1/n)$ are optimal for linear functions [18].

Linear functions are a class of pseudo-Boolean functions where one can prove useful, general results on the expected optimization time of the (1+1)-EA. It is natural to consider other, maybe larger function classes. Each pseudo-Boolean function $f \colon \{0,1\}^n \to \mathbb{R}$ has a unique representation as polynomial

$$f(x) = \sum_{A \subseteq \{1,\ldots,n\}} w_A \cdot \prod_{i \in A} x_i,$$

with $w_A \in \mathbb{R}$. The degree of the polynomial is the maximal cardinality of $A$, such that $w_A \neq 0$. Since the optimization of polynomials of degree 2 is already NP-hard, one cannot hope to find good upper bounds on function classes that are defined based only on the degree of the polynomial. In particular, Wegener and Witt [81] have explicitly defined a polynomial of degree 2, such that the expected optimization time of the (1+1)-EA is exponential. This is done based on a well-known function called Trap, introduced by Ackley [1]. For this function, the (1+1)-EA needs, with probability $1 - 2^{-\Omega(n)}$, at least $\Omega(n^n)$ steps to find the global optimum. The function Trap is represented by a polynomial of degree $n$. One can apply a reduction technique introduced by Rosenberg [54] resulting in a function Trap* that has properties similar to Trap and is a polynomial of degree 2. For Trap*, Wegener and Witt [81]

prove that the probability of a successful optimization within $2^{cn \log n}$ steps is $2^{-\Omega(n)}$, where $c$ is a small positive constant. Thus, also multistarts of the (1+1)-EA cannot decrease the expected optimization time to a polynomial.

Yet, interesting results on quadratic functions can be shown. For quadratic functions where all weights $w_A$ are not negative, Wegener and Witt [81] prove that the expected optimization time of the (1+1)-EA is $O(N \cdot n)$, where $N$ denotes the number of weights that are different from 0. Another way of structuring quadratic functions is to consider the degree of separability. A function $f: \{0,1\}^n \to \mathbb{R}$ is called separable with respect to a partition $(X_1, \ldots, X_k)$ of the set of variables $X = \{x_1, \ldots, x_n\}$ iff $f$ can be represented as $f(X) = f_1(X_1) + \cdots + f_k(X_k)$, where all $X_i$ are not empty. Wegener and Witt [81] present an upper bound of $O\left(2^m \cdot n^{m+1}\right)$ for the expected optimization time of the (1+1)-EA for such functions, where $m = \max\{|X_1|, \ldots, |X_k|\}$. Another special case of quadratic functions are functions that can be obtained as square of some linear function. Such functions can be optimized using multistarts of the (1+1)-EA very efficiently: Wegener and Witt [81] prove that the (1+1)-EA optimizes such functions within $O(n \log n)$ steps with probability at least $1/8 - \varepsilon$, where $\varepsilon > 0$ is some arbitrarily small constant.

Another interesting class of functions that is defined syntactically according to the representation as polynomial are monotone polynomials, i. e., polynomials with $w_A \geq 0$ for all $A$. For monotone monomials of degree $d$, i. e., polynomials of degree $d$ with $N = 1$, the (1+1)-EA has expected optimization time $\Theta\left(n \cdot 2^d/d\right)$ [79]. For monotone polynomials with a larger number of terms $N$, the expected optimization time increases. Wegener [79] proves an upper bound of $O\left(N \cdot n \cdot 2^d/d\right)$ for such polynomials where $d \leq \log n$ holds. However, it is conjectured that this upper bound is not tight and an upper bound of $O\left((n \log n) \cdot 2^d/d\right)$ can be proven. This is supported by the analysis of a special subclass of polynomials. Let $n = d \cdot m$. Consider monotone polynomials with $N = m$ nonvanishing terms, where all terms are of degree $d$ and depend on disjoint sets of variables. For such polynomials an upper bound of $O\left((n \log n) \cdot 2^d/d\right)$ is proven. The generalization to other monotone polynomials of degree $d$ is an open problem.

Another way to look at the problem of classification of objective functions (see Jansen [32] for an overview) is to consider the expected optimization time of the (1+1)-EA. Let, for $i \in \{0, 1, \ldots, n\}$, $F_i$ denote the set of functions $f : \{0,1\}^n \to \mathbb{R}$ such that the expected optimization time of the (1+1)-EA is $O\left(n^i\right)$. Droste, Jansen, and Wegener [18] present a family of functions $\text{Jump}_m$:

$$\text{Jump}_m(x) := \begin{cases} m + ||x|| & \text{if } ||x|| \leq n - m \text{ or } ||x|| = n \\ n - ||x|| & \text{otherwise} \end{cases}$$

For $x$ with $n - m < ||x|| < n$, $\text{Jump}_m$ equals $n - \text{OneMax}$, otherwise it equals $m + \text{OneMax}$. Since the global optimum is $1^n$, the part of the search space with $n - m < ||x|| < n$ is misleading and it depends on $m$ how difficult $\text{Jump}_m$ is to optimize. An upper bound can be proved via a general proof technique. For a pseudo-Boolean function $f\colon \{0,1\}^n \to \mathbb{R}$ and two disjoint, non empty sets $A, B \subseteq \{0,1\}^n$, we say $A <_f B$ iff $f(a) < f(b)$ for all $a \in A$ and $b \in B$. We call a partition $A_1, \ldots, A_l$ of $\{0,1\}^n$ an $f$-based partition iff $A_1 <_f \cdots <_f A_l$ holds and all $a \in A_l$ are $f$-optimal. For $a \in A_i$ with $i < l$ let $s(a)$ be the probability that a mutation of the $(1+1)$-EA leads from $a$ to some $b \in A_{i+1} \cup \cdots \cup A_l$. Finally, let $s_i$ denote the minimum of all $s(a)$ for $a \in A_i$. Obviously, $s_i^{-1}$ is an upper bound on the expected number of steps the $(1+1)$-EA stays in $A_i$. Since each set $A_i$ is left at most once, $s_1^{-1} + \cdots + s_{l-1}^{-1}$ is an upper bound on the expected optimization time. Given appropriate $f$-based partitions, this proof method often gives surprisingly tight results. For $\text{Jump}_m$, using $A_i := \{a \in \{0,1\}^n \mid \text{Jump}_m(a) = i\}$ (with $i \in \{1, \ldots, n+m\}$) as the $f$-based partition yields the upper bound $O(n^m + n \log n)$ and in fact the expected optimization time of the $(1+1)$-EA is $\Theta(n^m + n \log n)$ on $\text{Jump}_m$. Together with obvious observations on the expected optimization time on some easy functions, this proves that the sets $F_0, F_1, \ldots, F_n$ discussed above form a proper hierarchy.

We introduced the $(1+1)$-EA as simple EA that can be used as basis for a theoretical analysis of problems that are of general interest. One type of question that can be investigated using the $(1+1)$-EA as object of study is the behavior of EA on fitness landscapes that are thought to be typical. One example is plateaus, i.e., regions in the search space with constant fitness. One is interested in the expected time different variants of evolutionary algorithms spend in such flat regions of the search space. Jansen and Wegener [35] investigate two variants of the $(1+1)$-EA on such plateaus. The first variant is the $(1+1)$-EA defined as Algorithm 1. The second variant works in the same way with the exception that, in the selection step in line 3, the condition $f(y) \geq f(x)$ is replaced by $f(y) > f(x)$. We denote this variant as $(1+1)^*$-EA. The consequence of this difference is obvious: while the $(1+1)$-EA explores plateaus of constant fitness in a kind of random walk, the $(1+1)^*$-EA halts on the first point of such a plateau and stays until it finds a point with larger function value.

Jansen and Wegener [35] present two different functions that exemplify when the random exploration of plateaus is useful and when it is harmful. First, consider a function where the $(1+1)$-EA is led to a local optimum that has Hamming distance $n$ to the unique global optimum. There is only one path between this local and the global optimum, and all points on the path have the same function value. It is intuitively clear that the exploration of this small plateau eventually leads to the global optimum whereas waiting for one large mutation to lead there takes extremely long. In fact, it is proven that the $(1+1)$-EA has an expected optimization time $O(n^3)$, whereas the

(1+1)*-EA does not find the optimum of this function within $n^{n/2}$ steps with probability $1 - 2^{-\Omega(n)}$. Second, consider a function where the (1+1)-EA is led to a path leading to the unique global optimum. On the path there are many holes, i.e., points with small function value. A mutation of 2 bits simultaneously is needed to jump over such a point. At each such hole a small plateau similar to the one in the previous example starts, leading away from the path to some local optimum that has Hamming distance $\Omega(n)$ from the unique global optimum. Here, it is intuitively clear that the exploration of plateaus may lead to a local optimum that is hard to leave. In fact, it is proven that the (1+1)*-EA optimizes this function within $O\left(n^3\right)$ steps with probability $1 - 2^{-\Omega\left(n^{1/4}\right)}$, whereas the (1+1)-EA does not succeed in $n^{n/2}$ steps with probability $1 - n^{-\Omega\left(n^{1/2}\right)}$ [35].

One line of theoretical research of evolutionary algorithms investigates local performance measures that quantify the progress the EA makes in one single or at most a small number of steps. Two examples are quality gain, where the expected difference in the function value is considered, and progress rate, where the expected change in the Hamming distance to a global optimum is considered. Looking at the (1+1)-EA and the (1+1)*-EA on plateau functions, such as the ones described above, it becomes obvious that these local performance measures can be insufficient to distinguish between the two algorithms. Although the two algorithms behave quite differently on the functions, the quality gain suggests that the algorithms behave the same. Moreover, the progress rate predicts an advantage for the algorithm that is far worse by means of expected optimization time [35].

## 5.4   Multiobjective Evolutionary Algorithms

In the last section, we investigated evolutionary algorithms on single-objective optimization problems, i.e., we considered problems with a total order on the set of fitness values, which was always a subset of $\mathbb{R}$. However, we know from everyday life that we do not always have such a clear structure. It may depend on the customer whether an expensive high-quality computer is better than a cheap medium-quality one. Hence, there is good motivation to study search heuristics on functions $f\colon S \to V$, where $V$ is only partially ordered by the relation $\succeq$. Since we only consider finite search spaces $S$, $f(S) \subseteq V$ is also finite, and the existence of maximal (or minimal) search points $a$ is guaranteed. More precisely, $a$ is maximal if there is no $b \in S$ where $f(b) \succ f(a)$. Moreover for each $c \in S$, the existence of a maximal element $a$ where $f(a) \succeq f(c)$ is guaranteed.

What is the aim of optimization in the case of partially ordered fitness values? A population is called optimal and complete if it contains only maximal search points and all maximal search points. It is called optimal if it contains only maximal search points.

The best-known situation is the case called multiobjective optimization, i. e., the fitness of a search point is described by a fitness vector and the fitness function belongs to the class of functions $f\colon S \to \mathbb{R}^n$. The corresponding partial order is defined by $a \leq b$ iff $a_i \leq b_i$ for all $i \in \{1, \ldots, n\}$. The set of search points which are $\leq$-maximal is known as Pareto set and maximal points are called Pareto-optimal. The limit behavior of evolutionary algorithms in the case of multiobjective optimization has been analyzed by Rudolph [59] and Hanne [29].

However, multiobjective optimization is perhaps the most important case of optimization with partially ordered fitness values, but nevertheless it is only a special case of the much more general class of optimization problems with partially ordered fitness values. Rudolph [57] was the first to develop a theory on this large class of optimization problems with respect to evolutionary algorithms. This approach was advanced later [64, 68]. Before presenting this general theory we discuss other important situations that can be covered by our general framework [66].

The evaluation of fitness functions, also single-objective ones, is often numerically instable. Then the use of interval arithmetic (for an overview see Hansen [30]) is advisable because it enables an assessment of the numerical reliability of the fitness function. In this case $f(a)$ is a fitness interval and not a fitness value. As a consequence, evolutionary algorithms have to cope with an interval order (for an overview see Fishburn [23]). Let $a_1, a_2 \in S$ with fitness intervals $[s_1, t_1]$ and $[s_2, t_2]$, respectively. Then the following partial order on $S$ seems to be appropriate. We have $a_1 \preceq a_2$ iff $a_1 = a_2$ or $t_1 \leq s_2$. If $a_1 \preceq a_2$ and $a_1 \neq a_2$ we may exclude $a_1$ from our investigations if we keep $a_2$. If we want to keep all possibly best search points, we can replace $t_1 \leq s_2$ by $t_1 < s_2$ in the definition of the partial order.

If we have the opportunity to obtain numerically more stable outcomes, it will result in shorter fitness intervals and reduce the set of maximal elements. Hence, we get a sharper partial order leading to a set of maximal elements. This set of elements will still correspond to all optimal search points but perhaps to less nonoptimal search points that cannot be excluded from our considerations, since, due to numerical instability, we do not have enough information to exclude them from the set of possibly optimal search points. In many applications, including experiments and simulations, the numerical instability can be reduced by spending more time or money.

The second scenario we consider is the case of single-valued objective functions where the fitness evaluation is perturbed by bounded additive noise $Z$. The random variable $Z$ takes values in a bounded interval $[s, t]$ containing 0. Hence, the observed fitness value $\tilde{f}(x)$ contains the information that the true fitness value $f(x)$ is contained with probability 1 in the interval $[\tilde{f}(x) + s, \tilde{f}(x) + t]$. Then we can apply the same partial order on intervals as discussed above. A more general interval order has already been introduced by Guddat et al. [28].

Again, an optimal search point always leads to a maximal fitness interval and, if $s = -a$ and $t = a > 0$, then, as Rudolph [65] has shown, all maximal elements are $3a$-optimal. In this case, the intervals can be reduced by independent observations. Knowing with probability 1 that the true fitness value is contained in $I_j$, $1 \leq j \leq k$ implies that the true fitness value is contained with probability 1 in the intersection of these $I_j$. It is easy to compute for simple distributions of $Z$, such as the uniform distribution, the number of independent observations, which ensure that, with a large probability, the length of the fitness interval is smaller than a given bound. Altogether, the scenario of optimization with partially ordered fitness information is well motivated, although its special case of multiobjective optimization is by far the most important case.

Most of the results on single-objective optimization cannot be generalized directly to our general scenario. Rudolph [57] analyzed the limit behavior of a very general class of search heuristics with the special property of nonfixed population size. The only conditions are the following ones:

1. if $A(t)$ is the population at time step $t$, then each population has a positive probability to be created as $B(t + 1)$,
2. $A(t+1)$ contains the maximal elements from $A(t) \cup B(t+1)$ with respect to the considered partial order.

Then $A(t)$ converges with probability 1 to the set containing exactly all maximal elements of the search space.

However, the number of maximal elements may be exponentially large, and, in general it is not realistic to allow superpolynomial-sized populations. Rudolph and Agapie [68] have considered a search heuristic with populations whose size is bounded by a fixed parameter $m$. Now it is assumed that each population of size at most $m$ has a positive probability to be chosen as $B(t + 1)$. Let $B^*(t + 1)$ be the maximal elements of $B(t + 1)$. Then $A(t) \cup B^*(t + 1)$ contains the candidates for $A(t + 1)$. If $b \in B^*(t + 1)$ dominates some $a \in A(t)$ strictly, i. e., $f(b) \succ f(a)$, then we remove the dominated search points from $A(t)$ and add $b$ to $A(t)$. If $b$ is not comparable to all $a \in A(t)$, it is included in the initially empty set $C(t + 1)$. At the end of this procedure, $|A(t + 1)| \leq m$. We then obtain $A(t + 1)$ by adding $k := \min\{m - |A(t)|, |C(t+1)|\}$ arbitrary different search points from $C(t+1)$ to $A(t)$. With probability 1, the number of nonmaximal elements of the search space contained in $A(t)$ converges to 0 and the size of $A(t)$ converges with probability 1 to the minimum of the bound $m$ on the population size and the number of different maximal elements of the search space.

The population size of this algorithm is bounded above but not constant. Therefore, Rudolph [64] has designed a more complicated but still general framework of an evolutionary algorithm with a fixed size $m$. This algorithm guarantees that the number of nonmaximal elements in the search space contained in $A(t)$ converges to 0 with probability 1 if there are at least $m$ maximal search points.

All algorithms and results presented so far did not take into account that evolutionary algorithms often deploy mechanisms to adapt their search operators. Rudolph [59] noticed that successful mechanisms for single-objective evolutionary algorithms do not have the same positive effects in multiobjective optimization. Subsequent experiments by Laumanns, Rudolph, and Schwefel [43, 44] for multiobjective evolutionary algorithms with spatial structure and an underlying predator/prey selection model introduced in [42] identified some new and promising approaches to a self-adaptive control of the search operators. Theoretical results, however, are not yet available.

Weinert, Mehnen, and Rudolph [82] extended these spatially structured evolutionary algorithms by the dynamization of the neighborhood structure within the population. These algorithms are in an experimental state and they will have to prove their utility in future simulation studies.

We have reported several results about the limit behavior of evolutionary algorithms for multiobjective optimization and other problems. However, almost nothing is known about the expected time to reach the aim. Tinnefeld and Wegener [73] investigated search heuristics for shortest path problems. The black-box complexity of the shortest $s$-$t$-path problem is exponential (see Section 5.9). Therefore, they have investigated the single source shortest paths problem for graphs on $V = \{1, \ldots, n\}$ where $s = n$ is the source. A search point $x \in V^{n-1}$ describes for vertex $i$ by $x_i$ its predecessor on a path from $s$. This allows illegal search points, where the path backwards from $i$ does not reach $s$. However, this is taken into account by the fitness vector $f(x) \in (\mathbb{R} \cup \{\infty\})^{n-1}$ where the $i$th component equals the length of the path from $s$ to $i$ obtained by always choosing from vertex $i$ the predecessor described by $x$. This length equals $\infty$ if this procedure does not reach $s$. The algorithm is a simple (1+1) EA, accepting $x'$ only if $f(x') \leq f(x)$ (since the objective is minimization). Mutation changes $x_i$ with probability $1/n$ and then randomly chooses one of the vertices from $V \setminus \{i\}$. It has been shown that this simple evolutionary algorithm solves the single source shortest paths problem in an expected time of $O(n^3)$. Here we are in the special situation that the optimal fitness value is unique. Our result proves that it can be easier to solve $n - 1$ problems simultaneously than to solve a single one of these problems individually. The reason is that the fitness vector for the multiobjective problem reveals more information than the fitness value for a single problem.

Our results cannot hide the fact that the theory on evolutionary algorithms for multiobjective optimization is even less developed than the theory on evolutionary algorithms for single-objective functions.

## 5.5   Populations and Crossover

We have seen in Section 5.3 that the (1+1)-EA working with population size 1 and, therefore without crossover, it is surprisingly efficient in many

situations. People have suggested from the very beginning of the use of evolutionary algorithms that the use of populations and of the crossover operator is essential. However, to support this general claim with a simple example where the expected optimization time decreases significantly with the use of populations and crossover was an open problem for 35 years. Such examples can only be first steps towards a theory on populations and crossover. The examples themselves may be artificial, but analyzing the use of populations and crossover makes it necessary to develop new methods, and this is the essential part of this research. We start our investigations with the comparison of crossover-based and mutation-based evolutionary algorithms and finish with some results on the comparison of large populations and population size 1.

It would be unfair to base the investigation on a special or even problem-specific genetic algorithm. Therefore, we consider the following generic genetic algorithm with many free parameters:

- $s(n)$, the population size,
- $p_c(n)$, the probability of applying the crossover operator,
- the type of crossover,
- the selection scheme to choose parents (the only assumption is that, in the case of maximization, $f(x) \geq f(y)$ implies that it is not more likely to select $y$ than to select $x$).

We need some procedure to guarantee some diversity in the population. Hence, we investigate steady-state genetic algorithms (GA) (only one new individual is produced) and eliminate one of the individuals with the worst fitness. If the multiset representing the population has more than one worst individual, we eliminate one that has the largest number of copies in the population. This leads to the following algorithm:

## Algorithm 2 (Steady-state GA)
1. *Randomly choose the initial population.*
2. *With probability $p_c(n)$ use Step 3', otherwise use Step 3".*
3'. *Select two parents, apply the crossover operator to produce one offspring, and apply the mutation operator to produce the new individual.*
3". *Select one parent, apply the mutation operator to produce the new individual.*
4. *Add the new individual to the population and eliminate one individual from the population.*

*(Steps 2, 3, and 4 are repeated until a stopping criterion is fulfilled.)*

Many papers have discussed the fundamental role of the crossover operator; many experiments have been conducted; and arguments like the building block hypothesis have been used to explain the importance of crossover. This is an ideal example to show that only the rigorous analysis of the expected

optimization time can answer such questions. Mitchell, Forrest, and Holland [24, 46] tried to support their building block hypothesis with example functions that they called royal road functions because they were meant to be the royal roads allowing crossover to work with great success. The arguments based on the building block hypothesis work perfectly. However, the (1+1)-EA outperforms all genetic algorithms on the royal road functions, which has been shown by the original authors [47]. Another example function called hierarchical-iff (H-IFF) has been presented by Watson, Hornby, and Pollack [74, 75]. Here experiments fortify the conjecture that crossover supports the maximization of H-IFF. However, one may have some doubts. Experiments cannot distinguish between polynomial size and slowly increasing exponential size, and it is an open question whether H-IFF can be maximized in expected polynomial time by some evolutionary algorithm.

We have presented an example function without building blocks, the function $\text{Jump}_m$ for $m \leq n/3$ (see Section 5.3 for the definition of $\text{Jump}_m$). As long as $||x|| \leq n - m$, $\text{Jump}_m$ equals $\text{OneMax}$ (with an additional constant additive term). Since $m \leq n/2$, the initial population starts with overwhelming probability in this region (or the population has superpolynomial size implying superpolynomial optimization time). It is easy to find individuals with $n - m$ ones. The difficulty is to jump over the valley containing all $x$ where $n - m < ||x|| < n$. All mutation-based algorithms have an expected optimization time of $\Omega(n^m)$ independent of the probability of flipping bits.

In [38] we analyze genetic algorithms on $\text{Jump}_m$. After a short period of time all individuals have $n - m$ ones, and in order to produce the optimum $\mathbf{1}^n$ by uniform crossover, we need two individuals with no common $\mathbf{0}$-bit. Then these individuals are selected with a probability of at least $2/s(n)^2$; uniform crossover is successful with probability $2^{-2m}$; and mutation does not destroy $\mathbf{1}^n$ with a probability of approximately $1/e$. Hence, the expected optimization time would be polynomial if $s(n)$ is polynomial, $m = O(\log n)$, and we can guarantee the existence of two individuals with these properties.

Choosing strings with $m$ ones randomly ensures this diversity. However, it is much more likely that the population is concentrated somewhere. Crossover increases the concentration. Indeed, the stable situation for crossover is a population consisting of $s(n)$ copies of one individual. Mutation allows in such a situation the creation of new individuals. Nevertheless, even the stationary distribution on the set of populations is unknown. It is not the uniform distribution, since this distribution would lead to individuals with a large Hamming distance and mutation would create an individual close to an existing one. Discussions with experts on rapidly mixing Markov chains have led to the conclusion that this powerful theory cannot be applied here. Hence, the effect of mutation is estimated in the following way.

Independent of the initial population of individuals with $m$ ones, for each bit position, we prove that the algorithm accepting only individuals with $m$ ones has, after some time, with large probability for a long period of time, not

too many individuals with a **0**-bit at this position. Crossover is an operator which is not helpful to guarantee diversity. However, if $p_c(n)$ is small enough, we can cope with crossover assuming that it always has the worst effect. Let $m(n) = O(\log n)$. Simplifying the results from [38] leads to the first function, where it is proved that a population of polynomial size $s(n) = \Omega(m(\log n)^2)$ and a crossover probability of $c/mn$, $c > 0$ small enough, has the effect of the genetic algorithm having a polynomial expected optimization time, while all mutation-based algorithms have a superpolynomial expected optimization time of $n^{\Omega(\log n)}$. This is the first analysis that controls the hitchhiking effect caused by crossover and selection.

However, a trade-off of polynomial vs. exponential would be even better. Such a trade-off is not possible for Jump. We have learned from Jump that it is easier to cope with the effect of populations if the plateau of second-best individuals is smaller. In [37] we present a function, now called the real royal road function $R_m$ for 1-point crossover, since it has the properties that are expected from the royal road functions. It is a royal road for the crossover operator – without a clear building block structure. The function $R_m$ is obtained from Jump$_m$ by replacing the fitness value $||x||$ in the first line by $n \cdot ||x|| + b(x)$, where $b(x)$ is the length of the largest block in $x$ consisting of ones only. It is still at first the aim to increase the number of ones in $x$, but there is the secondary aim of concentrating the ones in one block. Hence, the number of second-best individuals equals $m + 1$. Having all of them in the population, we have a good chance of at least $1/s(n)^2$ to select $\mathbf{0}^m\mathbf{1}^{n-m}$ and $\mathbf{1}^{n-m}\mathbf{0}^m$ and a probability of at least $1/3$ (since $m \leq n/3$) to cut the strings in the common 1-block. Then 1-point crossover creates $\mathbf{1}^n$, and again there is a good chance that mutation does not destroy $\mathbf{1}^n$.

In the beginning, crossover may not support progress, but it has no bad consequences. The steps without crossover are enough to create a population containing all second-best individuals, provided that $s(n) \geq m+1$. To increase the number of ones in the string 1-bit mutations are helpful and to increase the block length without increasing the number of ones 2-bit mutations are helpful. Moreover, replications imply the elimination of bad individuals; and the selection rule to eliminate individuals with many copies implies that we obtain and keep all second-best individuals. Here we can choose $m = n/3$ leading to an exponential optimization time for mutation-based algorithms. The genetic algorithm where $p_c(n)$ is a constant in $(0, 1)$ and $s(n) = n$ has an expected optimization time of $O(n^4)$.

This function also improves the result for uniform crossover, since we now cover all reasonable values of $p_c(n)$. However, the trade-off is still between polynomial and superpolynomial, since we have to choose $m = O(\log n)$ to give uniform crossover a probability of $\Omega(1/poly(n))$ to create $\mathbf{1}^n$. In [37] a real royal road function for uniform crossover is also presented, leading to similar results for uniform crossover as have been discussed for $R_m$ and one-point crossover.

The functions mentioned can be optimized efficiently only with crossover, which makes it possible to create from two individuals one that is far from both of them. This is essential to jump over large valleys and to find a high mountain with a small base. Mutation has only a tiny chance of finding this mountain by jumping over the valley. However, if the valley is not too deep and rather flat, fitness-proportional selection may allow a walk through the valley. This idea has been made precise in [36]. For $s > 1$ let

$$\text{Jump}^*_m(x) := \begin{cases} s^{||x||} & \text{if } ||x|| \leq n - m \text{ or } ||x|| = n \\ s^{n-m} + n - m - ||x|| & \text{otherwise.} \end{cases}$$

The behavior of the $(1+1)$-EA on $\text{Jump}^*_m$ is essentially the same as on $\text{Jump}_m$. However, our generic algorithm with $p_c(n) = 0$ (no crossover), fitness-proportional selection of the individuals creating new individuals and the individual to be eliminated has a good chance that individuals with more than $n-m$ ones can create offspring with more ones and that these individuals survive long enough. Indeed, the expected optimization time for $s = \Omega(n^{\log n})$, $m = O(\log(n) \cdot \log \log(n))$ and the population size $n$ is polynomial while the $(1+1)$-EA needs superpolynomial time.

We have presented the first proofs that crossover (and even populations without crossover) can decrease the expected optimization time significantly. This opens the possibility of a rigorous analysis of more general evolutionary algorithms than the $(1+1)$-EA. In this section we have only presented special forms of the results of [36, 37, 38] in order to support its legibility.

## 5.6   Takeover Times and Related Key Measures

The analyses presented in the previous sections are mostly concerned with estimations of the expected optimization time. Thus, the evolutionary algorithm is considered as a whole and analyzed similarly to the way randomized algorithms are. An alternative well-established approach considers evolutionary algorithms as complex, dynamical systems and concentrates on the analysis of key aspects. In the field of engineering science such an approach is called dimensional analysis. The idea is to identify important dimensions of a complex system and to establish functional relationships between them. Applications of this idea to evolutionary algorithms typically lead to concentration on the three main operators employed in EA, i. e., recombination or crossover, mutation, and selection. If, for example, one concentrates on crossover, then results on the so-called mixing time are desired [51, 52]. When considering the effects of selection in isolation, the classical approach is to consider the takeover time as introduced by Goldberg and Deb [27].

The latter approach has the inherent difficulty that even exact knowledge about the properties of one operator may not be helpful when considering this operator in combination with other operators. However, the interest in the

determination of takeover times for different selection operators seems not to be weakened by this difficulty. The reasons probably lie in the fact that the underlying probabilistic model can be instantiated with different semantics (e. g., epidemics, forest fires, broadcasts in faulty networks, and others) so that the results concerning takeover times can lead to valuable results in other disciplines. We have participated in this kind of research by investigating takeover times and takeover probabilities for spatially unstructured as well as for spatially structured populations. Moreover, we propose a novel key measure that contains more information than the takeover time.

We consider a population of size $n$ that initially contains exactly one individual with best fitness. We are interested in the development of the number of copies of this individual under application of selection solely. We denote the number of such individuals in generation $t$ by $N_t$ and have $N_0 = 1$. We are interested in the first point in time $T$, where the best individual occupies the whole population, i. e., $N_T = n$. The takeover time is defined as the expected value of $T$. Obviously, depending on the selection method, $N_t$ may become 0 at some point of time with nonzero probability so that the takeover time is not well defined. This motivates the investigation of another measure, the takeover probability $P(N_t = n)$. It is, of course, possible to calculate this takeover probability numerically for different values of $t$ (see [5] and [71] for several generational and nongenerational selection methods). In order to derive analytical results, one considers the ultimate takeover probability $\lim_{t \to \infty} P(N_t = n)$. The random process of iterated selection can be modeled as a Markov chain with state space $\{0, 1, \ldots, n\}$, where 0 and $n$ are the only absorbing states. If one considers nongenerational selection methods, only transition probabilities $p_{i,j}$ with $|i - j| \leq 1$ are different from 0, and the ultimate takeover probability can be precisely and compactly stated [60]. Three special cases of particular interest are investigated. In noisy $s$-ary tournament selection $s$ randomly chosen individuals are ranked according to their fitness. With error probability $1 - \alpha$, $\alpha \in (0; 1)$, the second best individual in this set is chosen, while with probability $\alpha$ the best individual is chosen. The ultimate takeover probability equals $(r^n - r^{n-1})/(r^n - 1)$ with $r = (1 - \alpha)/\alpha$. In random replacement selection the better one of two randomly chosen individuals replaces a randomly chosen individual leading to an ultimate takeover probability of $(1/2) + 1/(2n)$. Finally, in kill tournament selection the better one of two randomly chosen individuals replaces the individual that is the worse one among two randomly chosen individuals. Here, the ultimate takeover probability can be proven to be in the interval $[1 - 1/n; 1 - 1/(5n)]$.

If the ultimate takeover probability is 1, it makes sense to investigate the takeover time. Nongenerational selection methods that never lose a best copy are particularly easy to analyze. This enables us to give precise values for the takeover time of $s$-ary tournament selection and replace worst selection.

If a selection method has an ultimate takeover probability different from 1, it can be extended to undo the extinction of the last copy of the best individual. For nongenerational selection methods with such an undo mechanism Rudolph [67] derives a compact and exact expression for the takeover time. Using this expression, the takeover times for many selection mechanisms can be derived. We give an overview of takeover times for many different well-known selection methods and refer to [67] for details. Note that $H_n$ denotes the $n$th Harmonic number.

| Selection Method | Takeover Time |
|---|---|
| 4-ary Tournament | $\leq \frac{1}{2} n H_{n-1}$ |
| Replace Worst | $\leq \frac{1}{2} n H_{2n-1}$ |
| Ternary Tournament | $\frac{2}{3} n H_{n-1}$ |
| Binary Tournament | $n H_{n-1}$ |
| Kill Tournament with Undoing | $\leq (n + \frac{1}{2}) H_{n-1}$ |
| Noisy Binary Tournament with $\alpha = \frac{1}{4}$ | $\leq 2 n H_{n-1}$ |
| Random Replacement with Undoing | $\frac{2n^2}{n+1} H_{n-1}$ |
| Noisy Binary Tournament with $\alpha = \frac{1}{2}$ | $n^2 H_{n-1}$ |

A population is said to be spatially structured if there is a neighborhood structure on the set of individuals $\{1, \ldots, n\}$ in the population given by an undirected graph $G = (\{1, \ldots, n\}, E)$, where $\{i, j\} \in E$ iff $i$ and $j$ are neighbors. Since we are interested in the takeover of the population by the best individual only, we can restrict ourselves to consider $V_i(t) \in \{0, 1\}$, where $V_i(t) = 1$ iff node $i$ contains a best individual in the $t$th generation. We have $N_t = \sum_{1 \leq i \leq n} V_i(t)$ and consider the Markov chain $\{(V_1(t), \ldots, V_n(t)) \mid t \geq 0\}$. We denote the takeover time, if the only best individual is initially at vertex $i$ by $E_i(T)$ and assume that a randomly chosen node contains the only best individual initially. Thus, obviously $E(T) = (\sum_{1 \leq i \leq n} E_i(T))/n$ holds. We denote $P(V_i(t+1) = 1 \mid V_i(t) = 0)$ as upgrade probability of node $i$ at generation $t$ and analogously denote $P(V_i(t+1) = 0 \mid V_i(t) = 1)$ as downgrade probability. If all downgrade probabilities are 0, we can prove that $E(T) \geq (\sum_{1 \leq i \leq n} \max\{\text{dist}(i, j) \mid 1 \leq j \leq n\})/n$ holds [62]. If all upgrade probabilities are 1, the inequality becomes an equality. For special graphs, such as rings or arrays much tighter results can be obtained, e. g., for vertex transitive graphs we can prove $E(T) = \text{diam}(G)$ under the same conditions. Let $G = (V, E)$ be an array with $V = \{1, \ldots, n\}$ and $E = \{\{i, i+1\} \mid 1 \leq i < n\}$. Then $(3n-3)/p \leq E(T) \leq (4n-3)/(4p)$ holds for the takeover time $E(T)$, where $p$ is the upgrade probability (identical for all nodes). If the neighborhood graph $G$ is a ring, the takeover time does not depend on the initial vertex, and we have $E(T) = n/(2p) - (1 - (-1)^{n-2} \cdot (p/(2-p))^{n-2})/4$ for the takeover time, where again $p$ is the upgrade probability [62].

As already discussed above, takeover time is a measure that is not very informative. Therefore, we propose a new measure for selection that hopefully

will turn out to be useful [63]. Let $B_T := \sum_{0 \leq t \leq T} N_t$ be the total number of best individuals before takeover at generation $T$. Then we define

$$\eta := 1 - \mathrm{E}(B_T)/(n \cdot \mathrm{E}(T)),$$

i.e., the mean fraction of nonbest individuals that were available for the generation of candidate solutions prior to takeover. Since we already know how to derive $\mathrm{E}(T)$, we can concentrate on $\mathrm{E}(B_T)$ here. Let $W_i$ be the number of time-steps $t$ where $N_t = i$, $i \in \{1, \ldots, n-1\}$, until takeover generation $T$. Then we have $\mathrm{E}(B_T) = \sum_{1 \leq i < n} i \cdot \mathrm{E}(W_i)$. For nongenerational selection methods that never lose a best copy, we have $\mathrm{E}(W_i) = 1/p_{i,i+1}$, so that $\eta$ can be easily obtained. Rudolph [63] describes some applications for concrete selection methods. As with any such measure, one has to be careful with the interpretation of the results. But since the classification number $\eta$ aggregates more information about the selection method than the takeover time alone, we hope it will turn out to be a helpful additional characterization.

## 5.7   Dynamization and Adaptation

In the previous sections we analyzed EA whose parameter settings were fixed to a given value, e.g., the mutation probability $1/n$ of the (1+1)-EA (Algorithm 1). Naturally, the search behavior of an EA is influenced by the setting of its parameters, e.g., population size, mutation, and crossover probability. It is typically the case that for an objective function, an EA will only be successful when the parameters are set to appropriate values. However, it is well known that finding good or even optimal settings is a hard problem on its own. As an example, we consider the choice of the mutation probability in detail and concentrate on the (1+1)-EA. The most often recommended choice for the mutation probability is $1/n$ [2]. For linear functions, we show that a mutation probability $\Theta(1/n)$ is optimal [18]. However, it is easy to come up with functions where the mutation probability $1/n$ is far from being optimal. For Ackley's `Trap` function [1] as well as the needle in the haystack functions, which have a flat fitness landscape with one single point that has maximal function value, setting the mutation probability to $1/2$ is superior to $1/n$. Since in both examples, the (1+1)-EA has exponential expected optimization time regardless of the mutation probability, these examples have no practical relevance. So, one is interested in examples where the expected optimization time can be reduced to some polynomial of small degree. Jansen and Wegener [33] construct a function called `PathToJump`, where the (1+1)-EA is led on a narrow path to a local optimum where $\log n$ bits have to flip simultaneously in order to reach a global optimum. If the mutation probability is substantially smaller than $(\log n)/n$, this final step takes very long. If, on the other hand, the mutation probability is substantially larger than $(\log n)/n$, it takes very long to find the local optimum. Thus, with mutation probability $c \cdot (\log n)/n$, the expected optimization time of the (1+1)-EA is

$O\left(n^{2+c}\ln^{-1}n + n^{c-\log c - \log\ln 2}\right)$, which is $O\left(n^{2.361}\right)$ with $c = 1/(4\ln 2)$. If the mutation probability is set to $\alpha(n)/n$ or $1/(\alpha(n)\cdot n)$ for some function $\alpha : \mathbb{N} \to \mathbb{R}$, where $\lim_{t\to\infty}\alpha(n) = 0$ holds, then the (1+1)-EA needs a super-polynomial number of steps to find a global optimum of PathToJump with a probability that converges to 1 [33].

While this result proves that the choice of the parameter settings is crucial for the success of evolutionary algorithms, it does not answer the question of how good parameter settings can be found. This is in fact a difficult problem. Thus, having evolutionary algorithms that find appropriate parameter settings for themselves and thereby solving the problem of finding good settings is obviously attractive. Bäck [3] distinguishes three different ways of choosing parameter settings. In a static setting the parameters are fixed at the beginning of a run and not changed throughout the run. Dynamic parameter control refers to a mechanism that changes the parameter settings during the run, whereas the actual setting can only depend on the number of steps the EA has taken so far. The most complex way is adaptive parameter control, where the control mechanism can take into account not only the number of steps but also all search points and their function values. Whereas for evolutionary algorithms, adaptive parameter control is the most usual nonstatic setting, it makes sense to investigate the structurally simpler dynamic control, too. Moreover, simulated annealing employs dynamic parameter control to change the temperature during the run. One can describe simulated annealing as a kind of evolutionary algorithm that is similar to the (1+1)-EA. The mutation is different: in each step, exactly one of the $n$ bits is chosen uniformly at random, and its value is changed. The selection is different too: the new child $y$ replaces the parent $x$ with probability $\min\left\{1, \alpha(t)^{f(y)-f(x)}\right\}$, where $\alpha : \mathbb{N} \to [1;\infty)$ is the so-called cooling schedule; the value $\alpha(t)^{-1}$ is referred to as temperature. If $\alpha(t)$ is fixed during the run, i.e., with static parameter setting, this algorithm is known as Metropolis algorithm. And it is an open problem to prove for a natural objective function that changing the temperature during the run can be necessary for successful optimization in expected polynomial time. Using methods similar to the ones used to analyze the (1+1)-EA, Droste, Jansen, and Wegener [16] were able to prove for two artificial example functions that simulated annealing can have polynomial expected optimization time, whereas the Metropolis algorithm has exponential expected optimization time even when $\alpha$ is chosen optimally. This can be done for a function $\alpha : \mathbb{N} \to [1;\infty)$ that is increasing as well as for a decreasing function $\alpha$.

A dynamic variant of the (1+1)-EA where the mutation probability is subject to dynamic parameter control was introduced by Jansen and Wegener [33]. The main idea is to try a reasonable number of different mutation probabilities from a reasonable range. Since setting the mutation probability to values smaller than $1/n$ implies that, on average, in many steps no bits flip at all, $1/n$ is taken to be a reasonable lower bound. Setting the muta-

tion probability to 1/2 implies that the child is chosen independently of the parent uniformly and at random. Thus, 1/2 is an upper bound on the range of sensible mutation probabilities. Initially, the mutation probability is set to $1/n$. After each step the mutation probability is doubled. If the new value exceeds 1/2, the mutation probability is set to $1/n$ again. Thus, $\log n$ different mutation probabilities of the form $2^i/n$ are used. Jansen and Wegener [33] prove that this dynamic (1+1)-EA optimizes the function PathToJump, on average, in $O\left(n^2 \log n\right)$ steps. Thus, it successfully overcomes the problem of finding a good mutation probability in this case.

A more detailed analysis of this dynamic (1+1)-EA is given by Droste, Jansen, and Wegener [16] and Jansen and Wegener [34]. One can divide a run of the algorithm in phases of length $\log n$, where in each phase, all different mutation probabilities are used. Considering such phases instead of single steps, all upper bounds for the (1+1)-EA proved with the help of $f$-based partitions carry over to the dynamic (1+1)-EA with an additional factor of $\log n$. This holds for example for OneMax and LO [16], where also corresponding lower bounds can be proved. For LO the optimization time is again sharply concentrated, the expected optimization time equals $\Theta\left(n^2 \log n\right)$ [16]. For OneMax the expected optimization time is $\Theta\left(n \log^2 n\right)$ [34]. Various other upper bounds on the expected optimization time of the dynamic (1+1)-EA for different classes of objective functions, e. g., unimodal functions, linear functions, and monotone polynomials of bounded degree can be found in [34] too.

Since the upper bound on the expected optimization time for PathToJump is slightly smaller for the dynamic (1+1)-EA than for its static counterpart, one could speculate that the dynamic mutation probability control might be advantageous in some cases, even when compared to optimal static settings. In fact, Jansen and Wegener [34] introduce a function that is, on average, optimized by the dynamic (1+1)-EA in $O\left(n^2 \log n\right)$ steps, whereas the (1+1)-EA needs an exponential number of steps with probability exponentially close to 1, even with optimal mutation probability. However, the opposite can also be true. In [34], an example function is presented that is optimized by the (1+1)-EA with standard static mutation probability $1/n$ in $O\left(n^2\right)$ steps with probability exponentially close to 1, while the dynamic (1+1)-EA needs with a probability exponentially close to 1 exponentially many steps.

As mentioned above, the most complex way to control parameter is to use adaptive parameter control, where the control mechanism may take into account all points in the search space encountered so far, together with the function values, in order to decide on the change of the parameter settings. One relatively simple and well-known mechanism is the 1/5 success rule introduced by Rechenberg [53] for the (1+1) evolution strategy (ES). This (1+1)-ES is similar to the (1+1)-EA but is designed for optimizing functions $f\colon \mathbb{R}^n \to \mathbb{R}$. Mutation is therefore implemented by adding some random vector according to some chosen distribution. One observes a prescribed number

of steps and measures the relative frequency of steps where the child $y$ replaces its parent $x$. If this quantity is larger than $1/5$, then the step size control parameter (typically the variance of the mutation distribution) is increased by some factor, if it is smaller than $1/5$, then this parameter is decreased by some factor. Schwefel [69] introduces a modification by replacing the prescribed factor by a lognormally distributed random variable and making the control parameter a part of the individual. This way, the control parameter is evolved together with the individual, a mechanism that is referred to as self-adaptation. Rudolph [55] proves that self-adaptation methods for which the control parameter can be bounded above and below by some positive constants guarantee that a global optimum is found with probability 1. However, for the $1/5$ success rule, no such bounds can be given. Thus, it is not obvious that one can be sure that eventually a global optimum will be found. In fact, Rudolph [61] proves for a example function that the (1+1)-ES with Gaussian mutations as well as with Cauchy mutations gets stuck in a local optimum with a probability that is bounded below by a positive constant. The idea is that once the local optimum is reached, the number of steps where the child $y$ replaces its parent $x$ become less frequent. Thus, by the $1/5$ success rule, the variance of the mutation distribution is more and more reduced, such that it becomes even less likely to leave the local optimum. This result remains valid if in each step, not only 1 but $\lambda \geq 1$ children are produced, i. e., it holds for a $(1 + \lambda)$-ES with Gaussian or Cauchy mutations [61]. Rudolph [65] is able to generalize this result further, proving it for a multivariant example function and showing that it holds for any mutation distribution with independent marginal distributions.

We argued above that dynamic settings of parameters may be necessary since there are objective functions that can be optimized with a dynamic parameter control within a small polynomial number of steps with probability close to 1, whereas no static parameter setting leads to efficient optimization. The objective function itself is not static during the optimization process but changes with time, which is another reason why dynamism may be necessary. Since this case differs drastically from the optimization of static functions, it makes sense to begin with the analysis of simplified algorithms and problems, as motivated and discussed in Section 5.3. Hence, we look at a very simple dynamic variant of OneMax. Maximizing OneMax is equivalent to minimizing the Hamming distance to the target bit-string $\mathbf{1}^n$. Using this interpretation, an almost canonical dynamic variant OneMax$_t$ of OneMax is to minimize the distance to a target bit-string $y_t$ that changes with time $t \in \mathbb{N}$:

$$\text{OneMax}_t(x) := n - H(x, y_t).$$

The (1+1)-EA for OneMax$_t$ works in the same way as for OneMax, except when using the fitness function OneMax$_t$ after $t$ mutations have been performed. Therefore, it is possible that the fitness of the actual element $x_t$ decreases with $t$, if the target bit-string $y_t$ moves away from $x_t$, and the

child $x_{t+1}$ cannot compensate for this. Certainly, the expected time until the (1+1)-EA finds an optimal $x_t$, i.e., $\texttt{OneMax}_t(x_t) = n$ depends on the process determining the change of $y_t$. Here, we assume that $y_{t+1}$ is generated by flipping one uniformly chosen bit of $y_t$ with probability $p'$ and setting $y_{t+1} = y_t$ with probability $1 - p'$. A similar process was examined in [72], but here, only the distribution of $x_t$ was computed numerically and compared to experiments.

In [9] we prove that for $p' = O((\log n)/n)$ the expected runtime of the (1+1)-EA on $\texttt{OneMax}_t$ is polynomial, but becomes superpolynomial for all substantially larger probabilities $p' = \omega((\log n)/n)$. For the polynomial upper bound we look at a slower process basically resulting from the (1+1)-EA by omitting all possibilities to increase the fitness of $x_t$ from one generation to the next by more than one. Then we can use standard methods to analyze this simpler process and show that even its expected runtime is polynomial.

For the superpolynomial lower bound for $p' = \omega((\log n)/n)$ we concentrate on the phase where the fitness of all individuals is at least $n - G$, and $G$ is essentially $\log n$. We have shown that this phase, which is necessary to reach the optimum, has a superpolynomial small probability (which is similar to a technique used by [52]). The main idea here is to upper bound the increase of fitness in one step and to show that this implies that the number of improvements has to be by a constant factor larger than its expected value, where all cases not considered can only have a superpolynomially small probability. Then, Chernoff bounds can be used to show that this is superpolynomially unlikely (see [9] for details).

All in all, we have proven a number of different theoretical results concerning dynamic EA or EA for dynamic problems, showing that our methods can also be carried over to nonstatic problems.

## 5.8   Metric-Based EA (MBEA) and an Application in GP

In the previous sections we have seen many theoretical results on EA for various pseudo-Boolean functions. Although the methods used for our analyses can often be generalized to other functions, they depend on the fact that we have a rather accurate knowledge of the fitness function. When designing EA, knowledge about the function to be optimized is often very limited. Hence, present theoretical results on EA can only hint at which operators to choose in order to get an efficient EA.

Furthermore, often the problem at hand is not given in a form suited to generic forms of EA, e.g., the input (of a graph) is either not given as a bit-string or a real vector. In this case, we can roughly distinguish two possibilities used in practice: either the inputs are transformed into one of the generic forms by a mapping function and standard crossover and mutation are used; or the representation of the inputs is not changed, but problem-specific

crossover and mutation operators are used. In the first case the so-called genotype set $G$ (the set of representations) differs essentially from the so-called phenotype set $P$ (the set of search points). Although it is trivial to find a mapping function $m : G \rightarrow P$, it can be a difficult problem to find one that preserves the good properties of the EA used. Therefore, the design of EA is often a very time-consuming, trial-and-error procedure.

In both cases the consequences of applying mutation or crossover on the represented elements of the phenotype space are often badly understood if the mapping or the problem-specific representation are not chosen carefully. This nullifies one of the most important advantages of EA, their intuitive understandability. For instance, mutation is believed most often to make only small changes. But when the mapping is not suitably chosen, a small change in the genotype space can imply a large change of the represented search point. Analogously, it is not clear that crossover creates search points that are similar to their parents.

In order to solve these problems and to present design guidelines for EA, we set up the metric-based EA [20, 21]. Its name stems from the fact that it is based on a metric $d_P : P \times P \rightarrow \mathbb{N}$ on the phenotype space. The fundamental role of $d_P$ derives from the fact that in the general MBEA concept $d_P$ is the only element that is directly influenced by the fitness function $f : P \rightarrow \mathbb{R}$. For $p_1, p_2 \in P$, the value $d_P(p_1, p_2)$ should be at least an upper bound for the difference of the fitness values $|f(p_1) - f(p_2)|$. Hence, two search points that are close to each other with respect to $d_P$ will have similar fitness. This leads to a smooth fitness landscape, which is necessary for the efficiency of every search heuristic relying on some form of locality of the fitness function [53].

Certainly, EA also respect locality, inasmuch as mutation is most often expected to do only small changes and crossover to generate search points similar to their parents. To formalize these properties with respect to the metric $d_P$ we need a related metric $d_G$ on $G$, where mutation and crossover work. Assuming that the mapping $m$ is bijective, we define $d_G$ by $d_G(g_1, g_2) := d_P(m(g_1), m(g_2))$.

One property intuitively required from mutation is that the probability of mutating an individual to another one should strictly decrease inversely proportional to the distance between them. This guarantees that mutation prefers small steps, which should be helpful if the fitness values of individuals with small distance are similar. Implicitly, this guideline is used in the design of most EA, but only incorporating a metric small steps made in the genotype space will lead to small steps of the fitness values. Other guidelines require that every individual can be reached by mutation and that all individuals with the same distance to the parent have the same probability of being reached (for the formal definition see [20, 21] and Section 6.4.2).

In many EA, crossover plays the role of concentrating information by generating individuals that share common properties with their parents. To

formally guarantee this behavior for MBEA, we include the guideline that the distance of a result of crossover to either of its parents should not be greater than the distance between the parents itself, while it should not generate individuals with a bias toward one of its parents. Hence, if the parents are similar to each other, crossover generates individuals that are similar to the parents. This should speed up the search process, as the fitness function is smooth with respect to the metric.

Although these guidelines defining an MBEA only formalize the intuitive role of mutation and crossover many researchers and practitioners on EA have defined, this is the first time that they have been explicitly and formally stated. (We do not give any guidelines on selection, as all common versions of selection fulfill its intuitive role: prefer those search points having a better fitness.) Theoretically showing that an MBEA outperforms another EA that is not fulfilling the MBEA guidelines for some fitness functions would prove the usefulness of this concept. But the difficulty in analyzing EA with crossover (see Section 5.5 for the only successful analysis made) restricts us to experimental comparisons.

Genetic programming (GP) [39] is an EA paradigm, where the goal is to find a program optimizing a given fitness function. Hence, the phenotype space consists of all functions between two sets $A$ and $B$ (in the following we assume $A = \{0,1\}^n$ and $B = \{0,1\}$). One of the most often used representation forms in GP is the S-expression: an S-expression is a tree where every inner node contains a function from a function set $F$ and every leaf contains a terminal from the terminal set $T$, including the input variables $x_1, \ldots, x_n$. Assuming that the function of every inner node has the same arity as the number of its successors, an S-expression is evaluated bottom-up.

We chose GP, as in most GP systems mutation and crossover fulfill their intuitive roles only in the genotype space. Typically, mutation randomly chooses a subtree and replaces it with a new random subtree, while crossover chooses two subtrees in the parents and swaps them. Hence, while the consequences on the S-expressions are easy to describe, the resulting change of the represented functions is very hard to understand and can be far reaching. Therefore, most GP systems do not fulfill the MBEA guidelines.

This becomes especially clear when we consider the problem of finding an unknown Boolean function $f^* : \{0,1\}^n \to \{0,1\}$, which can only be accessed by a fitness function $F$ returning the number of inputs where the valuated function agrees with $f^*$. In this case, an appropriate metric $d_P$ on the set of functions measures the number of inputs $x \in \{0,1\}^n$ where the two functions differ, i.e.,

$$d_P(g,h) := |\{x \in \{0,1\}^n \mid g(x) \neq h(x)\}|,$$

as the distance $|F(g) - F(h)|$ is exactly $d_P(g,h)$.

While creating a GP system fulfilling the MBEA guidelines for $d_P$, even if using S-expressions is easy, it is hard to create an efficient one, as S-expressions are a very general representation. For Boolean functions, ordered

binary decision diagrams (OBDD) [4] are a very efficient form of representation [76]. Hence, we use OBDD to represent the individuals of our GP system fulfilling the MBEA guidelines (therefore called MBGP system).

OBDD are directed acyclic graphs with one source and two sinks, where every inner node is labelled by one of the variables $x_1, \ldots, x_n$ and has two outgoing edges, called 0- and 1-edge, while the two sinks are labelled 0 and 1. To evaluate an OBDD for an input $a \in \{0,1\}^n$, we start at the source and follow a node labelled $x_i$ its $a_i$-edge until we reach a sink, whose label determines the function value. Every OBDD has a variable ordering $\pi$, requiring that an edge between two inner nodes has to point from a smaller variable to a larger one with respect to $\pi$. Only this guarantees that the evaluation merely needs time $O(n)$ and allows many efficient operators on OBDD with the same variable ordering, where the synthesis of two functions is the most important one for us. Furthermore, OBDD allow a unique minimal representation, which can be efficiently computed, leading to small memory requirements.

If $O_1$ and $O_2$ are two OBDD with respect to $\pi$ representing $f_1$ resp. $f_2$ and if $\otimes : \{0,1\}^2 \to \{0,1\}$ is a Boolean function, the synthesis of $O_1$ and $O_2$ with respect to $\otimes$ consists of computing the OBDD representing the function $f_1 \otimes f_2$. This can be done in time $O(|O_1| \cdot |O_2|)$ [76]. The synthesis of two OBDD allows the represented functions to be changed efficiently in a precisely defined way, making it the basis for mutation and crossover in our MBGP system. Although OBDD have been used in GP systems before (e.g., [84] or [8]), these systems do not make full, or any use of the synthesis operator.

Mutation of an OBDD $O_1$ in our MBGP system flips each of the $2^n$ output bits of the represented function independently, with probability $2^{-n}$. By first creating an OBDD representing the flipping output bits and then doing an EXOR synthesis with $O_1$, this can be done in expected time $O(n \cdot |O_1|)$. Hence, mutation works efficiently and also fulfills all MBEA guidelines. Crossover of two OBDD uses a modified $\oslash$-synthesis of the OBDD, where $\oslash$ returns one of its arguments with probability $1/2$, which guarantees that the resulting function can only differ from the parents for those inputs where the parents do not have the same output. As the order of the parents has no influence on the probability distribution of the resulting function, crossover also fulfills the MBEA guidelines.

We have compared this MBGP system with four GP systems using S-expressions [6, 7, 39, 40] and also an GP system using OBDD which does not fulfill the MBEA guidelines [8]. As a measure we used the computational effort [39], an estimate for the number of search points to be evaluated to find an optimum with 99 percent probability. Test functions to be found were the multiplexer and the parity functions for values of $n$, the number of inputs, between 3 and 12, where the optimal variable ordering leading to minimal OBDD was given to the MBGP system. Without going into details, the computational effort of the MBGP system for all test functions was, by at least a factor of 10, smaller than those of the other systems. Further

experiments showed that crossover essentially speeds up the optimization process. As the problem of finding a completely defined function is equivalent to maximizing OneMax, this emphasizes the lack of theoretical results for any EA with crossover on OneMax. All in all, the MBEA concept seems to be advantageous at least for the problem of finding Boolean functions considered here.

Although the variable orderings of the OBDD do not have any influence on the computational effort of the experiments made, it is crucial for time and space needs of the GP system, as there are functions where, for a random variable ordering, even approximations on $m$ inputs have only OBDD of size $\Omega(m/\log m)$ with probability converging on one [41]. Furthermore, the problem of computing a minimal variable ordering cannot be efficiently approximated up to a constant factor if NP $\neq$ P [70]. Hence, we have presented a GP system using OBDD, where the variable ordering used is not predetermined but part of the evolutionary search [10].

Because having OBDD with different variable orderings in the same population would require complicated and inefficient genetic operators, we have used a distributed approach. Here, we have four different populations, each having its own variable ordering, which is initially randomly chosen. On these populations, separate EA are working, trying to minimize the size of the OBDD, which are forced by the genetic operators to be consistent with the incomplete training set. Only every $M$ generations, are elements from one population transferred to another population, while the variable ordering of the populations is adjusted every $N$ generations using one of the well-known heuristics for this problem. The experimental data shows that using this distributed GP system (refer to [10] for further details), substantially smaller OBDD could be found than are found when using a random variable ordering, leading GP systems with OBDD one more step closer to practical applications.

## 5.9    Black-box Optimization

In the preceding sections, we presented many results on the analysis of evolutionary algorithms. It has been shown how EA can be analyzed rigorously. Upper bounds for EA and special problems as well as lower bounds for classes of EA on these problems have been obtained. The value of these lower bounds is limited to the class of algorithms just considered. One always may argue that perhaps some EA may be significantly better where a certain parameter is changed. General lower bounds would be better, but they are possible only under a complexity theoretical assumption, such as NP$\neq$P or RP$\neq$NP.

In [11] we present another approach to proving lower bounds. It is based on the fact that EA are considered to be robust and general randomized search heuristics. This implies, as argued in the introduction to this chapter, that the search heuristic may have problem-specific operators, but it works

independently of the current instance. More precisely, its knowledge of the current instance is limited to the $f$-values of all sampled search points. This leads to the black-box optimization scenario and the black-box complexity of problems.

Black-box algorithms know the problem and, initially, nothing about the instance. According to some probability distribution, they choose $a_1$ and learn the value of $f(a_1)$. The probability distribution to choose $a_t$ may depend on $a_1, f(a_1), \ldots, a_{t-1}, f(a_{t-1})$. The algorithm is only charged for samples, and its runtime is the smallest $t$, such that $a_t$ is $f$-optimal.

Black-box optimization differs from classical optimization. Since computation is for free, NP-equivalent problems can be polynomial-time solvable. We know that the maximization of degree-2 pseudo-Boolean polynomials (for the definition see Section 5.3) is NP-equivalent. However, sampling $f(a)$ for all $O(n^2)$ inputs $a$ with at most two ones offers enough information for the efficient computation of all parameters (coefficients) of the unknown polynomial (instance). Then we may solve the optimization problem without sampling. Although this takes exponential time, it is charged no cost. This makes black-box optimization in some situations unrealistically powerful.

In any case, black-box complexity is a lower bound on the runtime of all randomized search heuristics (not working with the parameters of the specific instance). Sometimes, as seen above, these bounds are too small to be significant. In such situations space-restricted black-box optimization may be a solution. Most randomized search heuristics store only few of the sampled search points (simulated annealing or the (1+1)-EA store only one point, and the population size of EA is typically rather small, often much smaller than $n$ for the search space $\{0, 1\}^n$). Lower bounds for space-restricted black-box optimization, which are larger than the best known lower bounds for the non-restricted case are not known and are a subject for future research. Therefore, we discuss only the non-restricted case.

Black-box optimization offers the possibility of proving lower bounds not based on complexity theoretical assumptions. The main lower-bound technique is Yao's Minimax principle [85]. If a problem has a finite set of instances of a fixed size and allows a finite set of deterministic algorithms, the minimal worst-case instance expected optimization time is lower bounded for each probability distribution on the inputs by the expected optimization time of an optimal deterministic algorithm. This implies the freedom to choose an appropriate probability distribution on the inputs. Then a deterministic black-box algorithm can be described by a decision tree. The root is labeled with the first query $a_1$ and is left by an edge for each possible answer. The edge with label $z$ leads to a node labeled with the second query if the $f$-value of $x_1$ equals $z$. The lower bounds even hold for nonuniform black-box algorithms.

A simple example is the class of needle in a haystack functions $f_a$ where $f_a(x) = 0$ if $x \neq a$, and $f_a(a) = 1$. An algorithm knowing the instance and,

therefore, the parameter $a$ can immediately produce the output $a$. Considering the uniform distribution on all $a$, we obtain a lower bound of $2^{n-1}+1/2$ on the black-box complexity of this class of functions. The decision tree of each black-box algorithm (not repeating former queries) is a degenerated tree with $2^n$ leaves. Its expected depth equals $2^{n-1} + 1/2$. A matching upper bound is easy to obtain. This simple result already has implications. People tend to claim that EA are inefficient on needle in a haystack functions. The truth is that the black-box optimization of these functions is a difficult problem and the expected optimization time of the (1+1)-EA is away from the lower bound only by a factor $2(1 - e^{-1}) \pm o(1)$, and therefore the (1+1)-EA is efficient.

The situation is different for so-called trap functions $\mathtt{Trap}_a$ where $\mathtt{Trap}_a(x)$ is $||x||$ if $||x|| \neq a$, and $\mathtt{Trap}_a(a) = 2n$. Again the black-box complexity equals $2^{n-1} + 1/2$, but most evolutionary algorithms are inefficient, since their expected optimization time is $2^{\Omega(n \log n)}$. The simple reason is that they follow the wrong hints given by the fitness function $\mathtt{Trap}_a$.

Black-box complexity is defined for problems and not for single functions. Hence, it is important to define an appropriate class of functions with the same properties as the example functions. This has been done in [11] for many of the example functions often discussed in EA papers, among them the linear functions $\mathtt{OneMax}$ and $\mathtt{BV}$ (binary value) and the function $\mathtt{LO}$ (leading ones). Among others the royal road functions (see Section 5.5) have been investigated. The base function equals for $n = dk$

$$\mathtt{RR}_d(x) := x_1 x_2 \cdots x_d + x_{d+1} x_{d+2} \cdots x_{2d} + \cdots + x_{(k-1)d+1} \cdots x_{kd}.$$

One corresponding class of functions considers all functions obtained from $\mathtt{RR}_d$ by replacing some $x_i$ with their negations $1 - x_i$. The task is to find the right bits, and the fitness only scores for each block where all bits are correct. In this situation we can prove a lower bound of size $\Omega(2^d + n/\log n)$ and an upper bound of size $O(\min\{2^d \cdot n, 2^d + \frac{n}{d} \cdot \log(n/d)\})$. Another possible generalization is to also consider all partitions of $\{1, \ldots, n\}$ into $k$ blocks of size $d$. Then we cannot get larger lower bounds and our upper bound is of size $O(\min\{2^d \cdot \log(n/d) + \frac{n^2}{d}, 2^d \cdot n \cdot \log(n/d)\})$. We can conclude from the lower bound that no randomized search heuristic (whether or not it uses crossover) can outperform the (1+1)-EA significantly for the class of royal road functions.

In the rest of this section we discuss one lower bound more intensively. A pseudo-Boolean function $f$ is called unimodal if the global maximum is unique and all other points have a Hamming neighbor with a larger $f$-value. People believed for a long time that evolutionary algorithms could maximize each unimodal function efficiently. Then long path functions were defined, which were assumed to be difficult for the (1+1)-EA. This was proved in [12]. However, some people believe that crossover might help on these special long path problems. Then other people could define other unimodal functions,

and so on. This discussion has now been finished by proving that the black-box complexity of the class of unimodal functions is exponential. Llewellyn, Tovey and Trick [45] prove such a bound for deterministic black-box algorithms, which is much simpler by using adversary arguments. We generalize the bound to randomized algorithms.

First, we choose a probability distribution on the set of unimodal pseudo-Boolean functions. A random path $R$ starts at $1^n$ and has a length of

$$l := l(n) := 2^{n^{1-\varepsilon}}, \varepsilon > 0 \text{ a constant.}$$

The successor of a point on the path is chosen randomly among all its Hamming neighbors. We observe that $R$ is typically not a simple path (the path may contain some points several times). Let $p = (p_0, \ldots, p_l)$ be the random path $R$. The corresponding random path function RP is defined as follows:

$$\text{RP}(x) := \begin{cases} \text{OneMax}(x) - n & \text{if } x \neq p_i \text{ for all } i \\ i & \text{if } x = p_i \text{ and } x \neq p_j \text{ for all } j > i. \end{cases}$$

For $x$ on $R$, let $i$ be the largest index, such that $x = p_i$. This implies $\text{RP}(x) = \text{RP}(p_i) < \text{RP}(p_{i+1})$. Each point $x$ outside $R$ has a Hamming neighbor $y$ with more ones and in any case $\text{RP}(x) < \text{RP}(y)$. Hence, RP is unimodal with a unique global maximum at $p_l$. In the following, we investigate deterministic black-box algorithms against RP, which describes a probability distribution on the set of all unimodal pseudo-Boolean functions.

Since RP has cycles with overwhelming probability, we also consider the so-called true path $T$ obtained from $R$ by omitting all cycles. We believe that it is hard to optimize $R$ since knowing the initial part of $R$ up to $p_i$, we cannot guess with non-negligible probability a point $p_j$ where $j - i = \Omega(n)$. However, after some queries, we know some points on the path together with their RP values and we know some points outside $R$. This knowledge implies that the path is no longer purely random. In order to cope with all these problems we strengthen the black box.

The current knowledge is described by the set $P$ of points known to lie outside $R$ and some index $j$, implying that we know $p_0, \ldots, p_j$ and that $\text{RP}(p_i) \leq \text{RP}(p_j) = j$ for all $i \leq j$. Based on this knowledge the next query point $p^*$ is chosen. Let $p$ be the $n$th successor of $p_j$ on $T$. We can stop our search successfully if $p^*$ is optimal or $\text{RP}(p^*) > \text{RP}(p)$. Otherwise, our knowledge is updated by replacing $j$ by $\text{RP}(p)$ (we obtain $n$ points on $T$ and all cycles in $R$ before $p$) and by inserting $p^*$ into $P$ if $p^* \notin R$. Although our knowledge restricts the possibilities of $R$ and $T$, this knowledge is not enough to guess some point on $T$ which is far behind $p_j$. It is tedious to prove this statement. The result is that the success probability even after $l/n^2$ steps is just $2^{-\Omega(n)}$. This example shows how the investigation of the black-box complexity of interesting classes $F$ of functions leads to lower bounds on the worst-case expected optimization time of each randomized search heuristic on $F$ getting information on the instance $f$ solely via sampling.

## Conclusions

We have presented several theoretical results on various aspects of EA. The central point is that we have obtained rigorously proven estimates of the expected optimization time of different EA on some important problems and on several artificial functions. The latter results have been obtained to show certain important properties of EA.

In order to clarify the discussion about the NFL theorem, we showed that, although differences between optimization strategies in complexity-restricted scenarios can occur, such scenarios always include objective functions almost surely implying exponential runtime. Therefore, we analyzed a very simple EA, the (1+1)-EA, for different syntactically defined classes of objective functions, such as linear or unimodal functions, with respect to its runtime. The methods and insights gained allow us to analyze more complicated EA that use populations or crossover. In particular, we solved the old problem of proving that crossover can significantly decrease the expected optimization time.

Furthermore, we presented results concerning the convergence of EA for multiple objective functions. Many of the problems encountered in practice are of this type, showing the importance of developing methods to analyze them. Other results deal with takeover times, a key measure for analyzing different selection operators. The MBEA guidelines presented are a first approach to giving the design process of EA a better foundation.

Moreover, we showed lower bounds on the black-box complexity of problems, which are lower bounds for all randomized search heuristics. Nevertheless, the collection of all results showing what a theory on EA can look like is even more important than the separate results.

## References

1. D. Ackley. *A Connectionist Machine for Genetic Hillclimbers.* Kluwer, Boston, MA, 1987.
2. T. Bäck. Optimal mutation rates in genetic search. In S. Forrest, editor, *Proceedings of the 5th International Conference on Genetic Algorithms (ICGA 93)*, pages 2–8. Morgan Kaufmann, San Francisco, CA, 1993.
3. T. Bäck. An overview of parameter control methods by self-adaptation in evolutionary algorithms. *Fundamenta Informaticae*, 35:51–66, 1998.
4. R. E. Bryant. Graph-based algorithms for Boolean function manipulation. *IEEE Transactions on Computers*, 35:677–691, 1986.
5. U. Chakraborty, K. Deb, and M. Chakraborty. Analysis of selection algorithms: A Markov chain approach. *Evolutionary Computation*, 4(2):133–167, 1996.
6. K. Chellapilla. Evolving computer programs without subtree crossover. *IEEE Transactions on Evolutionary Computation*, 1(3):209–216, 1997.
7. K. Chellapilla. A preliminary investigation into evolving modular programs without subtree crossover. In J. R. Koza, W. Banzhaf, K. Chellapilla, K. Deb, M. Dorigo, D. B. Fogel, M. H. Garzon, D. E. Goldberg, H. Iba, and R. Riolo,

editors, *Proceedings of the Third Genetic Programming Conference (GP 98)*, pages 23–31. Morgan Kaufmann, San Francisco, CA, 1998.

8. S. Droste. Efficient genetic programming for finding good generalizing Boolean functions. In J. R. Koza, K. Deb, M. Dorigo, D. B. Fogel, M. Garzon, H. Iba, and R. L. Riolo, editors, *Proceedings of the Second Genetic Programming Conference (GP 97)*, pages 82–87. Morgan Kaufmann, San Francisco, CA, 1997.

9. S. Droste. Analysis of an evolutionary algorithm for a dynamically changing objective function. In *The 2002 IEEE Congress on Evolutionary Computation (CEC 2002)*, 2002. (Accepted).

10. S. Droste, D. Heutelbeck, and I. Wegener. Distributed hybrid genetic programming for learning Boolean functions. In M. Schoenauer, editor, *Proceedings of the 6th Parallel Problem Solving from Nature (PPSN VI)*, volume 1917 of LNCS, pages 181–190. Springer, Berlin, 2000.

11. S. Droste, T. Jansen, K. Tinnefeld, and I. Wegener. A new framework for the valuation of algorithms for black-box optimization. Technical Report CI-118/01, Collaborative Research Center 531, University of Dortmund, 2001.

12. S. Droste, T. Jansen, and I. Wegener. On the optimization of unimodal functions with the $(1 + 1)$ evolutionary algorithm. In A. E. Eiben, T. Bäck, M. Schoenauer, and H.-P. Schwefel, editors, *Proceedings of the 5th Parallel Problem Solving from Nature (PPSN V)*, volume 1498 of LNCS, pages 13–22. Springer, Berlin, 1998.

13. S. Droste, T. Jansen, and I. Wegener. A rigorous complexity analysis of the $(1 + 1)$ evolutionary algorithm for linear functions with Boolean inputs. In *Proceedings of the Third IEEE International Conference on Evolutionary Computation (ICEC 98)*, pages 499–504. IEEE Press, Piscataway, NJ, 1998.

14. S. Droste, T. Jansen, and I. Wegener. A rigorous complexity analysis of the $(1 + 1)$ evolutionary algorithm for separable functions with Boolean inputs. *Evolutionary Computation*, 6(2):185–196, 1998.

15. S. Droste, T. Jansen, and I. Wegener. Perhaps not a free lunch but at least a free appetizer. In W. Banzhaf, J. Daida, A. E. Eiben, M. H. Garzon, V. Honavar, M. Jakiela, and R. E. Smith, editors, *Proceedings of the First Genetic and Evolutionary Computation Conference (GECCO 99)*, pages 833–839. Morgan Kaufmann, San Francisco, CA, 1999.

16. S. Droste, T. Jansen, and I. Wegener. Dynamic parameter control in simple evolutionary algorithms. In W. N. Martin and W. M. Spears, editors, *Proceedings of the Sixth Foundations of Genetic Algorithms Workshop (FOGA 6)*, pages 275–294. Morgan Kaufmann, San Francisco, CA, 2000.

17. S. Droste, T. Jansen, and I. Wegener. A natural and simple function which is hard for all evolutionary algorithms. In *Proceedings of Third Asia-Pacific Conference on Simulated Evolution and Learning (SEAL 2000)*, pages 2704–2709. IEEE Press, Piscataway, NJ, 2000.

18. S. Droste, T. Jansen, and I. Wegener. On the analysis of the $(1+1)$ evolutionary algorithm. *Theoretical Computer Science*, 2002. (Accepted).

19. S. Droste, T. Jansen, and I. Wegener. Optimization with randomized search heuristics – the (A)NFL theorem, realistic scenarios, and difficult functions. *Theoretical Computer Science*, 2002. (Accepted for publication in a special issue on natural computation).

20. S. Droste and D. Wiesmann. On representation and genetic operators in evolutionary algorithms. Technical Report CI-41/98, Collaborative Research Center 531, University of Dortmund, 1998.

21. S. Droste and D. Wiesmann. Metric based evolutionary algorithms. In R. Poli, W. Banzhaf, W. B. Langdon, J. F. Miller, P. Nordin, and T. C. Fogarty, editors, *Proceedings of the Third European Workshop on Genetic Programming (EuroGP 2000)*, volume 1802 of LNCS, pages 29–43. Springer, Berlin, 2000.

22. A. E. Eiben and G. Rudolph. Theory of evolutionary algorithms: A bird's eye view. *Theoretical Computer Science*, 229:3–9, 1999.

23. P. C. Fishburn. *Interval Orders and Interval Graphs: A Study of Partially Ordered Sets*. Wiley, New York, 1985.

24. S. Forrest and M. Mitchell. Relative building-block fitness and the building-block hypothesis. In *Proceedings of the 2nd Workshop on the Foundations of Genetic Algorithms (FOGA II)*, pages 109–126, 1993.

25. J. Garnier, L. Kallel, and M. Schoenauer. Rigorous hitting times for binary mutations. *Evolutionary Computation*, 7(2):173–203, 1999.

26. D. E. Goldberg. *Genetic Algorithms in Search, Optimization, and Machine Learning*. Addison-Wesley, Reading, MA, 1989.

27. D. E. Goldberg and K. Deb. A comparative analysis of selection schemes used in genetic algorithms. In G. J. E. Rawlins, editor, *Proceedings of the 1st Workshop on Foundations of Genetic Algorithms (FOGA I)*, pages 69–93. Morgan Kaufmann, San Mateo, CA, 1991.

28. J. Guddat, F. Guerra Vasquez, K. Tammer, and K. Wendler. *Multiobjective and Stochastic Optimization Based on Parametric Optimization*. Akademie-Verlag, Berlin, 1985.

29. T. Hanne. On the convergence of multiobjective evolutionary algorithms. *European Journal of Operational Research*, 117(3):553–564, 1999.

30. E. Hansen. *Global Optimization Using Interval Analysis*. M. Dekker, New York, 1992.

31. J. Horn, D. E. Goldberg, and K. Deb. Long path problems. In Y. Davidor, H.-P. Schwefel, and R. Männer, editors, *Proceedings of the 3rd Parallel Problem Solving From Nature (PPSN III)*, volume 866 of LNCS, pages 149–158. Springer, Berlin, 1994.

32. T. Jansen. On classifications of fitness functions. In L. Kallel, B. Naudts, and A. Rogers, editors, *Theoretical Aspects of Evolutionary Computing*, pages 371–386. Springer, Berlin, 2001.

33. T. Jansen and I. Wegener. On the choice of the mutation probability for the (1+1)-EA. In M. Schoenauer et al., editors, *Proceedings of the Sixth Parallel Problem Solving from Nature (PPSN VI)*, volume 1917 of LNCS, pages 89–98. Springer, Berlin, 2000.

34. T. Jansen and I. Wegener. On the choice of the mutation probability for the (1+1)-EA. Technical Report CI-92/00, Collaborative Research Center 531, University of Dortmund, 2000.

35. T. Jansen and I. Wegener. Evolutionary algorithms – how to cope with plateaus of constant fitness and when to reject strings of the same fitness. *IEEE Transactions on Evolutionary Computation*, 2001. (Accepted).

36. T. Jansen and I. Wegener. On the utility of populations in evolutionary algorithms. In L. Spector, E. D. Goodman, A. Wu, W. B. Langdon, H.-M. Voigt, M. Gen, S. Sen, M. Dorigo, S. Pezeshk, M. H. Garzon, and E. Burke, editors, *Proceedings of the Third Genetic and Evolutionary Computation Conference (GECCO 2001)*, pages 375–382. Morgan Kaufmann, San Francisco, CA, 2001.

37. T. Jansen and I. Wegener. Real royal road functions – where crossover is provably essential. In L. Spector, E. D. Goodman, A. Wu, W. B. Langdon, H.-M. Voigt, M. Gen, S. Sen, M. Dorigo, S. Pezeshk, M. H. Garzon, and E. Burke, editors, *Proceedings of the Genetic and Evolutionary Computation Conference (GECCO 2001)*, pages 1034–1041. Morgan Kaufmann, San Francisco, CA, 2001.
38. T. Jansen and I. Wegener. On the analysis of evolutionary algorithms – A proof that crossover really can help. *Algorithmica*, 2002. (Accepted).
39. J. R. Koza. *Genetic Programming*. MIT Press, Cambridge, MA, 1992.
40. J. R. Koza. *Genetic Programming II: Automatic Discovery of Reusable Programs*. MIT Press, Cambridge, MA, 1994.
41. M. Krause, P. Savický, and I. Wegener. Approximations by OBDDs and the variable ordering problem. In *International Colloquium on Automata, Languages, and Programming (ICALP 99)*, volume 1644 of LNCS, pages 493–502, 1999.
42. G. Laumanns, G. Rudolph, and H.-P. Schwefel. A spatial predator-prey approach to multi-objective optimization. In T. Bäck, A.E. Eiben, M. Schoenauer, and H.-P. Schwefel, editors, *Proceedings of the 5th Parallel Problem Solving From Nature (PPSN V)*, volume 1498 of LNCS, pages 241–249. Springer, Berlin, 1998.
43. M. Laumanns, G. Rudolph, and H.-P. Schwefel. Approximating the pareto set: Concepts, diversity issues, and performance assessment. Technical Report CI-72/99, Collaborative Research Center 531, University of Dortmund, 1999.
44. M. Laumanns, G. Rudolph, and H.-P. Schwefel. Mutation control and convergence in evolutionary multi-objective optimization. In R. Matousek and P. Osmera, editors, *Proceedings of the Seventh International Conference on Soft Computing (MENDEL 2001)*, pages 24–29. Brno University of Technology, Czech Republic, 2001.
45. D. C. Llewellyn, C. Tovey, and M. Trick. Local optimization on graphs. *Discrete Applied Mathematics*, 23:157–178, 1989.
46. M. Mitchell, S. Forrest, and J. H. Holland. The royal road function for genetic algorithms: Fitness landscapes and GA performance. In *Proceedings of the 1st European Conference on Artificial Life*, pages 245–254. MIT Press, Cambridge, MA, 1994.
47. M. Mitchell, J. H. Holland, and S. Forrest. When will a genetic algorithm outperform hill climbing? In J. Cowan, G. Tesauro, and J. Alspector, editors, *Advances in Neural Information Processing Systems*. Morgan Kaufmann, San Francisco, CA, 1994.
48. R. Motwani and P. Raghavan. *Randomized Algorithms*. Cambridge University Press, Cambridge, 1995.
49. H. Mühlenbein. How genetic algorithms really work. Mutation and hillclimbing. In R. Männer and R. Manderick, editors, *Proceedings of the 2nd Parallel Problem Solving from Nature (PPSN II)*, pages 15–25. North-Holland, Amsterdam, 1992.
50. C. H. Papadimitriou. *Computational Complexity*. Addison-Wesley, Reading, MA, 1994.
51. A. Prügel-Bennett. The mixing rate of different crossover operators. In W. N. Martin and W. M. Spears, editors, *Proceedings of the Sixth Foundations of Genetic Algorithms Workshop (FOGA VI)*, pages 261–274. Morgan Kaufmann, San Francisco, CA, 2000.

52. Y. Rabani, Y. Rabinovich, and A. Sinclair. A computational view of population genetics. *Random Structures & Algorithms*, 12, 1998.

53. I. Rechenberg. *Evolutionsstrategie: Optimierung technischer Systeme nach Prinzipien der biologischen Evolution*. Frommann-Holzboog Verlag, Stuttgart, 1973.

54. I. Rosenberg. Reduction of bivalent maximization to the quadratic case. *Cahiers du Centre d'Etudes de Recherche Operationnelle*, 17:71–74, 1975.

55. G. Rudolph. *Convergence Properties of Evolutionary Algorithms*. Dr. Kovač, Hamburg, 1997.

56. G. Rudolph. How mutation and selection solve long path-problems in polynomial expected time. *Evolutionary Computation*, 4(2):195–205, 1997.

57. G. Rudolph. Evolutionary search for minimal elements in partially ordered finite sets. In V. W. Porto, N. Saravanan, D. Waagen, and A. E. Eiben, editors, *Proceedings of the 7th Annual Conference on Evolutionary Programming*, volume 1447 of LNCS, pages 345–353. Springer, Berlin, 1998.

58. G. Rudolph. Finite Markov chain results in evolutionary computation: A tour d'horizon. *Fundamenta Informaticae*, 35(1-4):67–89, 1998.

59. G. Rudolph. On a multi–objective evolutionary algorithm and its convergence to the Pareto set. In *Proceedings of the 1998 IEEE International Conference on Evolutionary Computation*, pages 511–516. IEEE Press, Piscataway, NJ, 1998.

60. G. Rudolph. The fundamental matrix of the general random walk with absorbing boundaries. Technical Report CI-75, Collaborative Research Center 531, University of Dortmund, 1999.

61. G. Rudolph. Self-adaptation and global convergence: A counter-example. In *Proceedings of the IEEE Congress on Evolutionary Computation (CEC 99)*, pages 646–651, Volume 1. IEEE Press, Piscataway, NJ, 1999.

62. G. Rudolph. On takeover times in spatially structured populations: Array and ring. In K. K. Lai, O. Katai, M. Gen, and B. Lin, editors, *Proceedings of the Second Asia-Pacific Conference on Genetic Algorithms and Applications (APGA 2000)*, pages 144–151. Global-Link Publishing Company, Hong Kong, PR China, 2000.

63. G. Rudolph. Takeover times and probabilities of non-generational selection rules. In L. D. Whitley et al., editor, *Proceedings of the Second Genetic and Evolutionary Computation Conference (GECCO 2000)*, pages 903–910. Morgan Kaufmann, San Francisco, CA, 2000.

64. G. Rudolph. Evolutionary search under partially ordered finite sets. In M. F. Sebaaly, editor, *Proceedings of the International NAISO Congress on Information Science Innovations (ISI 2001)*, pages 818–822, Dubai, U.A.E., 2001. ICSC Academic Press.

65. G. Rudolph. A partial order approach to noisy fitness functions. In J.-H. Kim, B.-T. Zhang, G. Fogel, and I. Kuscu, editors, *Proceedings of the 2001 IEEE Congress on Evolutionary Computation (CEC 2001)*, pages 318–325. IEEE Press, Piscataway, NJ, 2001.

66. G. Rudolph. Some theoretical properties of evolutionary algorithms under partially ordered fitness values. In C. Fabian and I. Intorsureanu, editors, *Proceedings of the Evolutionary Algorithms Workshop (EAW 2001)*, pages 9–22, Bucharest, Romania, 2001.

67. G. Rudolph. Takeover times of noisy non-generational selection rules that undo extinction. In V. Kurkova et al., editors, *Proceedings of the Fifth International*

*Conference on Artificial Neural Networks and Genetic Algorithms (ICANNGA 2001)*, pages 268–271. Springer, Berlin, 2001.

68. G. Rudolph and A. Agapie. Convergence properties of some multi-objective evolutionary algorithms. In A. Zalzala et al., editors, *Proceedings of the 2000 IEEE Congress on Evolutionary Computation (CEC 2000)*, pages 1010–1016, Volume 2. IEEE Press, Piscataway, NJ, 2000.

69. H.-P. Schwefel. *Numerische Optimierung von Computer-Modellen mittels der Evolutionsstrategie*. Birkhäuser, Basel, 1977.

70. D. Sieling. The nonapproximability of OBDD minimization. *Information and Computation*, 2002. (To appear).

71. J. Smith and F. Vavak. Replacement strategies in steady state genetic algorithms: Static environments. In W. Banzhaf and C. Reeves, editors, *Proceedings of the Fifth Foundations of Genetic Algorithms Workshop (FOGA V)*, pages 219–233. Morgan Kaufmann, San Francisco, CA, 1999.

72. S. A. Stanhope and J. M. Daida. (1+1) genetic algorithm fitness dynamics in a changing environment. In P.J. Angeline et al., editor, *Proceedings of the Congress on Evolutionary Computation (CEC 99)*, pages 1851–1858, IEEE Press, Piscataway, NJ, 1999.

73. K. Tinnefeld and I. Wegener. Evolutionary algorithms for shortest path problems. Technical report, 2002. Preprint.

74. R. A. Watson, G. S. Hornby, and J. B. Pollack. Modelling building-block interdependency. In *Proceedings of the 5th Parallel Problem Solving from Nature (PPSN V)*, volume 1498 of LNCS, pages 97–106, 1998.

75. R. A. Watson and J. B. Pollack. Hierarchically-consistent test problems for genetic algorithms. In *Proceedings of Congress of Evolutionary Computation (CEC 99)*, pages 1406–1413, 1999.

76. I. Wegener. *Branching Programs and Binary Decision Diagrams – Theory and Applications*. SIAM, 2000.

77. I. Wegener. On the design and analysis of evolutionary algorithms. In *Workshop on Algorithm Engineering as a New Paradigm (RIMS)*, pages 36–47, 2000.

78. I. Wegener. On the expected runtime and the success probability of evolutionary algorithms (invited paper). In *26th Workshop on Graph-Theoretic Concepts in Computer Science (WG 2000)*, volume 1928 of LNCS, pages 1–10. Springer, Berlin, 2000.

79. I. Wegener. Theoretical aspects of evolutionary algorithms (invited paper). In *International Colloquium on Automata, Languages, and Programming (ICALP 2001)*, volume 2076 of LNCS, pages 64–78. Springer, Berlin, 2001.

80. I. Wegener. Methods for the analysis of evolutionary algorithms on pseudo-Boolean functions. In R. Sarker, X. Yao, and M. Mohammadian, editors, *Evolutionary Optimization*. Kluwer, 2002. (To appear).

81. I. Wegener and C. Witt. On the behaviour of the (1+1) evolutionary algorithm on quadratic pseudo-boolean functions. Technical Report CI-97/00, Collaborative Research Center 531, University of Dortmund, 2000.

82. K. Weinert, J. Mehnen, and G. Rudolph. Dynamic neighborhood structures in parallel evolution strategies. Technical Report CI-112/01, Collaborative Research Center 531, University of Dortmund, 2001.

83. D. H. Wolpert and W. G. Macready. No free lunch theorems for optimization. *IEEE Transactions on Evolutionary Computation*, 1(1):67–82, 1997.

84. M. Yanagiya. Efficient genetic programming based on binary decision diagrams. In *Proceedings of the First International IEEE Conference on Evolutionary Computation (ICEC 94)*, pages 234–239. IEEE Press, Piscataway, NJ, 1994.

85. A. C. Yao. Probabilistic computations: Towards a unified measure of complexity. In *Proceedings of the 17th IEEE Symposium on Foundations of Computer Science (FOCS 77)*, pages 222 – 227, 1977.

# 6 Design of Evolutionary Algorithms and Applications in Surface Reconstruction

Thomas Beielstein[1], Jörn Mehnen[2], Lutz Schönemann[1],
Hans–Paul Schwefel[1], Tobias Surmann[2], Klaus Weinert[2], and
Dirk Wiesmann[1]

[1] University of Dortmund, Department of Computer Science, Informatik XI,
   44221 Dortmund, Germany
   {beielstein, schoenemann, schwefel, wiesmann}@Ls11.cs.uni-dortmund.de
[2] University of Dortmund, Faculty of Mechanical Engineering, Institute of
   Machining Technology, 44221 Dortmund, Germany
   {mehnen, surmann, weinert}@isf.mb.uni-dortmund.de

**Summary.** Evolutionary algorithms are a class of direct search methods. They can be used whenever classical optimization methods do not yield satisfactory results. In the following we discuss the design of problem-specific evolutionary algorithms. We present a technique for the systematic integration of domain knowledge. Surface reconstruction by means of evolutionary algorithms serves as a practical example. For this problem the integration of domain knowledge is essential for a successful application of evolutionary algorithms.

## 6.1 Introduction

Evolutionary algorithms (EA) have been successfully applied to a wide range of practical problems. They are an accepted tool in the field of technical optimization [48]. Evolutionary algorithms are general search heuristics which are used whenever classical optimization methods do not yield satisfactory results.

In order to use the full potential of evolutionary algorithms, practical experiences as well as theoretical results indicate that EA need the integration of problem-specific knowledge. Due to its flexible structure, the evolutionary scheme provides a lot of starting points for the integration of domain knowledge. Until now it has not been possible to identify a suitable technique for the integration of domain knowledge for a given problem. Especially for applications with nonstandard representations, it is difficult to make design decisions about representations and variation operators.

In the following, we will discuss the design of problem-specific EA. The first six sections deal with the basics of these algorithms. Here, the fundamentals for a formal discussion and a technique for a systematic integration of domain knowledge is presented. A special focus will be on mutation operators for structure optimization. In the second half the practical problem of

evolutionary surface reconstruction will be discussed. In this field of application, the use of domain knowledge is essential to obtain EA that generate solutions that can be used in practice.

## 6.2  Design of Evolutionary Algorithms

Every conference in the field of EA gives birth to a variety of new variation operators and representations of the genotype space. These operators and representations are tailored for the optimization problems at hand. This perfectly reflects the fact that EA (like all algorithms) can only achieve above-average performance on a specific subset of problems. But fortunately, EA have shown to be very robust. They work satisfactorily even if they are not adapted to the problem very well. The robustness of EA leads to a great freedom in the design process. In the following, we discuss how one can use this freedom to obtain powerful EA.

Section 6.3 offers a short introduction to the framework of EA. Section 6.4 is devoted to a general discussion of the design of EA and to the integration of domain knowledge. As a concrete example, we look at the design of an evolutionary programming system in Section 6.5. Up to now, the focus will have been on the representation and the variation operators. To use a special kind of domain knowledge, a threshold selection operator is introduced in Section 6.6.

## 6.3  Evolutionary Algorithms: The General Framework

In order to locate an optimal point of an objective function $f : \mathcal{P} \to W$ ($W$ is at least partially ordered), an EA works with a multiset (population) of candidate solutions (individuals) that are modified by variation operators, such as mutation and recombination. A selection operator is used to filter out less promising solutions within the transition from one population to another. The implementation of an EA requires that the abstract elements of $\mathcal{P}$ are represented by a data structure, i.e., elements of a space $\mathcal{G}$. The set $\mathcal{P}$ is called the phenotype space and $\mathcal{G}$ the genotype space.

We assume that $\mu$ denotes the number of parents and $\lambda$ the number of offspring in one generation. In the context of EA, it is sufficient to store an individual in the form of its genotype, since the genotype-phenotype mapping $\mathrm{map} : \mathcal{G}^\lambda \to \mathcal{P}^\lambda$ is deterministic, and environmental influences are not taken into consideration. Hence, we can denote a population at generation $t$ by $Pop(t) \in \mathcal{G}^\mu$.

The mapping $\mathrm{map} : \mathcal{G}^\lambda \to \mathcal{P}^\lambda$ can be composed simply from $\lambda$ reduced mappings $\mathrm{map}_r : \mathcal{G} \to \mathcal{P}$, which represent the genotype-phenotype mapping for single individuals:

$$\mathrm{map}(g) = (\mathrm{map}_r(g_1), \ldots, \mathrm{map}_r(g_\lambda)), \quad g = (g_1, \ldots, g_\lambda) \in \mathcal{G}^\lambda$$

The mapping $\text{map}_r : \mathcal{G} \to \mathcal{P}$ determines the abstract element $\text{map}_r(x)$ of the search space being represented by $x \in \mathcal{G}$. Thus, the mapping $\text{map}_r$ describes the relation between genotype and phenotype space.

The recombination operator $\text{rec} : \mathcal{G}^\mu \times \Omega_r \to \mathcal{G}^\lambda$ generates $\lambda$ offspring of the parent population by mixing the parental genetic information. The probabilistic influence during recombination is described by the probability space $(\Omega_r, \text{Prob}_r)$, i.e., the outcome of the recombination depends on the random choice of $\omega_r \in \Omega_r$ according to $\text{Prob}_r$. The recombination is followed by mutating the $\lambda$ offspring according to $\text{mut} : \mathcal{G}^\lambda \times \Omega_m \to \mathcal{G}^\lambda$. Here, $(\Omega_m, \text{Prob}_m)$ is the underlying probability space.

The new population $Pop(t+1) \in \mathcal{G}^\mu$ is selected from the set of offspring of $Pop(t)$, where the selection of an individual is based directly or indirectly on the objective function $f : \mathcal{P} \to W$. The objective function assesses only the phenotype. This relationship is formalized by the selection operator

$$\text{sel} : \mathcal{G}^\lambda \times \mathcal{P}^\lambda \times \Omega_s \to \mathcal{G}^\mu$$

(with probability space $(\Omega_s, P_s)$). With the auxiliary function

$$\text{map}^* \; : \; \mathcal{G}^\lambda \to \mathcal{G}^\lambda \times \mathcal{P}^\lambda,$$
$$\text{map}^*(g) := (g, \text{map}(g)) \text{ for } g \in \mathcal{G}^\lambda$$

the equation

$$Pop(t+1) = \text{sel}(\text{map}^*(\text{mut}(\text{rec}(Pop(t), \omega_r), \omega_m)), \omega_s)$$

holds, where $\omega_r \in \Omega_r$, $\omega_m \in \Omega_m$ and $\omega_s \in \Omega_s$ are chosen randomly according to $\text{Prob}_r$, $\text{Prob}_m$, and $\text{Prob}_s$.

Thus, an EA can be described by the iterated use of the variation operators and the selection operator. Now, the fundamental design decision is: How do we have to choose $\mathcal{G}$, rec, mut, and map, if a certain search space $\mathcal{P}$ and an objective function $f : \mathcal{P} \to W$ are given?

## 6.4   Design of Evolutionary Algorithms and Integration of Domain Knowledge

There are various hypotheses about the working principles of EA [8, 35, 68]. All these hypotheses share the use of local observations or measures in order to explain the global behavior of EA. Therefore, the power of the hypotheses is limited. They can explain the behavior of an EA in one or a few generations only. But previous work mostly ignored the limitations of these hypotheses. In particular, there is a misinterpretation of the schema theorem leading to the building block hypothesis. The schema theorem says nothing about the macroscopic behavior of an EA. This has to be considered when a design approach based on the building block hypothesis is used.

Another design approach is based on the work of Davis and Michalewicz [21, 57]. Here, the idea is to choose a representation that is most suitable for the problem. The variation operators are adapted to the type of encoding. If a current heuristic for the problem is known, then the positive features of the current heuristic are incorporated in the EA. Often this approach leads to a successful application of EA. But this design approach is not very systematic and transparent.

Recently, dynamically changing the representation during the search has been suggested [49, 53, 67]. But all methods proposed use dynamical mappings from genotype to phenotype space only. The actual data structures and operators remain the same. Thus, the dynamical changing may be seen as a special kind of mutation [49]. The research in dynamic and self-adapting representations is still in its beginnings. It is an open question, whether an a priori choice of representation and operators can be replaced by a continuous change of the representation [19].

In fact, the designer of an EA has little more alternatives than to follow his or her favorite hypotheses, to integrate some domain knowledge if available, and finally, tune the algorithm through exhaustive testing. In most cases, the integration of domain knowledge is not straightforward. In practice, the design process becomes difficult and time consuming.

### 6.4.1   Representation, Recombination, and Mutation

The design of an EA includes the choice of a suitable representation and the construction of the variation operators. In the following, we explore whether some general recommendations can be made in this regard. We assume that a particular optimization problem is given. Every instance of the problem is given by a fitness function that can be used to evaluate the problem-specific objects. If the problem is to minimize a pseudo-Boolean function $f : \{0,1\}^n \to \mathbb{R}$, then the objects evaluated by the fitness function are binary strings of length $n$. But for real-world problems, things can be different. Take the synthesis of chemical plants as an example, where various processing units, interconnected by material streams, form a complex network, the structure of which has to be optimized. Now the fitness function has to evaluate a whole plant. In the case of a computer-based optimization, the design of a plant is evaluated by a numerical simulation software. The simulation software uses special data structures to represent the chemical plants.

For the design of an EA, this means that the representation of the phenotype space is given by the problem or the fitness function, respectively. The question is, on what kind of representation the variation operators should operate. One possibility is to choose for the genotype space the same representation (data structure) as for the phenotype space. Thus, the mapping will be the identity. In the case of pseudo-Boolean functions the bit-flip mutation and the standard crossover can be used.

But the design of variation operators for the chemical plant problem is not evident. They must be adapted to the representation given by the simulation model. Of course it is also possible to choose a bit-string representation for the genotype space. Then the standard variation operators could be used again. Now the problem is to design a complex mapping between bit-strings and the representation used by the simulation model. Very often, these mappings are difficult to understand. Therefore, it becomes difficult to predict how the variation of an individual affects the search in the phenotype space. Another problem can be the loss of information on the context of the genotypical atoms (genes) as implied by the coding. As an example, Groß et al. [37] compared a bit-string representation and a graph representation of heat exchanger network synthesis. The operators of the graph representation were able to use information on the connections between the streams and the heat exchangers. It was difficult to extract this kind of information from a bit-string representation and the corresponding mapping.

Another well known example is the optimization of functions $f : \mathbb{R}^n \to \mathbb{R}$ with genetic algorithms. A genetic algorithm with a standard mapping between bit-strings and rational numbers shows problems in convergence speed and reliability, in comparison to an evolution strategy, which can be used directly for continuous parameter optimization. One reason for the loss of performance is the fact that adjacent rational values are represented by bit-strings having a Hamming distance greater than one. To make a small step in the phenotype space, a large step in the genotype space is required. Thus, with a bit-flip mutation the approximation of a solution is difficult. A more general explanation is that the mapping reduces the strong causality between the elements of the genotype space and the corresponding function values [74]. Thus, to obtain a transparent process it seems to be necessary to adapt the genotype space and the variation operators to the problem.

There are different opinions regarding the importance and the capability of the varying representations and variation operators. On the one hand, the advantage of crossover together with a binary coding of the genotype space is emphasized. On the other hand recombination is regarded as dispensable. The discussion about the schema theorem has already been mentioned. In the following we look at the purely mutation-based approaches.

The advocates of purely mutation-based EA refer to empirical tests [29] and to the work of Fogel and Ghozeil [30]. Fogel and Ghozeil conclude from technical considerations, that no representation offers any intrinsic advantage over another. Furthermore, there is no intrinsic advantage for any particular two-parent operator over any particular one-parent operator. Therefore, recombination can be substituted by mutation. But one can show that, in general, both statements are wrong. The following discussion gives some insight into the problems.

We assume that an EA $A_1$ is working on the phenotype space $\mathcal{P}$ and the genotype space $\mathcal{G}_1$. The algorithm $A_1$ uses the mutation operator

$$\text{mut}' : \mathcal{G}_1 \times \Omega_{m'} \to \mathcal{G}_1$$

with

$$\text{Prob}_{m'}(\text{mut}'(u) = v) := \text{Prob}(\{\omega \in \Omega_{m'} | \text{mut}'(u, \omega) = v\}).$$

The connection between the genotype space and the phenotype space is established by the bijective mapping $\text{map}_m : \mathcal{G}_1 \to \mathcal{P}$. Thus, the variation operator of algorithm $A_1$ is working on the representation $\mathcal{G}_1$ of the original search space $\mathcal{P}$. Let $A_2$ be a second algorithm that uses another representation $\mathcal{G}_2$ for the genotype space. $A_2$ uses the variation operator $\text{var}' : \mathcal{G}_2 \times \Omega_{v'} \to \mathcal{G}_2$ and the bijective mapping $\text{map}_v : \mathcal{G}_2 \to \mathcal{P}$. If the mutation operator $\text{var}'$ satisfies the equation

$$\text{Prob}_{v'}(\text{var}'(u) = v) = \text{Prob}_{m'}(\text{mut}'(\text{map}_m^{-1}(\text{map}_v(u))) = \text{map}_m^{-1}(\text{map}_v(v))),$$

then the algorithms $A_1$ and $A_2$ show evidently the same search behavior in the phenotype space. In particular, the expected number of fitness-function evaluations is the same for $A_1$ and $A_2$. But to conclude from this observation that the choice of the representation has no influence on the performance of an EA is not permitted because the efficiency of the variation operators must not be ignored. Even if the operator $\text{mut}'$ is able to efficiently create a new point $v = \text{mut}'(u)$ in the length of $u \in \mathcal{G}$, the same does not necessarily hold for the operator $\text{var}'$. The EA $A_1$ and $A_2$ need the same number of fitness calculations, but $A_2$ possibly needs so much time to create a new search point that in the meantime, algorithm $A_1$ has terminated. In this respect the variation operators have to be supported by a suitable data structure. In addition to the efficiency of the operators the transparency is another important aspect. Some representations will meet the requirements of the problem better than others.

Until now the efficiency of the variation operators has not been taken into consideration. Mainly, representations and operators for which an efficient implementation was given naturally (e.g., bit-flip mutation or one-point crossover on bit-strings) were tested. This has also significantly influenced the further development of operators and representations. However, as soon as special requirements are set on the variation operators, an efficient implementation of the operators can turn out to be difficult. To rate the performance of an EA only by means of the fitness evaluations or the number of generations is not sufficient. For complex variation operators, the effort to create a new search point must be taken into account.

Recombination and mutation are the main variation operators in EA. Of course, if two search operators are used, each of them should fulfill a different task. Fogel and Ghozeil [30] showed that for EA operating on a finite representation, every search operator that is applied to two parents to create

a single offspring can be replaced by an equivalent search operator operating on a single parent and a specific index number. Thus, one might conclude that a recombination operator is unnecessary. But Fogel and Ghozeil ignored the fact that every operator has to meet specific requirements. In the next section, we present a set of requirements for the variation operators, such that mutation and recombination cannot be equivalent. It is not advisable to make general statements about the benefit of mutation and recombination without taking the problem (fitness function) into account.

## 6.4.2   Metric-Based Evolutionary Algorithms

In order to systematize the design of EA, we present some design guidelines in this section. The guidelines were originally proposed by Droste and Wiesmann [25]. The aim of the guidelines is to support a smooth motion of the search process through the fitness landscape [68, 73]. By defining two related distance measures within the genotype and phenotype spaces so that neighboring elements have similar fitness, problem-specific knowledge can be integrated. The guidelines for the variation operators are based on the distance measures and guarantee that the search process will behave in a controlled fashion. To motivate these guidelines, we highlight some desirable features of the variation operators and the representation in the following [26]. Since we want to run the EA on a computer, we concentrate on finite search spaces and discrete time. To make the discussion easier, we furthermore assume that the search space is $\mathcal{G} = \mathcal{P}$ (thus, map is the identity).

At first the reachability of every point in the search space is important for an EA. Every point $u \in \mathcal{G}$ could be the desired global optimum or a good local optimum. The property of visiting the global optimum with probability one is also a precondition if one wants to assure convergence of the EA. If any other point $v \in \mathcal{G}$ from each point $u \in \mathcal{G}$ can be reached with probability greater zero in one mutation step, then the EA visits the global optimum after finite time with probability one, regardless of the initialization [71].

If the selection operator guarantees that the best individual visited so far will survive with probability one, then the EA converges completely and in mean to the global optimum, regardless of the initialization. The property of reachability in one mutation step can be weakened if some other properties can be assured (see [71] for details).

The next question is where the variation operators should place a new search point. As long as no additional information is available, the operators should not prefer any direction, i.e., mutation and recombination should not generate a bias. Furthermore, they should support a gradual approximation of the optimum. For the design and the analysis of an EA, it may be advantageous for the operators to be clearly separated, i.e., have different search characteristics.

Let us assume that we have an EA with well specified variation operators and selection. Then the variation operators and the selection mechanism de-

termine the efficiency of the search process (how many points $u \in \mathcal{G}$ are visited on average until the optimum is found). On the other hand, the representation affects the time efficiency of the operators. Thus, from an algorithmic point of view, the representation of the phenotype space has to support an efficient calculation of the objective function, whereas the representation of the genotype space has to support an efficient implementation of the variation operators. Of course, both have to support an efficient implementation of the mapping-function map.

**Metric.** In the following, we assume that we have information about the phenotype space in a specific problem domain. This domain knowledge will be used to establish a distance measure on the phenotype space. This distance measure shows the similarity between two phenotypes with regard to a certain measure of quality, i.e., two elements of the phenotype space with small distance should have similar objective function values. Hence, the distance measure should be strongly connected to the objective function to be optimized. The available domain knowledge directly influences the choice of this distance measure and only indirectly the design of the recombination and mutation operators. Therefore, the performance of the resulting EA mainly depends on the suitability of the distance measure.

In order to fulfill the most basic properties that are reasonable for distance measures, we only allow *metrics* $d_{\mathcal{P}} : \mathcal{P} \times \mathcal{P} \rightarrow \mathbb{N}_0^+$ to measure the distance between two points of the phenotype space (we restrict ourselves here to metrics with values only in $\mathbb{N}_0^+$). It would be desirable for any three points $p_1, p_2, p_3 \in \mathcal{P}$ to show the following property with regard to fitness difference and distance:

$$|f(p_1) - f(p_2)| \leq |f(p_1) - f(p_3)|$$
$$\Leftrightarrow \qquad (6.1)$$
$$d_{\mathcal{P}}(p_1, p_2) \leq d_{\mathcal{P}}(p_1, p_3)$$

Because a metric like this will be very hard to find in practical applications, (6.1) is not a formal guideline as to what the correlation between fitness function and metric should look like. The success of finding a strong relationship is clearly bounded by the available domain knowledge. Therefore, we simply claim that there should be a strong correlation so that elements of $\mathcal{P}$ with small distance should have similar function values. We present an example in Section 6.5, where one such strong relationship is given. In the following, both the genotype space and the phenotype space are assumed to be finite spaces.

**Mapping.** Here we assume that the mapping map is bijective. Then the metric $d_{\mathcal{P}}$ automatically induces a metric $d_{\mathcal{G}} : \mathcal{G} \times \mathcal{G} \rightarrow \mathbb{N}_0^+$ on the genotype space using the following definition:

$$\forall g_1, g_2 \in \mathcal{G} : d_{\mathcal{G}}(g_1, g_2) := d_{\mathcal{P}}(\mathrm{map}_r(g_1), \mathrm{map}_r(g_2)) \qquad (6.2)$$

This guarantees, that the distance relations between three points translate from the genotype to the phenotype space and back.

**Recombination.** Without loss of generality, we look at the reduced recombination operator $rec' : \mathcal{G}^2 \times \Omega_{r'} \to \mathcal{G}$ with a finite probability space $(\Omega_{r'}, \mathrm{Prob}_{r'})$ instead of the general recombination operator $r$. Let

$$\mathrm{Prob}_{r'}(r'(u,v) = w) := \mathrm{Prob}_{r'}(\{\omega \in \Omega_{r'} \mid r'(u,v,\omega) = w\})$$

be the probability that two parents $u, v \in \mathcal{G}$ are recombined to an offspring $w \in \mathcal{G}$. The first guideline on the recombination operator states that the genotype distance $d_{\mathcal{G}}$ between any of the two parents and the offspring should be smaller than the distance between the parents themselves. Hence, children are not allowed to differ extremely from their parents (according to the metric):

**Guideline R 1** *The recombination $rec'$ should fulfill:*

$$\forall u, v \in \mathcal{G}, \omega \in \Omega_{r'} \text{ with } z = rec'(u,v,\omega) :$$
$$\max(d_{\mathcal{G}}(u,z), d_{\mathcal{G}}(v,z)) \le d_{\mathcal{G}}(u,v)$$

Moreover, the recombination operator should not generate an additional bias independent of the parents, i.e., the offspring should not tend to one of its parents to avoid any bias in the search process:

**Guideline R 2** *The recombination $rec'$ should fulfill:*

$$\forall u, v \in \mathcal{G}, \forall \alpha \ge 0 :$$
$$\mathrm{Prob}_{r'}(d_{\mathcal{G}}(u, rec'(u,v)) = \alpha) = \mathrm{Prob}_{r'}(d_{\mathcal{G}}(v, rec'(u,v)) = \alpha)$$

This guarantees that recombination produces an element having distance $\alpha$ from its first parent with the same probability as an element having distance $\alpha$ from its second parent. Hence, no parent is favored.

**Mutation.** The mutation operator $mut : \mathcal{G}^\lambda \times \Omega_m \to \mathcal{G}^\lambda$ operates on all $\lambda$ individuals at the same time. Since the mutation of a single individual does not depend on other individuals, the discussion is restricted to the reduced mutation operator $mut' : \mathcal{G} \times \Omega_{m'} \to \mathcal{G}$ with the finite probability space $(\Omega_{m'}, \mathrm{Prob}_{m'})$. With probability

$$\mathrm{Prob}_{m'}(mut'(u) = v) := \mathrm{Prob}_{m'}(\{\omega \in \Omega_{m'} \mid mut'(u,\omega) = v\}),$$

the mutation operator $mut'$ changes an element $u \in \mathcal{G}$ to a new element $v \in \mathcal{G}$. The first guideline states that from each point $g \in \mathcal{G}$, any other point $u \in \mathcal{G}$ can be reached in one mutation step. Hence, there can be no regions in the search space that are excluded from the search process:

**Guideline M 1** *The mutation mut' should fulfill:*

$$\forall u, v \in \mathcal{G} : \ Prob_{m'}(mut'(u) = v) > 0$$

Moreover, small mutations (with respect to $d_\mathcal{G}$) should occur with higher probability than large mutations because in a smooth fitness landscape, good elements are at least partially surrounded by other good elements:

**Guideline M 2** *The mutation mut' should fulfill:*

$$\forall u, v, w \in \mathcal{G} :$$
$$(d_\mathcal{G}(u, v) < d_\mathcal{G}(u, w)) \Rightarrow (Prob_{m'}(mut'(u) = v) > Prob_{m'}(mut'(u) = w))$$

For a given $u \in \mathcal{G}$, all genotypes $v \in \mathcal{G}$ that have the same distance to $u$ should have the same probability to be produced by mutation of $u$, guaranteeing that the mutation operator does not induce a bias on its own:

**Guideline M 3** *The mutation mut' should fulfill:*

$$\forall u, v, w \in \mathcal{G} :$$
$$(d_\mathcal{G}(u, v) = d_\mathcal{G}(u, w)) \Rightarrow (Prob_{m'}(mut'(u) = v) = Prob_{m'}(mut'(u) = w))$$

Since at the beginning of the search, there is no information on the position of the global optimum, the mutation should not push the search process in any particular direction. By working with a dynamic adaptation of the step size (e.g., self-adaptation of $n > 1$ step sizes), guideline M 3 might be violated. By learning an internal model of the topology, the preference of a certain local search direction can be useful. Because each individual can only learn a local model of the fitness function (especially in multimodal optimization), the topology learned can become useless when the individual leaves the subspace where the local model is valid. Whenever there is no longer any suitable information about the topology of the search space available, it should be guaranteed that a state can be reached in which M 3 holds again.

Since all guidelines are based on the definition of a suitable metric on the phenotype space, an EA which fulfills the guidelines H 1, R 1, R 2, M 1, M 2, and M 3 for a given metric and mapping is called metric-based evolutionary algorithm (MBEA). The metric-based design approach has already been successfully applied both to synthetic and to practical problems. Droste and Wiesmann [25, 26] developed a metric-based genetic programming (GP) system to learn Boolean functions (see Section 5.8). The metric-based system showed a significant increase in performance in comparison to standard GP systems [25]. The metric-based approach was also used to design a hybrid evolutionary search for single relevant fuzzy rules (see Section 3.2.3 and [75]) and to design an EA for the synthesis of chemical plants [27]. All applications comprise a structure optimization. In the next section we present an application in the field of evolutionary programming [100]. We show how one can obtain a semantical mutation operator from a pure syntactical (structure-based) mutation operator.

## 6.5   Structure Optimization and Evolutionary Programming

Within the field of evolutionary programming (EP), the evolution of finite automata has been studied since the 1960s [31, 29]. The starting point of the research was the question whether a simulated evolution on a population of contending algorithms is able to produce some kind of artificial intelligence. Intelligent behavior was considered to be the ability to predict an event in a given environment and to react to this event to meet a given goal. For the purpose of simplification, the environment was modeled as a sequence of symbols taken from a finite alphabet $\Sigma$. The algorithms were represented by Mealy automata reading the sequence of symbols. Every symbol the automaton reads activates a state transition and produces one output symbol from the finite alphabet $\Omega$. The task of the EP system was to evolve an automaton that correctly predicts, i.e., produces, the next symbol to appear in the environment on the basis of the sequence of symbols it has observed previously. Thus, the number of wrong predictions was minimized.

An EP system works on the graph representation of an automaton [31]. An automaton is represented as a directed graph. The nodes represent the states of the automaton, and the edges correspond to the state transitions. Every edge is labeled by an input and an output symbol. Five different mutations are derived from the graph description: change of an output symbol, change of a state transition, addition of a state, deletion of a state, and change of the initial state. The mutation operator selects with equal probability a certain mode of mutation and applies it to an individual. Depending on the mode of mutation the nodes and edges are selected with equal probability. The number of mutations per offspring is chosen with respect to a probability distribution [29]. A recombination operator was proposed but not implemented.

The graph representation of an automaton and the resulting five modes of mutation have two advantages. Every single mode of mutation can be performed efficiently. Provided that the graph is stored as an adjacency list, every change of an output symbol and every mutation of a state transition needs only linear time in the number $|Q|$ of nodes. To add or to delete a state needs quadratic time. The change of the initial state can be done in constant time. Since the deletion of a state and the change of the initial state are only allowed when the parent automaton has more than one state, every mutation leads to a feasible representation of an automaton, but the resulting automaton is not necessarily minimal. In particular, there can be nodes and even whole subgraphs that are not reachable from the initial state.

A potential drawback of the mutation may be that every single mode of mutation is based solely on the structure of an automaton. Thus, the standard mutation is a purely syntactical operator. The size of a mutation, e.g., the length of a mutation step, is directly related to the complexity of the structural modification. A mutation that deletes a state and changes a state transition has greater influence on the structure than a mutation that

only changes a symbol of the output alphabet. Thus, the impact on the input/output behavior is not considered here. Even the influence of two mutations of the same mode may vary significantly (see Section 6.5.2). Moreover, it is difficult to find a suitable distance measure (metric), which measures the structural difference of two automata. Especially for the gradual approximation of a solution in a large search space, it is important that mutation prefers small steps in the search space (regarding a suitable distance measure). By using EP to evolve programs in the form of symbolic expressions it has been observed that preferring mutations with a small effect has some advantages [18].

In the following, we use the metric-based approach to design an EP system. In the next section we define two related distance measures within the genotype and the phenotype space so that neighboring elements have similar fitness. The metric allows us to reason about the distance of individuals and the size of mutation steps in a formal framework. The size of a mutation is correlated with the change in the fitness value and is not directly based on the structural modifications within the representation of an individual. To simplify our consideration we focus on deterministic finite automata (DFA) in the following [42]. For each point in time a DFA is in a state $q$ of a finite nonempty set of states $Q$. In every step, the automaton reads a symbol $w_i \in \Sigma$ and changes its state according to the mapping $\delta : Q \times \Sigma \to Q$. The automaton starts in the initial state $q_o \in Q$. An input string $w \in \Sigma^*$ is said to be accepted, iff the automaton is in a final state after reading $w \in \Sigma^*$, where $\Sigma^*$ denotes the set of all finite strings of symbols from the alphabet $\Sigma$. Thus, a DFA $A$ can be described as a system $A = (Q, \Sigma, q_o, \delta, F)$. Where $F \subseteq Q$ is the set of final (accepting) states. The set of all strings accepted by $A$ is denoted as the regular language $L(A)$. The language $L^n \subset \Sigma^n$ consists of strings of a fixed length $n$ only. Thus, $L^n(A)$ is the set of all strings of fixed length $n$ accepted by the DFA $A$. In general the restriction to decision problems is not too strong [34].

### 6.5.1   Fitness and Distance of DFA

Let $S \subseteq \Sigma^*$ be a finite subset and $T = \{(w, f(w)) | w \in S\}$ a training set. Based on the training set a DFA has to be found which accepts the language $L(A) := f^{-1}(1)$ for a function $f : \Sigma^* \to \{0, 1\}$. The EP system has to evolve DFA which will generalize from the training set to $L(A)$.

For simplicity, we restrict the problem space in two ways. At first, we only consider languages with strings of fixed length $n$. Furthermore, the search will be based on the complete training set $T_v$. For a function $f : \Sigma^n \to \{0, 1\}$ and the training set $T_v = \{(w, f(w)) \mid w \in \Sigma^n\}$ a DFA has to be found which accepts the language $L^n(A) := f^{-1}(1)$. Thus, $A$ must achieve:

$$\forall (w, 1) \in T \text{ is } w \in L^n(A) \text{ and } \forall (w, 0) \in T \text{ is } w \notin L^n(A)$$

The effects of the restrictions will be discussed later. The fitness function $F(A) := |\{(w, f(w)) \in T_v \mid f(w) = 1 \Leftrightarrow w \in L^n(A)\}|$ counts the number of strings the DFA $A$ classifies correctly.

Now, how can we measure the similarity of two DFA $A$ and $B$? The distance $d^n$ of $A$ and $B$ should be the number of strings on which $A$ and $B$ disagree:

$$d^n(A, B) = |L^n(A)| + |L^n(B)| - 2|L^n(A) \cap L^n(B)|$$

The maximum difference in fitness values of two DFA is $d^n(A, B)$. Note that the fitness calculation is based on $T_v$. The distance measure $d^n : \Sigma^n \times \Sigma^n \to \mathbb{N}$ is a metric. Obviously, this distance measure cannot distinguish between two structurally different DFA accepting the same language. To clarify the problems of structure optimization we first arrive in an intermediate step to the so-called weighted mutation operator. Then the metric-based mutation operator is presented.

### 6.5.2   Proposals for EP Mutation Operators

**Weighted Mutation.** The first proposal for a new mutation operator is motivated by the observation that the fitness calculation can provide more information than the pure fitness value. To compute the fitness of a given DFA for every word in the training set, a path beginning at the initial state has to be traversed.

In order to assess the influence of a mutation event every node (state) is assigned a weight index with initial value 0. Every time a node is visited during fitness calculation the weight index is incremented by 1. After fitness calculation on the complete training set, the weights give an upper bound for a change in the fitness value caused by mutation.

Let us consider the following example: Let $\Sigma^4 = \{0, 1\}^4$ and let $L^4 = \{0000, 0011, 0101, 0110, 1001, 1010, 1100, 1111\}$ be the language of all strings with an even number of 0s and an even number of 1s of length 4. Figure 6.1 shows the graph representation (state diagram) of an automaton with the corresponding weights after fitness calculation on the complete training set $T_v$. In three cases the DFA makes the wrong decision on $T_v$.

Now, let us discuss the impact of different mutations. The state $q_2$ has a relatively high weight of 15. If the state $q_2$ is deleted by a mutation event, then the fitness changes by at most 15. By comparison, the deletion of state $q_5$ can change the fitness by the value of 3 at most. By ranking the states according to their weights, states with a lower weight can be mutated with higher probability than states with a higher weight. State transitions can be mutated likewise. Transitions beginning in a state with a lower weight will be mutated with higher probability than transitions beginning in a state with a higher weight. The insertion of new states will take place with higher probability between states with a lower weight.

This approach allows the definition of a reasonable probability distribution on mutation events for every single mode of mutation. But it is not obvious how the different modes of mutation should be weighted among themselves. For example, should transitions be mutated with higher probability than states? Should we mutate states with a low weight more often than transitions beginning in a state with a high weight? Additionally, even with respect to $T_v$ the upper bound may turn out to be a bad estimate for the real change in fitness. For example, an improvement and a decline of the fitness may cancel out each other. These observations show once again the problems of variation operators purely based on the structure, even when additional information is available.

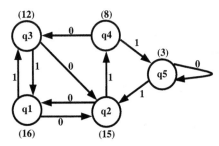

**Fig. 6.1.** DFA with weighted states (weights in parenthesis) after fitness evaluation on the complete training set $T_4$ for the language $L^4$ of all strings with an even number of 0's and an even number of 1's. State $q_1$ is initial and final state.

**Metric-based Mutation.** In order to overcome the deficiencies described above, we will now design a mutation operator in accordance with the metric-based guidelines, which use the metric $d^n$ defined in Section 6.5.1. The mutation is based on efficient synthesis operations for DFA presented in [24]. In order to obtain a bijective mapping the mutation operates on reduced DFA only.

The first step in mutating the DFA $A$ to a DFA $B$ is to randomly choose a step size $K$ with $0 \le K \le |\Sigma^n|$. The mutation operator will choose a subset $M^n \subseteq \Sigma^n$ with $|M^n| = K$. Every word $w \in \Sigma^n$ is selected with the same probability $p$ for the set $M^n$. Thus we have

$$\text{Prob}(K = k) = p^k \cdot (1 - p)^{|\Sigma^n| - k}.$$

To obtain a fast selection of the set $M^n$ we assume that every word can be addressed by an index $i \in \{0, \ldots, |\Sigma^n| - 1\}$. If the word at position $i$ is already selected for the set $M^n$, then the word with index $(i + l)$ will be chosen next with probability $p \cdot (1 - p)^{l-1}$. The random variable $L$ with

$$\text{Prob}(L = l) = p \cdot (1 - p)^{l-1}$$

describes the relative position of the next word to be chosen. Thus, $L$ is geometrically distributed with parameter $p$. If $p = 1/|\Sigma^n|$, then the expected size of the set $M^n$ is 1.

The set $M^n$ is split in two sets $X^n$ and $Y^n$ with

$$\forall x \in X^n : x \in L^n(A), \quad \forall y \in Y^n : y \notin L^n(A), \text{ and } X^n \cup Y^n = M^n.$$

No $x \in X^n$ should be accepted by the DFA $B$. The DFA $B$ should only accept all $y \in Y^n$. On every other input string, $A$ and $B$ should agree. Thus, $d^n(A, B) = K$ holds. For the partitioning in the sets $X^n$ and $Y^n$, the DFA $A$ has to be tested $K$ times (cost: $K \cdot n$). In order to obtain $B$, two DFA $A_X$ and $A_Y$ are constructed with

$$L^n(A_X) = \Sigma^n \setminus X^n \text{ and } L^n(A_Y) = Y^n.$$

With this, we get $B$ as $L^n(B) = (L^n(A) \cap L^n(A_X)) \cup L^n(A_Y)$.

$A_X$ and $A_Y$ are constructed as follows: For every $x^i \in X^n = \{x^1, \ldots, x^{|X^n|}\}$ we construct an automaton $A_{x^i}$ that accepts only the string $x^i$, thus $L^n(A_{x^i}) = \{x^i\}$. This automaton has $n + 2$ states. Figure 6.2 shows the structure of a DFA accepting the string $a = a_1 \ldots a_n$ only. Thus, we have

$$L^n(A_X) = \overline{L^n(A_{x^1}) \cup \cdots \cup L^n(A_{x^{|X^n|}})}.$$

For every $y^i \in Y^n = \{y^1, \ldots, y^{|Y^n|}\}$ we construct an automaton $A_{y^i}$ only accepting the string $y^i$ as well. With $L^n(A_{y^i}) = \{y^i\}$ we have

$$L^n(A_Y) = L^n(A_{y^1}) \cup \cdots \cup L^n(A_{y^{|Y^n|}}).$$

After each synthesis operation the resulting DFA is minimized. An EP system using this mutation operator is called a metric-based evolutionary programming (MBEP) system.

**Theorem 1.** *The constructed mutation operator fulfills the guidelines M 1, M 2 and M 3.*

*Proof.* The guideline M 1 is fulfilled because for every step size $k \in \{0, \ldots, |\Sigma^n|\}$ every subset $M^n \subseteq \Sigma^n$ with $|M^n| = k$ has a probability $p^k \cdot (1-p)^{|\Sigma^n|-k} > 0$ to be chosen. According to the design of the operator, every language $L^n \subseteq \Sigma^n$ can be generated. The guideline M 2 is fulfilled because for $k_1 < k_2$ it is guaranteed that $P(K = k_1) > P(K = k_2)$, and all subsets $M^n \subseteq \Sigma^n$ with $|M^n| = K$ have an equal probability of being chosen. This implies that guideline M 3 is also fulfilled.   $\square$

## 6.5.3   Experiments and Discussion

A direct comparison between EP and MBEP is not possible. Due to its design the MBEP system searches for languages with strings of fixed length $n$. An

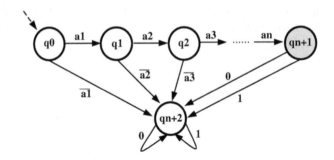

**Fig. 6.2.** The structure of a DFA on $\Sigma^n = \{0,1\}^n$. The DFA only accepts the string $a = a_1 \ldots a_n$. The final state is hatched and $q_0$ is the initial state.

EP system can operate on strings of arbitrary length. A (1+1)-MBEP system was tested on two different languages. The first language $L^n_{\text{even}}$ consists of all strings of length $n$ with an even number of 0s and an even number of 1s. The second language $L^n_{\text{fel}}$ consists of all strings of length $n$ where the last symbol equals the first. The initial starting point was chosen by random selection of an element from the set of all languages with strings of length $n$ with equal probability. We used a constant $p = 1/2^n$, but a dynamic adaptation of the parameter $p$ is also possible. The number of generations (mutations) until the language was found the first time was averaged over 50 independent runs (Table 6.1). One has to keep in mind that the time needed for a mutation depends on the length $n$ of the strings, the step size $K$, and the size of the DFA. The mutation operator is efficient in these sizes but more time-consuming than standard EP mutation (see Sections 6.5 and 6.5.2 and [25]). For 500 mutations, the MBEP system needed for $n = 4$, $n = 6$, and $n = 8$, about 1, 4, and 16 seconds, respectively (on a Sparc Ultra 10/300). It is not surprising that the evolution processes for both languages need similar amounts of time.

To explain this observation, recall that a function $f : \Sigma^n \rightarrow \{0,1\}$ can be described as a list of all $f(x)$, $x \in \Sigma^n$, and therefore as a bit-string. Assume a bit-string of length $|\Sigma^n|$ is given for every DFA. At the $i$th position the bit-string holds a 1 if the DFA makes the right decision for the $i$th string. Otherwise this bit equals 0. Basically, the fitness function counts the number of 1s in a bit-string. Thus, the fitness function equals the counting-ones problem [70] on a string of length $|\Sigma|^n$. Since the MBEP mutation operator is based on the metric $d^n$, the difficulty to be found is the same for all languages $L^n$.

The MBEP system is subject to substantial restrictions. The system can only work on regular languages with strings of fixed length $n$. But this restriction can be weakened. Prior to a mutation step a string length could be chosen with respect to a probability distribution. Then the mutation oper-

**Table 6.1.** Number of generations averaged over 50 independent runs until the $(1+1)$-MBEP system found the language the first time.

| Language | Runs | Generations | Language | Runs | Generations |
|---|---|---|---|---|---|
| $L_{\text{even}}^4$ | 50 | 118.09 | $L_{\text{fel}}^4$ | 50 | 133.18 |
| $L_{\text{even}}^6$ | 50 | 725.06 | $L_{\text{fel}}^6$ | 50 | 696.34 |
| $L_{\text{even}}^8$ | 50 | 3956.78 | $L_{\text{fel}}^8$ | 50 | 4056.26 |

ator operates only on strings of the chosen length. The restriction that the MBEP system can only work on the complete training set is much stronger. In its current implementation the system has no generalization ability. Due to the construction of the DFA (see Figure 6.2), cycles over final states cannot occur. Strings that are too long or too short are kept in a nonaccepting state. This problem may be solved by setting transitions starting in state $q_{n+1}$ (see Figure 6.2) randomly to states in $\{q_0, \ldots, q_{n+1}\}$. Additionally, states in $\{q_0, \ldots, q_{n+1}\}$ have to be final states with a certain probability. Unfortunately, first experiments show that these modifications may result in very large DFA. The size of a DFA depends on the size of the incomplete training set. If the training set is too small, the DFA may become too large.

## 6.6  An Analysis of Threshold Selection in Noisy Environments

In many real-world optimization problems, the fitness function evaluation is disturbed by noise. Evolution strategies (ES) can generally cope with noisy fitness function values [4, 14]. It has been proposed that the plus strategy can find better solutions by keeping over-valued function values, thus preventing inferior offspring with fitness inflated by noise from being accepted. The plus strategy builds an implicit barrier around the current best population. We propose to make this barrier-building process explicit and to employ a threshold value $\tau$ to be used in a selection operator for noisy fitness functions:

> *threshold selection (TS) accepts a new individual if its apparent fitness is better than that of the parent by at least the margin $\tau$.*

Analytical and empirical investigations on the sphere model and on the S-ring model (see below) were carried out [13, 51].

### 6.6.1  Elevator Group Control and the S-Ring Model

The elevator group control task is a real-time optimization problem of allocating elevator cars to passengers requesting service. Although it has been investigated for many decades, it is still an open research problem [50]. The

main difficulties lie in the stochastic nature of passenger arrivals, and in the combinatorial explosion of the number of system states. There are many proposed control methods, but because of the many differences in buildings, traffic patterns, and elevator movements, the published results cannot be compared directly, and their critical evaluation is difficult.

We investigated a closely related artificial problem, the S-ring model. The S-ring model is based on the Kac ring (see, e.g., [36]), and has properties that makes it useful for an analysis of an elevator group control simulation:

- it can be solved exactly for small problem sizes, while still exhibiting nontrivial dynamics;
- it retains interesting properties of the elevator system;
- it has a problem complexity similar to that of the elevator problem.

Since the S-ring model is easily reproducible, other researchers should be able to compare their work with the results given here.

By comparing different optimization techniques experimentally (stochastic approximation, Q-learning, and evolution strategies), it has been found that a (1+1)-ES outperforms the other methods on the S-ring model.

A common problem of the S-ring model and the elevator control group problem is the occurrence of noisy fitness function evaluations. Frequently, the probability of generating a real improvement is very small. However, a considerable proportion of generated solutions may *appear* to be an improvement due to noisy fitness function evaluation. This effect can lead to a decrease in the convergence speed, or even to divergence.

A recent study [5] of the $(1 + 1)$-ES on a noisy sphere has shown that in the course of the search, candidate solutions that are retained tend to be overvalued on average if the fitness of a candidate solution is evaluated and stored only upon its creation rather than every time its numerical value is used. This is intuitively clear, as those candidate solutions that are perceived to be fitter than they actually are, due to noise, are more likely to survive in comparisons with other candidate solutions than those that are undervalued. The systematic over valuation of the parental fitness implicitly introduces a threshold into the selection process. Only those candidate solutions that exceed the perceived parental fitness by at least a certain amount are retained. For the noisy sphere model, it has also been shown that this overvaluation can in fact be useful: while it decreases success probabilities, it increases the expected fitness gain $Q$ per generation, as it mostly leads to the rejection of candidate solutions that would otherwise be accepted merely due to noise rather than to a real advantage in fitness.

We investigated the effect of such a threshold for a real-world optimization problem. However, instead of implicitly implementing the threshold simply by not reevaluating parental fitness, we choose to gain greater control over it by explicitly setting its value and reevaluating the parental fitness in every generation. This becomes especially important for real-world optimization

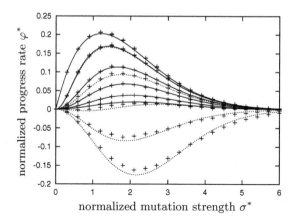

**Fig. 6.3.** Normalized progress rate $\varphi^*$ as a function of normalized mutation strength $\sigma^*$. The solid curves represent results for optimally chosen threshold parameter $\tau^* = \sigma_\epsilon^{*2}$, the dashed curves for $\tau^* = 0$. The respective curves correspond to, from top to bottom, normalized noise strengths $\sigma_\epsilon^* = 0.0,\ 0.4,\ 0.8,\ 1.2,\ 1.6$, and $2.0$. The crosses visualize data measured in ES runs.

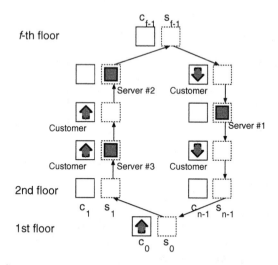

**Fig. 6.4.** The S-ring model as an elevator system. $c_i$ denotes the $i$th customer, whereas $s_j$ denotes the $j$th elevator car. We are seeking an optimal policy $\pi$, that defines the relation $c_i \Leftrightarrow s_i$, in order to minimize the number of waiting customers.

problems, when the target system is not necessarily stationary. In such a case reevaluation is needed for tracking and TS may be recommended.

The resulting algorithm can be sketched as follows:

**Algorithm 1 ((1+1)-ES with TS)**

**Step 1:** *choose one initial parent* $\mathbf{y}^{(g)}$
**Step 2:** repeat
    **Step 3:** *generate a mutation vector* $\mathbf{z}^{(g)}$
    **Step 4:** *determine the (noisy) fitness values* $\tilde{Q}(\mathbf{y}^{(g)} + \mathbf{z}^{(g)})$ *and* $\tilde{Q}(\mathbf{y}^{(g)})$
    **Step 5:** *accept* $\mathbf{y}^{(g)} + \mathbf{z}^{(g)}$ *as* $\mathbf{y}^{(g+1)}$ *if its noisy fitness exceeds that of*
        $\mathbf{y}^{(g)}$ *by at least a margin of* $\tau$; *otherwise retain* $\mathbf{y}^{(g)}$
until *some stopping criterion is satisfied.*

### 6.6.2 Results

A (1+1)-ES with TS was introduced and analyzed for the sphere model. We determined the optimal normalized threshold and performed numerical evaluations that may support the conjecture that TS can improve the performance of ES in noisy environments. The following situation was analyzed: The goal is a minimization of a function $Q(\mathbf{y})$, $\mathbf{y} \in \mathbb{R}^N$, disturbed by additive Gaussian noise $\epsilon \sim \mathcal{N}(0, \sigma_\epsilon^2)$. Let $\tilde{Q}(\mathbf{y})$ be the actually measured fitness (noisy fitness), i.e.,

$$\tilde{Q}(\mathbf{y}) := Q(\mathbf{y}) + \sigma_\epsilon \mathcal{N}(0, 1). \tag{6.3}$$

The selection scheme of the $(1 + 1)$-ES with TS reads

$$\left.\begin{aligned}
\mathbf{y}^{(g+1)} &:= &\mathbf{y}^{(g)} + \mathbf{z}^{(g)} \\
&\Leftrightarrow &\tilde{Q}(\mathbf{y}^{(g)} + \mathbf{z}^{(g)}) \leq \tilde{Q}(\mathbf{y}^{(g)}) - \tau^{(g)} \\
\mathbf{y}^{(g+1)} &:= &\mathbf{y}^{(g)} \\
&\Leftrightarrow &\tilde{Q}(\mathbf{y}^{(g)} + \mathbf{z}^{(g)}) > \tilde{Q}(\mathbf{y}^{(g)}) - \tau^{(g)},
\end{aligned}\right\} \tag{6.4}$$

where

$$\mathbf{z}^{(g)} \sim \sigma \mathcal{N}(\mathbf{0}, \mathbf{E}) \tag{6.5}$$

is the mutation vector with normally distributed random components of standard deviation $\sigma$, and $\tau \geq 0$ is the threshold value. At each generation the parental fitness is measured anew (even when the offspring of generation $g$ did not replace the parent at $g + 1$).

A spherically symmetric fitness function $Q(\mathbf{y})$ is assumed with the location of the minimum at $\hat{\mathbf{y}}$, i.e.,

$$Q(\mathbf{y}) = Q(\|\mathbf{y} - \hat{\mathbf{y}}\|) = Q(R). \tag{6.6}$$

$R := \|\mathbf{y} - \hat{\mathbf{y}}\|$ is the distance to the optimum. Without loss of generality, the optimum is assumed at $\mathbf{0}$. As a special fitness model, we consider the case

$$Q(R) = Q_0 + cR^\alpha. \tag{6.7}$$

If one is able to estimate $\sigma_\epsilon$ and the fitness distance (to the optimal fitness value $Q_0$), the optimal threshold $\hat{\tau}$ can be used in the $(1+1)$-ES:

$$\hat{\tau} = \frac{\sigma_\epsilon^2 N}{\alpha(Q - Q_0)} \tag{6.8}$$

Numerical evaluations showed that TS requires a modified adaptation rule for the mutation rate. The results from the sphere model were transferred to the S-ring model that was introduced as a simulation model to investigate an elevator group control task – a complex real-world optimization task with noisy fitness function values. We used thresholding to handle the noise because reevaluations of the fitness function are too costly. Experimental results on the S-ring model were successful.

### 6.6.3  Further Research

The analytical results from the sphere are consistent with experimental results on the S-ring model, but many details remain unknown, so that the exploration of the following topics will remain for future research:

- robustness of the $\tau$ parameter,
- dynamical adaptation of the $\tau$ parameter for real-world optimization problems, and
- performance of multi-membered evolution strategies, e.g., $(\mu/\mu_I, \lambda)$-ES.

In practical optimization work, useful values of $\tau$ might be determined according to some statistical testing procedure from observed process statistics, somewhat similar to the recommendation of Stagge [79]. Currently, experiments continue with an ES algorithm on the S-ring model to achieve deeper insight into the dynamic behavior of TS [10].

## 6.7  Evolutionary Surface Reconstruction

The intention of surface reconstruction – sometimes also called reverse engineering – is to represent the geometry of an arbitrary physical object in a computer-processable way. Practical applications of this technique cover areas such as medicine, physics, and mechanical engineering as well as the reconstruction of sculptures or the design of flacons in the perfume industry. In all of these fields the physical geometry of a body or model is given, and a visual or physical copy of the original object has to be made. In surface reconstruction many different steps are necessary in order to obtain a highly realistic representation. Each step implies the solution of challenging technical as well as mathematical problems. Sophisticated techniques of computer science as well as mathematics or statistics, combined with the practical knowledge of a mechanical engineer, are needed to establish a well dovetailed process chain.

Surface reconstruction implies the treatment of problems of optimal filtering, approximation, and structural design. Additionally, the mathematical results have to be prepared for further processing steps such as visualization, manipulation in CAD/CAM systems, and manufacturing. Generally, surface reconstruction for applications in mechanical engineering can be divided into the following process steps:

- digitization
  - magneto-electronic scanning
  - optical scanning
  - tactile scanning
- point selection (filtering)
  - surface depending schemes
  - generalizing schemes
- surface generation using
  - triangulation (interpolation or dynamic approximation)
  - nonuniform rational B-spline (NURBS) surface approximations
  - constructive solid geometries (CSG) and hybrids
- representation and application in CAD/CAM systems

In the following, all steps of surface reconstruction are treated. There is a major focus on the issues of point selection and surface generation. A more comprehensive and more formal discussion can be found in the references.

The computation of realistic surface representations implies difficult optimization problems. Discrete approximation, structural design, and pattern recognition are mathematically as well as computationally very complex and require high sophisticated techniques. Here, evolutionary algorithms are used because they proved their capability and efficiency in many practical applications. In the following, a taxonomic approach to the application of EA for surface reconstruction is presented. The general concept of evolutionary algorithms is described in [6, 15, 73] and the more theory-oriented chapters of this book. Depending on the surface reconstruction problem to be solved, different probabilistic algorithms such as evolution strategies, genetic programming, ES GP hybrids, and simulated annealing have been used. A comparison with deterministic approaches is given when possible.

A new parallelizing concept for evolutionary algorithms will be introduced. Therefore, a short more theory-oriented section is presented, where a new modeling scheme for dynamic population structures is introduced. First results on the application of parallel evolution for surface reconstruction with CSG and NURBS structures are presented.

The process chain is closed for each surface model by the manufacturing of prototypic hardware models. The completion of the reconstruction loop demonstrates the possibilities and difficulties of the different approaches in a very practical context.

## 6.8   Scanning Techniques

Scanning techniques are used to measure the surface of an object in order to generate a digital representation of its geometry. The result of a digitization process is a set of discrete 3D point samples that can be recorded using different measuring techniques. According to the digitization method, the 3D point set is regularly structured (e. g., equally spaced), weakly structured (e. g., scan lines), or unstructured.

The decision on a specific scanning system depends on practical demands, (e. g., costs, time, size or material of the workpiece, precision, etc.) as well as on the existing hardware and software environment. For the setup of a complete surface reconstruction system the design of the complete process chain has to be considered. In order to establish a flexible surface reconstruction system, standard CAD interfaces such as STL [78], IGES [66], STEP [80], or VDA-FS [84] should be used. In the context of this work IGES and STL proved to be the best concepts.

One can distinguish between non-tactile and tactile scanning techniques. Non-tactile scanners record the object geometry by electromagnetic, optical or acoustic methods. This type of scanner is able to measure surfaces that are soft, sticky, or should not be touched or destroyed. There exists a large variety of non-tactile systems such as flying spot scanners, CCD cameras using phase shift and Moiré techniques, satellite radar for geographic purposes, computer tomography (CT), and nuclear magnetic resonance (NMR). Ultrasonic methods are used to measure layer thicknesses and in medical applications [55].

In the field of mechanical engineering, tactile sensor systems are prevalent. This has historic as well as practical reasons. The handling of tactile scanners is often quite similar to that of CNC machines. Actually, copymilling machines are the early predecessors of modern surface reconstruction systems. Modern systems can be divided into point-by-point recording or tactile scanning systems. The first approach is slow but very precise ($0.5\,\mu m$ to $10\,\mu m$) and is often used for quality assurance purposes [33]. Tactile scanning systems use a touch probe, which is guided mechanically along fixed paths (digitization strategy). These paths should be chosen according to the structure of the surface to be measured. The optimal scanning paths need not be identical to the optimal milling strategy (this is the handicap of copymilling machines). Therefore, both strategies should be considered and optimized separately. The result of a tactile scanning process is a large set (up to millions of points) of weakly structured 3D samples. Touch probe radius compensation techniques reduce the offset, which is due to the radius of the probe [95]. Typical tactile scanning configurations reach qualities between $10\mu m$ to $50\mu m$ [33].

Most optical and tactile scanning systems generate discrete surface descriptions that do not show undercuts. For many practical applications, such as die and mold manufacturing, this is not a restriction because the tools

cannot be manufactured nor applied having undercuts. Therefore, here only surfaces that can be described by functions from $\mathbb{R}^2$ to $\mathbb{R}$ will be considered.

In the context of this work, a tactile scanning machine (Renishaw Cyclone) was used. Comprehensive tests of the machine yielded nearly normal distributed errors with mean zero and a standard deviation of $\pm 20\,\mu m$. A lot of surface reconstruction applications accept errors that are in this tolerance [33, 55].

## 6.9   Point Selection Schemes

Both, optical and tactile scanning processes usually generate large point sets. These sets may contain thousands to several million samples. The aim of surface reconstruction is to find efficient, exact, and easy to use CAD descriptions on the basis of discrete 3D point data. The effort needed for surface reconstruction depends on the number, the density, and the quality of the measured samples as well as on the accuracy desired. Often the number of points is much too high to be processed efficiently in a CAD system. With respect to the surface model many points are redundant and can be eliminated from the original point set without a loss of quality.

The sets of points can be reduced due to structural or by geometrical criteria. Structural selection intends to generate sparse point distributions that are sufficient for good surface-fitting purposes. It is assumed that the geometry of the original object is correctly captured by the point set. Geometric criteria select point for smoothing or error reduction purposes.

Two ways to reduce point data sets exist: A generalizing approach uses selection schemes that generate point distributions which can be used for a large variety of different surface models. The redundancy of the point set may not be optimal for special surface types but fast and good enough for a lot of practical applications. This class of point-selection schemes, called surface model independent, uses heuristics and global mathematical properties and are often motivated by linear or higher-order approximation theory. Point selection schemes can also be interpreted as filters that let only the relevant points pass. Filters (e. g., Wavelets [54] or new string methods [20]) can be used to reduce measuring errors or to change the geometrical shape of a surface.

Schemes that are surface model dependent exploit special properties of the mathematical surface models. Approximation theory gives hints for necessary sampling point distributions with respect to the tolerable approximation error [40, 58]. The advantage of this approach lies in its ability – for certain metrics – to find the minimum of relevant points. The disadvantage is that the optimal surface has to be known before the points can be reduced.

Both approaches are used for surface reconstruction and are discussed in the following sections. Due to the high practical relevance, a major focus lies

on the generalizing approaches. An extensive description of the algorithms can be found in [55, 56, 87].

### 6.9.1  Generalizing Selection Schemes

Generalizing point selection algorithms can be divided into scan line oriented and multiple schemes. Cord deviation filters are used in postprocessing in tactile digitizing systems. These filters belong to the scan line oriented methods because they use the time series structure of the sampling points. Generalizations of these line oriented methods can process higher dimensional point sets with irregular structure. These methods are called multiple schemes. Cord deviation filters often use linear – seldom nonlinear – functions to approximate a sampling point set with a given tolerance $\epsilon \geq 0$. The functions have to be chosen piecewise and in such a way that each section approximates as many points as possible. This is already an optimization problem by itself and will be solved here with approximation theoretic and heuristic approaches. Depending on the mathematical norm used for distance measurement and the type of function used for the piecewise approximating sections there exist whole classes of point reduction schemes.

A widely used scan line oriented point selection scheme is the tangential distance strategy (TDS) developed by Aretz [3]. This strategy uses a linear model to select relevant points in point sets with time series characteristics. All points that fit into a cylinder with radius $\epsilon > 0$ except the starting point and the last point are deleted from a point set. This procedure is iterated for all points of each scan line. The distance from the points to each line can be measured by an arbitrary norm, which is usually the orthogonal distance. Variants of TDS are the adaptive tangential distance strategy using an orthogonal (ATDSO) and $z$-axis (ATDSZ) parallel distance measures [55, 87]. These strategies are more flexible than the classic TDS. A multiple generalization is the 2DTDS. This algorithm approximates a 3D point set with linear plane patches. A recursive flood fill method cuts out areas of points that fit into a cuboid with height $2\epsilon$. The points on the border of the upper and the lower plane of the cuboid, which cut through the 3D surface, are selected to be so-called relevant points while all other points are eliminated. Univariate linear regression (ULREG) [55, 86] is a method that uses statistical regression to approximate areas of points along a scan line. Points that have an absolute correlation coefficient $r$ close to one can be approximated by a line. The quality of the regression is given by $\gamma_u = 1 - |r_u| \geq 0$. Analogous to the chordal distance schemes, all samples are tested point by point as long as $\gamma$ does not become greater than a certain threshold $t_u$. One or more points can be chosen as representatives for the redundant points. The choice of these representative points is open to the user. A good choice is a sample from the point set near the center of gravity. A multiple approach of the ULREG is the multiple linear regression (MLREG) [55, 87]. This method uses planes that approximate areas of the surface. A multiple regression coefficient $r_m$ is

used to classify points that represent a planar area. A QuadTree method [32] divides the set of sampling points recursively into smaller sets until these sets meet the demands of a good planar approximation, i. e., $\gamma_m = 1 - |r_m| \leq t_m$, where $t_m$ is an external threshold. The center of gravity can be used as a representative for the deleted redundant points.

### 6.9.2    Surface Dependent Schemes

In the following, two approaches for surface dependent point selection will be introduced. First, the iterative method uses all points of a point set and approximates or interpolates these points. In the case of piecewise interpolation, e. g., by triangulation, all points are deleted from sections that fulfill certain criteria, e. g. the angular criterion of neighboring triangles is less than a given threshold value, and the distance of the new triangulation is less than the allowed tolerance. An example of this strategy is the deterministic triangulation-based point selection scheme (DTPSS) [86, 87]. The second approach is based on dynamic triangulations. A more elaborate discussion of dynamic triangulations follows in a later section.

### 6.9.3    A Comparison of Point Selection Schemes

The generalizing point selection algorithms developed and analyzed here in order to find better and more efficient reconstruction algorithms generate very different point distributions when applied to real-world surfaces. In order to illustrate this behavior, a surface of the hood of a car was digitized. The original point set contains 61, 893 points. The reduced point sets all have nearly the same numbers of samples ( 20 percent $\approx$ 12,000 points).

The linear strategies TDS, ATDSO and ATDSZ are fast (in the test application they needed about one second) and yield quite similar point distributions. Using the same tolerance value, the ATDSO reduces the most points, followed by the ATDSZ and the TDS.

For smoothly bended surfaces the multiple 2DTDS strategy is a better choice than any other linear strategy [87]. In the example, the sharp edges along the fender forced the strategy to use more points to yield qualitatively good point distributions.

The ULREG and the MLREG strategy are more robust against outliers than the TDS strategies and have a scalable smoothing effect. This is due to their basically statistical character. Peaks of smooth curves and surfaces are rendered very exactly while small ripples are eliminated (depending on the tolerance). Figure 6.5 shows that this strategy renders contours very well and that the reduction factor in planar areas is also relatively high. In the application to the samples from the hood of a car the MLREG performed best.

Figure 6.5 shows the different results of the point selection schemes.

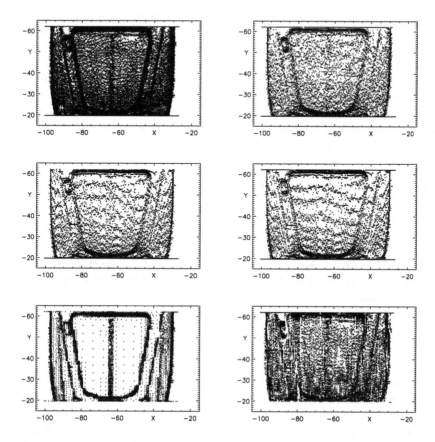

**Fig. 6.5.** (a) Point sets (original, (b) TDS, (c) ATDSO, (d) ATDSZ, (e) MLREG, (f) 2DTDS. Read from upper left to lower right picture.

The quality of point selection schemes cannot be measured without the application of a surface reconstruction method to the reduced point sets. Triangulations as well as NURBS surfaces (see also Sections 6.10 and 6.11) were used to test the selection schemes on various artificially generated data as well as real-world data [55, 56, 87]. The linear strategies are fast but generate point distributions that are influenced by the orientation of the scan lines. The multiple strategies do not exhibit that problem. They are more time consuming (ca. 10 seconds in the applications) but often yield point sets that can be used better for surface reconstructions. Especially MLREG accumulates samples in areas with high curvature while in planar areas the points are regularly distributed. Especially the last point is important for NURBS surface reconstructions. With increasing tolerance $\epsilon$, the number of relevant points decreases hyperbolically for each strategy. Tests with special geometries show that an optimal threshold value for $\epsilon$ exists that corresponds

to efficiently processible point distributions and densities. Reconstructions using triangulations need a lot more points than NURBS surface descriptions. Typical reduction factors, i.e., number of relevant points in relation to the number of irrelevant points, for triangulations lies between 70 percent and 90 percent. Typical reduction factors for NURBS surfaces are between 90 percent and 95 percent.

The reduction factor has a strong impact on the calculation time of a surface reconstruction. Therefore, a well designed preprocessing of the sampling data is highly appropriate.

## 6.10   Surface Reconstruction Using Triangulations

Triangulation is perhaps the most often used surface model in computer graphics. These piecewise linear models are very flexible and can be described in a simple and very compact way. Therefore, they are often used in CAD/CAM systems. All international CAD data-exchange standards such as STL [78], VDA-FS [84], IGES [66], and STEP [80] are able to describe triangulations.

A triangulation, also called piecewise linear interpolation [33], $C^0$ approximation [83] or faceted surface [44, 69], consist of a set of vertices

$$\mathbf{V} = \{v_i \in \mathbb{R}^n | i = 1, \ldots, m\}, m \in \mathbb{N}$$

and a set of edges $\mathbf{E} = \{e_j \in \mathbb{R}^n | j = 1, \ldots, r\}, r \in \mathbb{N}$. Here, only 3D triangulations are considered, i.e., $n = 3$. Every vertex is connected to another vertex via an edge. Three interconnected vertices form one triangle. Triangles must not intersect with each other. In the case of tactile sample points, where no undercuts can be recorded, all triangulations can be projected orthogonally into the $xy$ plane without any intersecting edges.

The structure of the mesh $\mathbf{E}$ of a triangulation is not unique [91]. The number of possible triangulations of a convex set of points increases with the factorial of the number of vertices [33, 55]. The aim of 3D surface reconstruction is to find that specific mesh structure which describes the original surface as realistically as possible. Finding an optimal 3D triangulation can become a very complex algorithmic task. Depending on the structure of the points and the demands on the reconstruction, the problems can have linear complexity (triangulation of equally spaced points) or can even be NP-complete [1, 28].

Experience shows that triangulations with more than 300,000 vertices cannot be processed efficiently with common commercial CAD systems. Therefore, the point sets have to be reduced by point selection schemes (see previous section). Triangulation of reduced point sets is difficult because the resulting triangulation still has to resemble the original surface as realistically as possible.

Two basic approaches to reconstructing surfaces with triangulations exist. The first one is to interpolate the digitized point data and to find an optimal

mesh of edges. The second approach is called adaptive mesh optimization [43, 81] or approximating triangulation [94]. Here, both approaches will be described briefly. Extensive discussion of the optimization of interpolating triangulations can be found in [55, 56, 86, 87, 90, 91, 92].

### 6.10.1   Interpolating Triangulations

For interpolating triangulations the set of vertices is identical to the set of sample points. The discrete character of the sample points does not provide any information on the smoothness or any other characteristics of the original surface. Therefore, for reconstruction algorithms, assumptions on the surface structure have to be made. The task of finding an optimal mesh in the huge set of possible mesh structures can be reduced to an optimization problem. The complexity of this task depends on the criteria used and the encoding of the problem structure. Evolutionary algorithms are very flexible and robust tools, which have often proven to be very efficient when applied to real-world problems. The modularity of an EA allows the use of different criteria as well as different problem encodings.

There are two basic problems that have to be solved before an evolutionary algorithm can be used. First, a fitness function that characterizes the optimum of a problem must be defined. This may seem simple, but in practice, fitness functions should be designed with care. Functions with large plateaus and small slopes as well as deep, narrow, and bended valleys should be avoided. The dimension of a problem has a direct influence on the time needed for the search and should be kept as small as possible. As a rule of thumb, a dimension of over 300 may be called large. The application of several criteria yields multiobjective optimization tasks at once [23]. Second, the structure of the problem is defined by its encoding. Typically, evolutionary algorithms use vectors of parameters or data structures such as lists or trees with variable dimensions. The encoding can be as decisive for the problem complexity as the definition of the fitness function [68]. Mappings to standard representations such as vectors of real values or bit-strings help employ well tested evolutionary algorithms. The knowledge of the behavior of these algorithms simplifies the often nontrivial adjustment of the parameter settings of these search strategies [46]. Theoretical discussions about computational intelligence methods are focused in other chapters of this book.

Here, the encoding of a triangulation is motivated by the definition of Inventor (compatible with virtual reality markup language (VRML)). This standard uses an indexed list of 3D coordinates and a list of edges. Six different fitness criteria that describe the smoothness of a triangulation have been tested [55, 91, 92]: minimum total length of the edges, minimum surface area, minimum maximal/maximum minimal angle in each triangle, minimum angle between normals, and minimum sum of total absolute curvatures. The TAC describes the 3D curvature of a triangulation around a vertex with respect to its adjacent vertices [83]. In order to find an optimal triangulation, the

concept of an evolution strategy and simulated annealing were applied. Here, only mutation operators are used because the recombination of new regular triangulations is very time consuming. Furthermore, one might expect that recombination is helpful during the startup of an ES. During the fine tuning phase it might more likely destroy good meshes and yield small triangles, which are often a bad choice for smooth triangulations. The mutation operator flips randomly chosen edges of the triangular mesh. The number of edge-flip operations follows a decreasing temperature curve (SA), deterministically defined hyperbolic function (ES) and a concept of self adaptive step sizes (ES) [55]. Delaunay triangulations are used to form the startup population [22, 41]. Typically, these meshes form only a rough approximation of the optimal structure of the triangulation. The evolutionary search algorithms optimize the structure according to the fitness criterion used. The total absolute curvature excelled in comparison to the other criteria. This fitness function supports the evolution of long triangles that allow the smooth approximation of cylindric surfaces as well as short triangles that optimally fit to spheric surfaces.

initial triangulation                                    optimized triangulation

**Fig. 6.6.** Interpolative triangulation of the hood of a car (on the left initial triangulation according to Delaunay and on the right optimized solution using the TAC criterion).

The evolutionary optimization of triangulations was applied to various real-world surfaces. Figure 6.6 shows the initial Delaunay triangulation and the solution after the optimization process. The original touch probe radius compensated 87,618 samples have been reduced to 4021 points. The differences of the mesh structures become visible in the cylindric regions around the fender, and the rib in the middle of the hood. Here, the waffle like structure is replaced with a smooth surface by using long triangles. The triangulation has

been converted from Inventor into the stereo lithography (STL) file format and passed to a CAD/CAM system.

Figure 6.6 shows the results of milling simulations [95]. The quality of the physically milled workpieces and the simulations are the same. The total time needed for the complete reconstruction process starting with digitizing (ca. 1 hour), point selection and optimization of the triangles (ca. 4 hours), and finally generating of the NC-paths (ca. 4 hours) and milling the surface (ca. 1 hour) sums up to about 10 hours [55].

## 6.10.2   Approximating Triangulations

Approximating triangulation, also called mesh optimization, fits a triangulation $T = (\mathbf{V}, \mathbf{E})$ of vertices $\mathbf{V}$ and edges $\mathbf{E}$ into a set of sample points $\mathbf{P} = \{p_i \in \mathbb{R}^3 | i = 1, \ldots, n\}$. The number of vertices is a lot lower (about 10 percent or less) than the number of samples in $\mathbf{P}$. In contrast to interpolating triangulations, the vertices of $T$ can be changed, while the structure of the edges is not varied. More general approaches may change the structure of the triangulation as well as the position of the vertices.

The basic idea of mesh optimization is to find the position of the vertices $\mathbf{V}$ of a 3D triangulation that fits optimally into a given set of samples $\mathbf{P}$. The solutions give important indicators for the design of efficient deterministic point selection schemes because the positions of the vertices of the optimized mesh will lie in the neighborhood of the relevant points which should be used for optimal interpolating triangulations. As long as there is no explicit mathematical theorem about the optimal choice of the relevant points for 3D triangulations, only approximative approaches can be used.

The problem of approximating triangulations is algorithmically very complex. A general, optimal, and fast solution still does not exist. Several authors use dynamic graphs for triangulations [2, 43, 81]. A discussion comparing their work with the system introduced here can be found in [99].

The dynamic triangulation algorithm introduced here [99] uses an evolution strategy to adapt the position of the vertices in order to find a triangulation that approximates a given set of samples with a minimum distance. An evolution strategy is applied because deterministic strategies follow a fixed, predefined scheme to search for an optimum. The ES just uses a fitness criterion, which does not introduce any explicit and possibly misleading knowledge of an expert.

The fitness function used here characterizes a good approximation by the sum of deviations of the sample points to the triangles measured in parallel to the $z$-axis. Additionally, in some experiments a multiobjective approach was tested. In these cases the values from the TAC criterion [83] and the point-to-triangle deviations were aggregated to a weighted sum.

In order to use a standard ES, a special encoding was used to represent a complete 3D triangulation as a vector of real values. Efficiency considerations

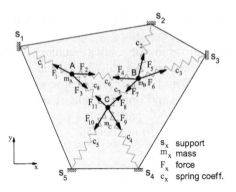

**Fig. 6.7.** System of springs representing a triangulation.

lead to the usage of a system of springs (see Figure 6.7). This idea was mo-
tivated by the deterministic approach of Hoppe [43]. Masses are dispensable
because the mechanical system is fully defined by set of spring coefficients.[1]
The structure of the mesh of springs, i.e. the length of the edges in the $xy$
plane, is uniquely characterized by one vector of spring coefficients and the
position of the support points. The position of the vertices is efficiently calcu-
lated via a linear equation system. The $z$ component of the vertices is directly
adapted by the evolution strategy.

A well-elaborated and easy to use evolution strategy called EASE [76] was
applied. This tool was developed in the course of the project B3 (Schwefel)
and permits the introduction of arbitrary fitness functions into an ES (written
in C) via a comfortable interface (written in Tcl/Tk). The usage of a standard
ES for application purposes is strongly recommended because the fitness
functions are usually already very difficult to understand, and errors due
to the unexpected behavior of not well understood nonstandard optimization
strategies should be avoided.

The optimization starts with a population of triangulations with planar
equidistantly distributed vertices. This structure seems to be the best for
many applications. The $xy$ coordinates of the border points of the triangula-
tion are not varied during the optimization process. Any other value of the
vector (individual) was adapted by a $(30, 100)$-ES [73] with multiple variable
step sizes. The recommendations of Kursawe [46] and [15] to use intermediate
recombination for the object variables and discrete or intermediate recombi-
nation of the strategy parameter proved to be a good choice.

---

[1] If masses had been introduced, the solution of differential equations would have
been necessary to calculate an equilibrium state of the net. This approach would
have had become too time consuming. First experiments with masses showed
that the vertices behave like a set of magnetic bodies, which tend to arrange in
inflexible suboptimal structures.

**Fig. 6.8.** Approximating triangulation of a step structure (generations 100, 1000, 5000, and 30000). Read from upper left to lower right picture.

Figure 6.8 shows an exemplarily selected test geometry which was reconstructed by evolutionary optimized approximative triangulation. The surface was defined manually and contained 2500 points. A triangulation with $10 \times 10$ vertices was used. The geometry was selected because it shows all problems, such as steep steps or sharp edges, which are very difficult to reconstruct using interpolating triangulations.

Due to the fact that the algorithm is able to place the vertices on the border of the steps, very sharp edges can also be exactly represented. In addition, long triangles are evolved when cylindric surfaces have to be approximated (not shown) [94]. These mesh structures resemble smooth surfaces, as results from interpolating triangulations already show.

The algorithm works in two phases. In the first step, the $z$-coordinates are adapted by the ES. During this phase the rough shape of the surface is found. During the fine tuning phase, the solution of the more complex problem of optimal spring coefficient adaptation is solved. The final structure of the surface is generated during this step. The algorithm tends to aggregate points in areas with high curvature while in planar areas only few points are used.

The experiments show that the evolution of good reconstructions using approximative triangulations can become very time consuming. Especially

the high number of dimensions needed to generate smooth triangulations and the strongly interdepending objective parameters increase the complexity of the reconstruction problem. The run time of the approximating triangulation increases cubicly with the number of vertices. The approximation quality increases with higher vertex numbers until the effect of oversampling appears, i. e., an increase in the number of vertices does not yield a significant improvement of the approximation quality. Depending on the surface structure a special effect may appear that inhibits the evolutionary process. In plateau-shaped areas differences between generations become very small. In this case the fitness function shows a plateau, which implies long evolution times. A solution to this problem was to apply the TAC criterion temporarily, which urged the points to move into areas with higher curvature. Another approach is to use redundancy [85] in the genome, but this has not been tested explicitly.

In order to compare the results of the EA with the solutions of a deterministic optimization method, the SIMPLEX algorithm [60], which is known to be quite robust [73], has been used. Due to the high multimodality of the approximation problem, this hill-climbing strategy often yielded fast but very insufficient solutions. Of course the solutions depend on the starting point of the search and the structure of the surface. The EA proved to be more robust than the SIMPLEX strategy. The fitness of these reconstructions can be several powers of ten better than deterministic solutions.

## 6.11   Surface Reconstruction Using Nonuniform Rational B-Splines

Triangulations and nonuniform rational B-splines are the most commonly used mathematical models to describe free form surfaces in CAD systems. The advantages of NURBS are their smoothness, their compact mathematical description, the intuitive manipulation, as well as the ability to combine NURBS patches to larger patches. NURBS are mainly used to describe surfaces, such as turbine blades or parts of the bodywork on cars, that excel by high smoothness. Typically, in practical applications complex surfaces are combinations of several singular NURBS patches. Therefore, it is necessary to know the basic behavior of single patches, especially when they are used in combination with evolutionary algorithms.

NURBS surfaces [65] are parametric bivariate tensor products that map the $\mathbb{R}^2$ to $\mathbb{R}^3$. The definition of a NURBS surface of order $(p, q)$ comprises the following components:

- the control point net with $n \times m$ vertices
  $\mathbf{P} = \{\mathbf{P}_{i,j} \in \mathbb{R}^3, i = 1, ..., n, j = 1, ..., m\}$,

- the knot vector $U$ and $V$, where

$$\mathbf{U} = (\underbrace{0, \ldots, 0}_{p+1}, u_{p+1}, \ldots, u_{r-p-1}, \underbrace{1, \ldots, 1}_{p+1})^T,$$

$$\mathbf{V} = (\underbrace{0, \ldots, 0}_{q+1}, v_{q+1}, \ldots, v_{s-q-1}, \underbrace{1, \ldots, 1}_{q+1})^T,$$

and $r = n + p + 1$ and $s = m + q + 1$,

- the weight vector

$$\mathbf{W} = \{w_{i,j} \in \mathbb{R}^+, i = 1, ..., n, j = 1, ..., m\}.$$

The parametric surface function is defined on the domain $(u, v) \in [0, 1] \times [0, 1]$ an expressed by

$$\mathbf{S}(u, v) = \frac{\sum_{i=0}^{n} \sum_{j=0}^{m} N_{i,p}(u) N_{j,q}(v) w_{i,j} \mathbf{P}_{i,j}}{\sum_{i=0}^{n} \sum_{j=0}^{m} N_{i,p}(u) N_{j,q}(v) w_{i,j}}.$$

$N_{i,p}(u) : \mathbb{R} \to \mathbb{R}$ ($N_{j,q}(v)$ are defined analogously) are the $i$th basis functions of order $p$ for the parameter $u \in \mathbb{R}$ computed on a knot vector $(u_0, ..., u_m)^T$ [65].

A NURBS patch can be manipulated locally by a variation of the 3D positions of a single control point. Due to the fact that the structure of the net of control points directly corresponds to the 3D geometry of the NURBS surface, any variation of the control net structure directly influences the shape of the NURBS surface. The variation of the positions of the control points, the values of the weights, and the structure of the knot vectors $U$ and $V$ all define the structure of a NURBS surface. The size of the area influenced by one weighted control point depends on the degrees $p$ and $q$. The higher the degree the larger the area concerned. Degrees of $p = 1$ and $q = 1$ correspond to piecewise linear models that can roughly be compared with approximating triangulations. Typically, degrees of $p = 3$ and $q = 3$ are used. In practical applications degrees higher than 5 are uncommon.

The general problem of discrete NURBS surface approximation is not easy to solve using deterministic schemes [47, 65]. [52] and [59] use evolutionary approaches for one-dimensional NURBS curves. NURBS surface interpolations should not be used because the number of sampling points is often so high that the corresponding matrices become too large. One of the advantages of NURBS surfaces compared to triangulations is that they produce smooth surfaces. NURBS show deep pass filtering properties and allow the use of smaller numbers of sampling points than needed for triangulations. Point selection schemes, such as ULREG or 2DTDS, seem to be most appropriate.

In the context of this work several different fitness functions have been tested to characterize good NURBS surface reconstructions. Experience shows that the best function seems to be a multiobjective combination of a $z$ axis parallel point-to-surface distance criterion and the sum of angles between normal vectors on the NURBS surface. Both criteria are aggregated in a weighted sum and must be minimized. The tensor product properties of NURBS surfaces imply a relatively time consuming evaluation of the distance criterion

because the calculations imply optimizations steps by themself. A new and efficiently designed distance function increases the evaluation time by a factor of five in comparison with early approaches [88].

NURBS surfaces can directly be encoded by mapping the vectors $\mathbf{P}$, $\mathbf{U}$ and $\mathbf{V}$ to one common vector $\mathbf{I} \in \mathbb{R}^\gamma$, $\gamma = 4nm + (n+p+m+q)$. This encoding can be used directly by an evolution strategy. Here the recommendations from theory supported the decisions for the initial parameter adjustments (e. g., [46]).

Practical considerations motivate a new three step approach: During the first step the $z$ coordinates of the control points are adjusted by the evolutionary process. The degrees of $p$ and $q$ are one. The result is a rough, piecewise linear, discrete surface approximation of the sampling points. In a second step the degrees are increased to $p = 3$ and $q = 3$ and the optimization is restarted from the linear solution. The resulting NURBS surface becomes smoother and the quality of the surface increases. In a third step all other parameters are open to the optimization process in order to find the global solution in the complete space of NURBS surfaces.

This stepwise approach has the advantage that undercuts can be prevented. Especially during the first phases of the reconstruction process, this effect is very likely to appear. NURBS are parametric functions that can represent nearly arbitrary 3D geometries, such as spheres and cylinders. In the application introduced here, undercuts are not allowed because tactile digitizing processes cannot scan objects with these shapes.

An example of the reconstruction process of the hood of a car is shown in Figure 6.9. Here, only the left half of the surface was reconstructed because the original surface is axially symmetric. The initial population consisted of horizontal and plane NURBS surfaces. The control net contained $40 \times 7$ points and 1600 sampling points were used.

**Fig. 6.9.** NURBS surface reconstruction of the hood of a car (from left to right: generation 100, 500, 1000).

The evolutionary reconstruction process took about 3 hours on a Pentium PC with 450 MHz. The average deviation is 0.05 $mm$. Experience shows that

the calculation of CAM NC-paths using NURBS is a lot faster than using triangulations. As a rule of thumb, accelerations of 50 times (compared with triangulations) are possible.

The NURBS reconstruction is part of a hybrid system, which is also able to reconstruct CSG objects (see the next section) using the same data structure [56, 96, 97, 98, 99]. The result of the reconstruction process can be imported via IGES into a standard CAD system (see Figure 6.10).

**Fig. 6.10.** NURBS surface reconstruction of the complete hood of a car displayed in a CAD system.

In order to solve the problem of oscillation, an additional deterministic pre-processing step may be introduced. This algorithm should locally change the density of the control net. In the actual application uniformly distributed control points are used. A well-balanced adaptation mechanism that inserts or deletes control points should lead to even better results. First tests showed that EA using variable dimensions prefer the insertion of control points rather than deleting points. This is due to the fact that insertion can be done without any change in the surface structure, while deletion usually implies impairment of an individual.

The application of a multiobjective fitness function, which also takes into account the distance together with the angles between the normal vectors, increased the smoothness of the NURBS surface and reduced the reconstruction time. The smoothing effect on sharp edges can be slightly reduced by adding more sample points in areas with high curvature. This increases the fitness locally and hence, improves the reconstruction quality.

Depending on the surface structure, deterministic approaches are also basically able to yield quite good approximations for small and comparatively simple structured NURBS patches (e. g., few and equally spaced control points, all weights equal one, etc.). The problem for deterministic strategies is again the multimodality of the fitness function. A hybrid algorithm com-

bining fast deterministic search strategies with the global search power of evolution strategies together with a preprocessing phase for the positioning of sample point and control point is matter of future research.

A more elaborate discussion of NURBS surface reconstructions can be found in [55, 56, 86, 87, 88, 89, 90, 97, 98].

## 6.12    A Parallel Reconstruction System Using CSG-NURBS Hybrids

Objects can either be described by boundary surfaces (e. g., NURBS surfaces, triangulations, etc.) or combinations of volumes [78]. Constructive solid geometries are the basic data structure for many volumetric 3D CAD systems. Typically, CSG objects are made of mathematically simple solid geometries, such as spheres, ellipsoids, cones, boxes, cylinders, tori, etc. These elements are combined by binary relations: intersection, union, and subtraction. Nearly all arbitrarily complex geometries can be formed by recursive application of CSG relations. This recursive description suggests the application of trees as a basic data structure. The integration of free-form surfaces into CSG is possible [38] and expands this concept to CSG hybrids. CAD data exchange formats, such as IGES and STEP, already support hybrid models [66, 80], although only few modern CAD systems permit the use of both descriptions simultaneously.

The applications of CSG together with evolutionary algorithms focus mainly on the evolutionary design of new objects rather than the reconstruction of existing elements. The spectrum of evolutionary design reaches from the setup of hydroelectric power plants [62] to design studies for cars [11, 12, 64] and evolutionary art [82].

Surface reconstruction tries to invert the production-process chain by generating a CAD description from the surface data of a physical object. Here, a parallel evolutionary NURBS-CSG surface reconstruction system [99], which is based on the CSG reconstruction system described in [45] is introduced. In the new hybrid system NURBS surfaces and CSG objects use one common data structure [56, 97, 98]. All NURBS surface reconstructions described below used this hybrid system. The reconstruction of hybrid surfaces by simultaneously using CSG and NURBS during the evolutionary process is possible but very difficult and a subject of current research. Until now the parallel reconstruction of either CSG or NURBS surfaces by the one reconstruction system is possible. A step-by-step approach, where first NURBS are used to describe free form surfaces and then complete CSG structures are added in a second step, is typical for real-world CAD construction.

For each surface type, two separate multiobjective fitness functions were used in the CSG-NURBS hybrid system. The discussion of the NURBS criteria can be found in the previous paragraph. For the reconstruction of CSG

structures, four criteria were aggregated to form one weighted sum. The fitness function compares all criteria of the point set with the properties of the CSG structure. The distance as well as the sum of angles between normals, the differences of curvatures, and the number of objects have to be minimized to fulfill the properties of a good CSG reconstruction [99].

The hybrid system makes use of the data structure of a binary tree. This encoding is very useful for comparing the CSG structure with the sampling point set because efficient algorithms from computer graphics can be used [32]. Each CSG element contains its own real-valued parameters (e. g., length, radius, etc.). Therefore, the data structure contains the CSG relations as well as vectors of real values.

**Fig. 6.11.** Reconstruction of a hash-shaped structure.

Due to the fact that the size of the tree, i. e., the structure of the object to be reconstructed, is a priori not known, a variable dimensional problem appears which can be solved by genetic programming. Additionally, an evolution strategy is used to adapt the real value parameters of the CSG elements [9]. Due to the complexity of the variable dimensional multiobjective and multimodal fitness function, the application of evolutionary algorithms seems to be necessary. Tests using a deterministic strategy – a Hough transformation [63] that is well known in graphical pattern recognition – were only successful for very simply structured geometries [55].

The ability of this ES/GP hybrid to reconstruct CSG structures has been proved and discussed in several publications [56, 97, 98]. Figure 6.11 shows the evolution of the reconstruction of a hash-shaped structure composed of four cylinders. Analyses with objects where subtractions are necessary have also been successful, although a global tendency of the algorithm to join new

elements to the structure rather than to subtract objects from a geometry should be remarked [55].

A parallel approach using a blackboard communication model [16] for data exchange was introduced to improve the convergence probability of the evolutionary algorithm [99]. The motivation to use parallel populations is based on the observation that a single population was able to reconstruct substructures of an object a lot faster than the complete structure as a whole. The combination of several populations which exchange solutions at a common market place allows a safer and faster evolution. Furthermore, the algorithm can be implemented as a distributed system which runs robustly in a LAN of common PC.

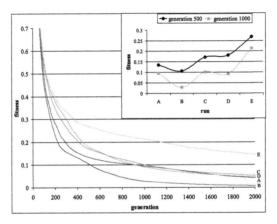

**Fig. 6.12.** Fitness progression for parallel evolutionary CSG reconstructions for $\nu_A = \frac{1}{5}, \nu_B = \frac{1}{20}, \nu_C = \frac{1}{50}, \nu_D = \frac{1}{100}, \nu_E = \frac{1}{1000}$.

Figure 6.12 shows the fitness progression of five different experiments.[2] An object made of two cylinders with four spheres at their ends combined in a cross structure was to be reconstructed. Five Pentium PC were connected by a 100 MBit LAN. Typically, running times of 30 minutes were needed for 4000 generations. The parameter $\nu_I, I \in \{A, B, C, D, E\}$, the reciprocal number of generations after which 10 percent of the individuals of each population were copied to the blackboard, was varied for the experiments. On the average, after 1500 generations the parallel algorithm found a solution. The corresponding panmictic algorithm needed more time to find a similar solution and had a smaller all over convergence probability. One can see from the small graph in Figure 6.12, which shows with respect to the migration frequency the fitness of a population after 500 and 1000 generations that a frequency of $\nu_B = \frac{1}{20}$ performed best. This was not expected because experiments with ES using distributed populations (island models) showed that

---

[2] 10 runs for each experiment. The results have been averaged.

exchanging only very few individuals during each generation performed best on artificial fitness functions (e. g., Rastrigin's function) [93]. This observation may be explained by the two extreme cases: A high frequency yields high data exchange, and the evolution resembles a panmictic scheme. A very low migration frequency implies the evolution of solutions in isolated populations. The benefit of data exchange is not used, and the genetic variability as well as the probability to generate better recombinations during a generation gets low. In the case of evolutionary CSG surface reconstruction, a moderate migration rate seemed to be the best choice.

## 6.13   Parallel Evolution Strategies

The positive experience of using parallel evolutionary algorithms in surface reconstruction motivated a deeper analysis of this concept. Parallelization is a straightforward approach to reducing computation time. Especially for evolutionary algorithms, which are inherently parallel, synergetic effects can also be expected. A more theoretical analysis was necessary to get a closer look at the dynamic processes that works behind the communication within a population. Applying the successful concept of self-adaptation of evolution strategies to the adaptation of the exchange of genetic data between individuals was the driving idea behind the implementation of an evolution strategy with dynamically changing neighborhood structures [93]. The communication model introduced next is based on a diffusion model approach. The results of these experiments also influenced the setup of the simpler blackboard models (see earlier chapters).

In parallel evolutionary optimization, three different kinds of static population structures can be distinguished: the panmictic model, where a population that is not explicitly structured is used [73]; the multipopulation model, where individuals migrate or move between a set of essentially isolated panmictic populations [72]; and neighborhood models, where each individual in a population is interconnected with an other individual via static relations [77].

Halpern introduced an evolutionary algorithm with dynamic population structures [39]. It is based on a structurally dynamic cellular automaton. Here, one individual is assigned to one corresponding cell of the cellular automaton. The individuals exchange their genetic data using a network that interconnects the cells with each other. An adjacency vector describes the net. During the optimization, this vector is changed according to deterministic rules. Halpern's approach has many similarities to an evolution strategy running on a diffusion model. Great differences can be seen in the way mutation, recombination, and selection are used. As in genetic algorithms, an explicit step size adaptation is not used by Halpern. The mutation rate is kept at a predefined fixed level and the step sizes are limited to small intervals.

The new approach only adopts Halpern's rules for changing the interconnections between the individuals. Alternative probabilistic connection rules have also been introduced. Essentially, the rules define the coupling and decoupling of conditions between individuals within a population. The deterministic rules decouple an individual's neighbors that have the lowest fitness while new connections to the so-called next nearest neighbors with high fitness values are established. The nondeterministic rules mimic the adaptation scheme of an ES. Individuals are decoupled from each other whenever a recombination with that specific neighbor yields an offspring with inferior fitness. At the same time, new connections are established with arbitrary elements of the population whenever the number of connections becomes lower than a certain threshold. Recombination and mutation have been implemented according to the scheme of standard ES [15]. The highly multimodal fitness function of Rastrigin [73] has been used to show the convergence probabilities of the parallel algorithm. Recommendations with regard to the parameter settings for ES have been taken from [46]. The software has been implemented on a parallel computer SGI Origin 2000 with 16 R12000 processors using a shared memory concept. Using message passing interface (MPI) [61] it is also possible to run the implementation on a distributed cluster of LINUX or Solaris workstations.

Depending on the parameter settings for the coupling and decoupling rules, the parallel ES showed the typical kinds of behavior of dynamic population growths. Stable, oscillating, or even chaotic population dynamics were observed. Populations with a long-term stable behavior were selected for the analyses.

A dynamic ES with probabilistic neighborhood adaptation rules that change the communication structure after pauses of several generations produced the best results, i.e., runs with high convergence probability. Restructuring the communication of a population after every generation cycle introduced too many turbulences into the genetic pool. This has a negative effect on the evolutionary process. The probabilistic adaptation process of the neighborhood of the individuals generated communication nets, which successively adapted to increasingly better structures. The best solutions showed dynamically changing substructures, such as islands with close connections. These islands had comparatively loose contact with other islands. The complete structure was dynamically changing but showed on the average a stable evolutionary behavior.

Generally, the speed of convergence (measured per generation cycle) of the parallel approaches was similar to that of a standard panmictic ES. Typically, the convergence probability was increased by parallelization. A comparison of parallel ES and parallel GP (a rule of thumb based on the experience with surface reconstruction) shows a higher potential of parallel GP.

The relative speed up and the relative efficiency[3] of the dynamic ES measured on the parallel computer are shown in Figure 6.13.

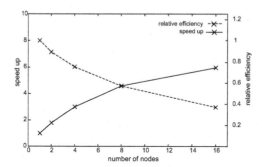

**Fig. 6.13.** Relative speed up and relative efficiency.

Due to the fact that the individuals are arranged in a matrix-shaped cellular automaton, the structure of the population can be subdivided according to the number of processors. With an increasing number of processors (nodes) the speed up also increased. Because of the rising amount of communication, the relative efficiency becomes smaller. In the tests of ES with dynamic and static communication structures, an optimal compromise between speed and efficiency was reached using 8 processors. An explicit use of the shared memory ability of the system, of course, would have increased this number. The experiments show that the application of the multiprocessor machine significantly reduced the calculation time and helped to improve the convergence behavior of evolutionary algorithms in the experiments.

## 6.14    Conclusions

The demand for efficient surface reconstructions that can be used in practical applications is high, as discussions with partners from the industry show. Practical applications often imply changing demands, which have to be met with flexible surface reconstruction techniques. The complex problems that have to be solved by each element of the process chain are very challenging. Evolutionary surface reconstruction is a new approach that may help to find good solutions to this class of problems.

---

[3] The execution time of a parallel program is the length of the time interval from the start of the first process to the end of the last process [17]. The relative speed up $s(p)$ is the quotient of the time $t_1$ needed by one processor to execute the parallel program to time $t_p$ needed by $p$ processors, i. e., $s(p) = t_1/t_p$. The relative efficiency $e(p)$ is the quotient of the relative speed up $s(p)$ and the number of processors $p$. This quotient $e(p) = s(p)/p$ expresses the effective gain of the parallelization.

Evolutionary algorithms show a very complex behavior that makes it difficult to provide a handbook of *how* to use evolutionary algorithms. Standard EA approaches and their variations, such as hybrid combinations and parallel approaches, open a large field of still unsolved problems. The practical application of EA necessitates experience, some sure instinct, and basically a new view to problem solving. We discussed an algorithmic approach to systematize the problem-specific design of EA.

# References

1. A. Aggarwal and S. Suri. Surface approximations and geometric partitions. In *Proceedings Fifth Symposium on Discrete Algorithms*, pages 24–33. Arlington, VA, 1994.
2. M.-E. Algorri and F. Schmitt. Surface Reconstruction from Unstructured 3D Data. *Computer Graphics Forum*, 15(1):47–60, 1993.
3. R. Aretz. Datenreduktion und mathematische Modellbildung. *CAD-CAM-Report*, 1:65–78, 1992.
4. D. V. Arnold. Evolution strategies in noisy environments – A survey of existing work. In L. Kallel, B. Naudts, and A. Rogers, editors, *Theoretical Aspects of Evolutionary Computing*, Natural Computing, pages 241–252. Springer, Berlin, 2001.
5. D. V. Arnold and H.-G. Beyer. Local performance of the (1+1)-ES in a noisy environment. Technical Report CI-80/00, Collaborative Research Center 531, University of Dortmund, 2000.
6. T. Bäck. *Evolutionary Algorithms in Theory and Practice*. Oxford University Press, New York, 1996.
7. T. Bäck, editor. *Seventh International Conference Genetic Algorithms (ICGA '97), East Lansing, MI, July 19–23, 1997*. Morgan Kaufmann, San Francisco, CA, 1997.
8. T. Bäck, D. B. Fogel, and Z. Michalewicz, editors. *Handbook of Evolutionary Computation*. Oxford University Press, New York, und Institute of Physics Publishing, Bristol, 1997.
9. W. Banzhaf, K. Weinert, R. Keller, and J. Mehnen. Surface Reconstruction from 3D Point Data with a Genetic Programming/Evolution Strategy hybrid. Technical Report CI-44/98, Collaborative Research Center 531, University of Dortmund, 1998.
10. T. Beielstein and S. Markon. Threshold selection, hypothesis tests, and DOE methods. Technical Report CI-121/01, Collaborative Research Center 531, University of Dortmund, 2001.
11. J. B. Bentley. Generic evolutionary design of solid objects using a genetic algorithm. PhD thesis, University of Huddersfield, UK, 1996.
12. P. J. Bentley. *Evolutionary Design by Computers*. Morgan Kaufmann, San Mateo, CA, May 1999.
13. H.-G. Beyer. Towards a theory of evolution strategies: Some asymptotical results from the $(1\overset{+}{,}\lambda)$-theory. *Evolutionary Computation*, 1(2):165–188, 1993.
14. H.-G. Beyer. Evolutionary algorithms in noisy environments: Theoretical issues and guidelines for practice. *CMAME (Computer methods in applied mechanics and engineering)*, 186:239–267, 2000.

15. H.-G. Beyer. *The Theory of Evolution Strategies*. Springer, Heidelberg, 2001.
16. W. Brenner, R. Zarnekow, and H. Wittig. *Intelligente Software Agenten*. Springer, Berlin, 1998.
17. H. Burkhart. *Parallele Programmierung*. P. Rechenberg und G. Pomberger (editors), Hanser, Munich, 1997. Informatik–Handbuch.
18. K. Chellapilla. Evolving computer programs without subtree crossover. *IEEE Transactions on Evolutionary Computation*, 1(3):209–216, 1997.
19. J. C. Culberson. On the futility of blind search: An algorithmic view of the "no free lunch". *Evolutionary Computation*, 6(2):109–127, 1998.
20. P. L. Davies and A. Kovac. Local extremes, runs, strings and multiresolution. *Annals of Statistics*, 29:1–65, 2001.
21. L. Davis. *Handbook of Genetic Algorithms*. Van Nostrand Reinhold, New York, 1991.
22. M. de Berg, M. van Kreveld, M. Overmars, and O. Schwarzkopf. *Computational Geometry: Algorithms and Applications*. Springer, Berlin, 1997.
23. K. Deb. *Multi-Objective Optimization using Evolutionary Algorithms*. Wiley, New York, 2001.
24. P. J. Denning, J. B. Dennis, and J. E. Qualitz. *Machines, Languages, and Computation*. Prentice-Hall, Englewood Cliffs, 1979.
25. S. Droste and D. Wiesmann. Metric based evolutionary algorithms. In R. Poli, W. Banzhaf, W. B. Langdon, J. F. Miller, P. Nordin, and T. C. Fogarty, editors, *Genetic Programming, Proceedings of EuroGP'2000, Edinburgh, April 15–16, 2000*, volume 1802 of *LNCS*, pages 29–43. Springer, Berlin, 2000.
26. S. Droste and D. Wiesmann. On the design of problem-specific evolutionary algorithms. In A. Ghosh and S. Tsutsui, editors, *Advances in Evolutionary Computing*. Springer, Berlin, 2002. (In print).
27. M. Emmerich, M. Grötzner, and M. Schütz. Design of graph-based evolutionary algorithms: A case study of chemical process networks. *Evolutionary Computation*, 9(3):329–354, 2001.
28. S. Fekete and W. Pulleyblank. Area optimization of simple polygons. In *Ninth Annual Symposium on Computational Geometry*, pages 173–182. ACM, San Diego, CA, 1993.
29. D. B. Fogel. *Evolutionary Computation: Toward a New Philosophy of Machine Intelligence*. IEEE Press, New York, 1995.
30. D. B. Fogel and A. Ghozeil. A note on representation and variation operators. *IEEE Transactions on Evolutionary Computation*, 1(2):159–161, 1997.
31. L. J. Fogel, A. J. Owens, and M. J. Walsh. *Artificial Intelligence through Simulated Evolution*. Wiley, New York, 1966.
32. D. Foley, A. van Dam, S. Feiner, and J. Hughes. *Computer Graphics*. The Systems Programming Series. Addison-Wesley, Menlo Park, CA, 2nd edition, 1992.
33. J. Friedhoff. Aufbereitung von 3D-Digitalisierdaten für den Werkzeug-, Formen- und Modellbau. PhD thesis, University of Dortmund, 1996.
34. M. R. Garey and D. S. Johnson. *Computers and Intractability: A Guide to the Theory of NP-Completeness*. W.H. Freeman, New York, 1979.
35. D. E. Goldberg. *Genetic Algorithms in Search, Optimization and Machine Learning*. Addison-Wesley, Reading, MA, 1989.
36. H. Gould. *Thermal and Statistical Physics Simulations*. John Wiley, New York, 1995.

37. B. Groß, U. Hammel, A. Meyer, P. Maldaner, P. Roosen, and M. Schütz. Optimization of heat exchanger networks by means of evolution strategies. In H.-M. Voigt, W. Ebeling, I. Rechenberg, and H.-P. Schwefel, editors, *Fourth International Conference Parallel Problem Solving from Nature (PPSN IV), Berlin, September 22-26, 1996*, volume 1141 of LNCS, pages 1002–1011. Springer, Berlin, 1996.

38. B. Guo and J. Menon. Local shape control for free-form solids in exact CSG representations. *Computer-Aided Design*, 28(6/7):483–493, 1996.

39. P. Halpern. Evolutionary algorithms on a self-organized dynamic lattice. Technical report, University of the Sciences in Philadelphia, PA, 1999.

40. G. Hämmerlin and K.-H. Hoffmann. *Numerische Mathematik*, volume 7 of *Grundwissen Mathematik*. Springer, Berlin, 1989.

41. C. Hao and J. Bishop. Delaunay triangulation for curved surfaces. In *Proceedings Sixth International Meshing Roundtable*, pages 115–127. Sandia National Laboratories, CA, 1997.

42. J. E. Hopcroft and J. D. Ullman. *Introduction to Automata Theory, Languages and Computation*. Addison-Wesley, Reading, MA, 1979.

43. H. Hoppe, T. DeRose, T. Duchamp, J. McDonald, and W. Stuetzle. Mesh optimization. In *Proceedings Computer Graphics SIGGRAPH '93*, Computer Graphics Annual Conference Series, pages 21–26. ACM SIGGRAPH, Anaheim, NY, 1993.

44. S. Karbacher. *Rekonstruktion und Modellierung von Flächen aus Tiefenbildern*. PhD thesis, University of Erlangen, Aachen, 1997.

45. R. E. Keller, J. Mehnen, W. Banzhaf, and K. Weinert. Surface reconstruction from 3D point data with a genetic programming/evolution strategy hybrid. In L. Spector, W. Langdon, U.-M. O'Reilly, and P. Angeline, editors, *Advances in Genetic Programming*, volume 3, chapter 2, pages 41–65. MIT Press, Cambridge, MA, 1999.

46. F. Kursawe. Grundlegende empirische Untersuchungen der Parameter von Evolutionsstrategien – Metastrategien. PhD thesis, University of Dortmund, 1999.

47. P. Laurent-Gengoux and M. Mekhilef. Optimization of a NURBS Representation. *Computer-Aided Design*, 25(11):699–710, 1993.

48. A. M. Law and W. D. Kelton. *Simulation Modeling and Analysis*. McGraw-Hill, New York, 3rd edition, 2001.

49. J. Ludvig, J. Hesser, and R. Männer. Tackling the representation problem by stochastic averaging. In Bäck [7], pages 196–203.

50. S. Markon. Studies on applications of neural networks in the elevator system. PhD thesis, Kyoto University, 1995.

51. S. Markon, D. V. Arnold, T. Bäck, T. Beielstein, and H.-G. Beyer. Thresholding – a selection operator for noisy ES. In *Proceedings Conference on Evolutionary Computation (CEC 2001)*. Seoul, Korea, May 27–30, 2001.

52. A. Márkus, G. Renner, and J. Váncza. Genetic Algorithms in Free Form Curve Design. In M. Daehlen, T. Lyche, and L. L. Schumaker, editors, *Mathematical Methods for Curves and Surfaces*, pages 343–354. Vanderbilt University Press, Nashville, TN, 1995.

53. K. E. Mathias and L. D. Whitley. Changing representation during search: A comparative study of delta coding. *Evolutionary Computation*, 2(3):249–278, 1994.

54. A. Le Mehaute, C. Rabut, and L. Schumaker. *Surface Fitting and Multiresolution Methods*. Vanderbilt University Press, Nashville, TN, 1997.
55. J. Mehnen. *Evolutionäre Flächenrekonstruktion*. PhD thesis, University of Dortmund, 2000.
56. J. Mehnen, J. Meng, and T. Surmann. Methoden der Flächenrekonstruktion. In K. Weinert, editor, *Spanende Fertigung*, volume 3, pages 489–498. Vulkan, Essen, 2001.
57. Z. Michalewicz. *Genetic Algorithms + Data Structures = Evolution Programs*. Springer, Berlin, 1996.
58. M. Müller. *Approximationstheorie*. Studienbuch für Studierende der Mathematik, Physik und Informatik. Akademische Gesellschaft Wiesbaden, Germany, 1978.
59. R. Mussa, R. Roy, and G. Jared. Surface optimisation within a CAD/CAM environment using genetic algorithms. In R. Roy, T. Furuhashi, and P. K. Chawdhry, editors, *Advances in Soft Computing – Engineering Design and Manufacturing*, pages 119–131. Springer, London, 1998.
60. J.A. Nelder and R. Mead. A simplex method for function minimization. *Computer Journal*, 7:308–313, 1965.
61. P. S. Pacheco. *Parallel Programming with MPI*. Morgan Kaufmann, San Francisco, CA, 1997.
62. I. Parmee. The development of dual-agent strategy for efficient search across whole system engineering design hierarchies. In H.-P. Schwefel, editor, *Parallel Problem Solving From Nature IV*, pages 523–532. Springer, Berlin, 1996.
63. D. Paulus and J. Hornegger. *Pattern Recognition and Image Processing in C++*. Advanced Studies in Computer Science. Vieweg, Wiesbaden, 1995.
64. D. T. Pham and Y. Yang. A genetic algorithm based preliminary design system. *Journal of Automobile Engineers*, 207(D2):127–133, 1993.
65. L. Piegl and W. Tiller. *The NURBS book*. Springer, Berlin, 1997.
66. US PRO. *IGES 5.2 An American National Standard*. U.S. Product Data Association, Gaithersburg, MD, 1993.
67. S. B. Rana and L. D. Whitley. Bit representation with a twist. In Bäck [7], pages 188–195.
68. I. Rechenberg. *Evolutionsstrategie '94*. Frommann–Holzboog, Stuttgart, 1994.
69. S. Roth-Koch. *Merkmalsbasierte Definition von Freiformgeometrien auf der Basis räumlicher Punktwolken*. PhD thesis, University of Stuttgart, 1996.
70. G. Rudolph. *Convergence Properties of Evolutionary Algorithms*. Dr. Kovač, Hamburg, 1997.
71. G. Rudolph. Finite Markov chain results in evolutionary computation: A tour d'horizon. *Fundamenta Informaticae*, 35(1-4):67–89, 1998.
72. G. Rudolph. Evolution strategies. In D. B. Fogel T. Bäck and Z. Michalewicz, editors, *Evolutionary Computation 1 - Basic Algorithms and Operators*, pages 81–88. Institute of Physics Publishing, Bristol, UK, 2000.
73. H.-P. Schwefel. *Evolution and Optimum Seeking*. Wiley, New York, 1995.
74. B. Sendhoff, M. Kreutz, and W. von Seelen. A condition for the genotype-phenotype mapping: Causality. In Bäck [7], pages 73–80.
75. T. Slawinski, A. Krone, U. Hammel, D. Wiesmann, and P. Krause. A hybrid evolutionary search concept for data-based generation of relevant fuzzy rules in high dimensional spaces. In *Proceedings Eighth International Conference on Fuzzy Systems (FUZZ-IEEE'99), Seoul, Korea, Aug. 22-25, 1999*, pages 1431–1437. IEEE Press, Piscataway, NJ, 1999.

76. J. Sprave. Ease - Evolutionary Algorithms Scripting Environment. Technical Report CI-55/99, Collaborative Research Center, University of Dortmund, 1999.
77. J. Sprave. Ein einheitliches Modell für Populationsstrukturen in evolutionären Algorithmen. PhD thesis, University of Dortmund, 1999.
78. G. Spur. *Die Genauigkeit von Maschinen*. Carl Hanser, Wien, 1996.
79. P. Stagge. Averaging efficiently in the presence of noise. In A. E. Eiben, T. Bäck, M. Schoenauer, and H.-P. Schwefel, editors, *Fifth International Conference Parallel Problem Solving from Nature (PPSN V)*, Amsterdam, Sep. 27–30, 1998, volume 1498 of LNCS, pages 188–197. Springer, Berlin, 1998.
80. STEP. *ISO 10303-42*, volume 42. International Organization for Standardization, Switzerland, 1994.
81. D. Terzopoulos and M. Vasilescu. Sampling and reconstruction with adaptive meshes. In *Proceedings IEEE Conference Computer Vision and Pattern Recognition*, pages 70–75, 1991.
82. S. Todd and W. Latham. *Evolutionary Art and Computers*. Academic Press, Winchester, UK, 1992.
83. R. van Damme and L. Alboul. Polyhedral metrics in surface reconstruction: Tight triangulations. Memorandum no. 1275, University of Twente, Netherlands, 1994.
84. Verband der Automobilindustrie e.V. *VDA-Flächenschnittstellen, Version 2.0*. Verband der Automobilindustrie e.V., Frankfurt, 1986.
85. K. Weicker and N. Weicker. Burden and benefits of redundancy. In W. Martin and W. Spears, editors, *Foundations of Genetic Algorithms 6*, pages 313–333. Morgan Kaufmann, San Francisco, CA, 2001.
86. K. Weinert and J. Mehnen. A Comparison of Point Data Selection Schemes for Evolutionary Surface Reconstructions. Technical Report CI-88/00, Collaborative Research Center 531, University of Dortmund, 2000.
87. K. Weinert and J. Mehnen. Comparison of selection schemes for discrete point data for evolutionary optimized triangulations and NURBS surface reconstructions. In F. Nagdy and F. Kurfess, editors, *ISA 2000, Intelligent Systems & Applications, CI'2000*, pages 104–110. University of Wollongong, Australia, ICSC, International Computer Science Conventions, December 2000.
88. K. Weinert and J. Mehnen. Discrete NURBS-surface approximations using an evolutionary strategy. Technical Report CI-87/00, Collaborative Research Center 531, University of Dortmund, 2000.
89. K. Weinert and J. Mehnen. NURBS-surface approximation of discrete 3D-point data by means of evolutionary algorithms. In R. Teti, editor, *CIRP International Seminar on Intelligent Computation in Manufacturing Engineering, ICME 2000*, pages 263–268. Capri, Italy, 2000.
90. K. Weinert and J. Mehnen. Reverse Engineering im Werkzeug- und Formenbau. In D. Kochan, editor, *3D-Erfahrungsforum*, pages 153–160. ART e.V., Dresden, 2000.
91. K. Weinert, J. Mehnen, F. Albersmann, and P. Drerup. New solutions for surface reconstructions from discrete point data by means of computational intelligence. Technical Report CI-22/98, Collaborative Research Center 531, University of Dortmund, 1998.
92. K. Weinert, J. Mehnen, F. Albersmann, and P. Drerup. New solutions for surface reconstructions from discrete point data by means of computational

intelligence. In R. Teti, editor, *ICME '98, CIRP International Seminar on Intelligent Computation in Manufacturing Engineering, Conference Report*, pages 431–438. Capri, Italy, 1998.

93. K. Weinert, J. Mehnen, and G. Rudolph. Dynamic neighborhood structures in parallel evolution strategies. Technical Report CI-112/01, Collaborative Research Center 531, University of Dortmund, 2001.

94. K. Weinert, J. Mehnen, and M. Schneider. Evolutionary optimization of approximating triangulations for surface reconstruction from unstructured 3D data. In M. Grana, editor, *The Fourth International Workshop on Frontiers in Evolutionary Algorithms, FEA 2002*. Research Triangle Park, Durham, NC, March 8–13, 2002. (Accepted).

95. K. Weinert, H. Müller, F. Albersmann, and J. Mehnen. Efficient rasterized offset computation of tactile digitization. In *Proceedings of the ISMTII '98*, pages 258–264. Miskolc, Hungary, Sep. 2–4, 1998.

96. K. Weinert and M. Stautner. Reconstruction of particle flow mechanisms with symbolic regression via genetic programming. In L. Spector, E. Goodman, A. Wu, and W.B. Langdon, editors, *Proceedings of the Genetic and Evolutionary Computation Conference, GECCO-2001*, pages 1439–1446. Morgan Kaufmann, San Francisco, CA, June 2001.

97. K. Weinert, T. Surmann, and J. Mehnen. Evolutionary surface reconstruction using CSG-NURBS-hybrids. Technical Report CI-114/01, Collaborative Research Center 531, University of Dortmund, 2001.

98. K. Weinert, T. Surmann, and J. Mehnen. Evolutionary surface reconstruction using CSG-NURBS-hybrids. In L. Spector, E. Goodman, A. Wu, and W. B. Langdon, editors, *Proceedings of the Genetic and Evolutionary Computation Conference, GECCO-2001*, page 1456. Morgan Kaufmann, San Francisco, CA, June 2001.

99. K. Weinert, T. Surmann, and J. Mehnen. Parallel surface reconstruction. In *EuroGP 2002*, Kinsdale, Ireland, 2002. EvoNet. (Accepted).

100. D. Wiesmann. Mutation operators for the evolution of finite automata. Technical Report CI-82/00, Collaborative Research Center 531, University of Dortmund, 2000.

# 7 Genetic Programming and Its Application in Machining Technology

Wolfgang Banzhaf[1], Markus Brameier[1], Marc Stautner[2], and Klaus Weinert[2]

[1] University of Dortmund, Department of Computer Science, Informatik XI, 44221 Dortmund, Germany
{banzhaf, brameier}@Ls11.cs.uni-dortmund.de
[2] University of Dortmund, Faculty of Mechanical Engineering, Institute of Machining Technology, 44221 Dortmund, Germany
{stautner, weinert}@isf.mb.uni-dortmund.de

**Summary.** Genetic programming (GP) denotes a variant of evolutionary algorithms that breeds solutions to problems in the form of computer programs. In recent years genetic programming has become increasingly important for real-world applications, including engineering tasks in particular. This contribution integrates both further development of the GP paradigm and its applications to challenging problems in machining technology. Different variants of program representations are investigated. While problem-independent methods are introduced for a linear representation, problem-specific adaptations are conducted with the traditional tree structure.

## 7.1 Introduction

In the first part of this chapter we focus on advanced concepts of linear GP. Linear genetic programming (LGP) denotes a variant of GP, where programs of an imperative programming language or a machine language are evolved. We analyze data flow within linear genetic programs and exploit structural aspects of the linear representation for detecting non-effective code. Extracting this code from programs before fitness calculation leads to a significant acceleration of runtime. The information about whether an instruction is effective or not is further used for designing efficient variation operators. Here we present a variant of linear GP that is based exclusively on mutations. Program solutions developed with this approach have not only been found to be more successful but less complex in size. Furthermore, a structural distance metric is defined that reveals causal connections between changes of the linear genotype and changes of the phenotype (fitness). Using this distance information allows variation step size and diversity to be controlled explicitly on the level of effective code.

The second part deals with the application to problems of machining technology. Here modeling general correlations out of data given only by observation is a difficult task. A lot of knowledge about the examined process is needed

to extract the correct coherences out of the given data. Especially when this process has to be done by hand. To support the scientist in this problem computational intelligence (CI) methods, here symbolic regression through genetic programming, are a well known possibility to help.

## 7.2   Linear Genetic Programming

Genetic programming [2, 16] applies an evolutionary algorithm that subjects computer programs to evolution. In contrast to traditional GP which uses expressions of a functional programming language, linear GP evolves programs of an imperative language. While in the former case the program structure is a tree, in the latter case the representation is linear, which derives from the sequence of imperative instructions such a genetic program is composed of.

Originally, linear genetic programming was introduced with a binary machine language so that the genetic programs can be executed directly without passing a time-consuming interpretation step first [21]. Apart from this speed advantage, we investigate more general characteristics of linear representations in this chapter. One principle difference compared to tree representations is that unused code parts occur and remain within linear genetic programs (see Section 7.2.1).

In our linear GP approach [4], an individual program is represented by a variable-length sequence of simple C instructions. The instruction set may include operations such as $r_i = r_j + c$, function calls $r_i = f(r_j)$, or branches $if(r_j > r_k)$ that may skip the succeeding non-branch instruction. All instructions operate on one or two indexed variables (registers) $r_j$ or constants $c$ from predefined sets and all instructions, but branches assign the result to a destination register $r_i$. The individual LGP approach has been extended to the evolution of program teams in [6].

### 7.2.1   Effective and Non-effective Instructions

In linear genetic programs two types of code can be distinguished on the structural level, effective and non-effective instructions. An instruction is defined as effective at its program position if it manipulates an effective register. A register, in turn, is effective at a certain position if a manipulation of its content can effect the behavior, i.e., the output(s), of the program. Otherwise, the register as well as the manipulating instruction are non-effective for that position. Such non-effective instructions do not contribute to the data flow in a program and are referred to as structural introns or data flow introns.

In the following excerpt of a linear genetic program, only the instructions that are marked with an exclamation point can have an influence on the final output, which is held in register r[0] here.

**Example 1**: (*Linear genetic program*)

```
void gp(r)
  double r[4];
{
    . . .

    r[3] = r[1] - 3;
    r[1] = r[2] * r[1];
!   r[3] = r[1] / r[0];
    r[0] = r[1] - 1;
    r[1] = r[2] * r[0];
    r[1] = r[0] * r[1];
!   r[0] = r[2] + r[2];
    r[2] = pow(r[1], r[0]);
!   r[2] = r[0] + r[3];
!   r[0] = r[3] - 1;
!   r[1] = r[2] - r[0];
!   r[3] = pow(r[0], 2);
!   r[2] = r[2] + r[1];
    r[0] = r[1] + 9;
    r[0] = r[1] / r[3];
!   r[0] = r[2] * r[2];
!   r[2] = r[1] * r[3];
!   r[0] = r[0] + r[2];
}
```

The structural non-effective code in linear individuals is not directly dependent on the set of instruction operators. However, the more registers that are provided, the easier the creation of this code will become for evolution. If only one register is available, this type of code cannot occur at all, and finding a solution is very difficult.

Semantic (or operational) intron instructions manipulate effective registers without affecting program semantics. In order to restrict the rate of semantic introns and thus, to keep the structural effective length of programs small, an instruction set can be chosen with a minimal tendency to create these introns. Only if the creation of structural introns is easier than the creation of semantic introns, semantic non-effective code can be expected to occur less frequently.

## 7.2.2   A GP Algorithm

Algorithm 1 shows the specific evolutionary algorithm used in our LGP approach. In such a steady-state EA, newly created individuals replace existing individuals in the same population, while typically, tournament selection is applied.

**Algorithm 1**: (*LGP algorithm*)

1. Initialize a population of individual programs randomly.

2. Select $2 \times n$ individuals from the population without replacement.

3. Perform two fitness tournaments of size $n$.

4. Reproduce the winners of the two tournaments by replacing the two losers with temporary copies of the winners.

5. Modify the two winners by one or more variation operators, including mutation and recombination, with certain probabilities.

6. Evaluate the fitness of the offspring.

7. If the currently best-fit individual is replaced by one of the offspring then validate the new best program.

8. Repeat steps 2–8 until the maximum number of generations has been reached.

9. Test the generalization performance of the program with minimum validation error.

10. Both the best program during training and the best program during validation define the output of the algorithm.

The fitness (training error) of an individual program $p$ is often computed by an error function, e.g., the sum of errors between the predicted and the desired program outputs for a given set of $n$ training examples (also called fitness cases):

$$fitness(p) = \sum_{k=1}^{n} |p(i_k) - o_k| \tag{7.1}$$

Better fitness means smaller error and the minimum fitness is zero then. The data domain from which the input-output examples $(i_k, o_k)$ are drawn defines the problem that should be solved or approximated by GP.

The generalization ability of LGP solutions is checked during training by calculating the validation error of the currently best-fit program. This requires the use of unknown data (validation data), which is sampled from the same data space as the training data. Finally, among all the best individuals emerging over a run the one with minimum validation error (point of best generalization) is tested on a test data set again once after training is over.

## 7.3  Removal of Non-effective Code

The most time-consuming step in GP is the evaluation of individuals. Usually this step includes multiple executions of a genetic program. In [21] execution

Individual

**Fig. 7.1.** Intron elimination in LGP. Only effective code (black) is executed.

time has been reduced by evolving binary machine programs that do not require interpretation. In this section we introduce another possibility to speed up linear GP.

The imperative program structure in linear GP permits non-effective instructions to be identified efficiently. This in turn allows the effective code to be extracted from a program and to be copied to a temporary program buffer once before the fitness of the program is calculated (see Figure 7.1). By only executing the effective program when testing each fitness case, evaluation can be accelerated significantly [4]. It is important to realize that the representation of the individuals in the population is not affected by this action. In this way, no potential genetic material gets lost.

Algorithm 2 detects all structural introns in a linear genetic program. Note that whether or not a branch instruction is an intron only depends on the status of the operation that directly follows. Copying all marked instructions at the end forms the effective program.

**Algorithm 2**: (*Detection of non-effective instructions*)

1. Let the set $R_{eff}$ always contain all registers that are effective at the current program position.
   $R_{eff} := \{\, r_i \mid r_i \text{ is output register}\,\}$
   Start at the last program instruction and move backwards.

2. Mark the next operation with destination register $r_i \in R$.
   If such an instruction is not found, $\rightarrow$ 5.

3. If the operation directly follows a branch or a sequence of branches, mark these instructions too, otherwise remove $r_i$ from $R_{eff}$.

4. Insert the operand register(s) of new marked instructions in $R_{eff}$ if not already contained. $\rightarrow$ 2.

5. Stop. All unmarked instructions are introns.

The algorithm needs linear runtime $O(n)$ at worst, with $n$ being the maximum length of the linear genetic program. Actually, detecting and removing

the non-effective code from a program only requires about the same time as calculating one fitness case.

By omitting the execution of non-effective instructions during program interpretation a large amount of computation time can be saved. A good estimate of the overall acceleration in runtime is the factor

$$\alpha_{acc} = \frac{1}{1 - p_{intron}} \qquad (7.2)$$

with $p_{intron}$ denotes the average percentage of intron code in a genetic program and $1 - p_{intron}$ the respective percentage of effective code. In [4] an average intron rate of about 80 percent is documented for several classification problems, which corresponds to an average decrease in runtime by a factor 5. In general, the proportion of effective and non-effective instructions in linear programs depends on both the problem and the configuration of the LGP system, including the provided program components (see Section 7.2) and the choice of variation operators (see Section 7.5) in particular.

## 7.4   Graph Interpretation

The imperative representation of a linear GP program can be transformed into a semantically equivalent functional program graph. The following Algorithm 3 achieves this for an effective linear program. Thus, this method requires the deletion of intron instructions by using Algorithm 2, first. The directed structure of the resulting graph better reflects functional connections and data flow in linear genetic programs. The graph is contiguous for two reasons: first, by not transforming structural intron instructions and, second, by assuming that there is only one output register. Finally, the graph is acyclic if instructions, such as loops or backward jumps do not occur in the program. Instructions, such as conditional branches result in graph structures that are not always visited during execution. Both rather special cases of programming concepts, iterations and branches, shall be excluded from the following discussion for simplicity.

**Algorithm 3**: (*Transformation of an effective linear program into a graph*)

1. Start at the last instruction in the program. Its destination register $r_{dest}$ is the output register of the program and labels the start node of the graph (sink node at this stage).

2. Go to the sink node in the graph that holds the destination register $r_{dest}$ of the current instruction as a label.

3. Assign the instruction operator to this node.

4. For each operand register $r_{op}$ (and constant) of the current instruction repeat steps 5–7:

5. If there is no sink node with label $r_{op}$, create it.

6. Connect $r_{dest}$ and $r_{op}$ by a directed edge. ($r_{dest}$ becomes inner node.)

7. If not all operations are commutative label this edge with 1 (2) if $r_{op}$ is the first (second) operand.

8. If the end of program is reached, $\rightarrow$ 10.

9. Go to the next preceding instruction in program. $\rightarrow$ 2.

10. Stop. Delete all register labels from inner nodes.

Each program instruction may be interpreted as a small subtree of depth one. In the effective program graph resulting from Algorithm 3, each inner node represents an operator and has as many outgoing edges as there are operands in the corresponding instruction, i.e., one or two.[1]

Sink nodes, i.e., nodes without any outgoing edges, are labeled with register identifiers or constants. Sink nodes that represent a constant are only created once during the calculation of the graph. Only sink nodes that represent a variable (register) are replaced regularly by operator nodes in the course of the algorithm. These are the only points at which the graph may grow. Since loops are not considered the only successors of sink nodes may become other sink nodes or new nodes. After each iteration of the algorithm all nonconstant sink nodes correspond exactly to the effective registers at the current program position. Because the number of effective registers is limited by the total number of registers, of course, the number of sink nodes and the maximum width of the graph respectively are limited as well. As a result, the program graph is supposed to grow in depth. The depth is restricted by the length of the imperative program because each imperative instruction corresponds to exactly one inner node in the graph. For those reasons the graph structure may be referred to as linear, like its imperative equivalent.

The actual width of a program graph indicates the number of parallel calculation paths in a linear genetic program. It can be determined by the maximum number of registers that are effective at a program position. It is important for the performance of linear GP to provide enough registers for calculations (calculation registers), in addition to the registers that hold the input data. This is especially true if input dimension is rather low. If the number of registers is not sufficient there are too many conflicts through overwriting of register information during calculation. The more registers are provided, however, the more independent program paths may develop.

It also follows from the above observations that the runtime of Algorithm 3 is $O(k \cdot n)$ with $n$ is the number of effective instructions and $k$ is the number of registers. For problems with a rather low input dimension, i.e., with a small number of input registers, runtime is almost linear in $n$.

The program listed in Example 1 corresponds exactly to the graph in Figure 7.2 after applying Algorithm 3. Both, the imperative representation

---

[1] The maximum number of operands in an instruction is two.

and the functional representation consist of effective code here that is free from unused instructions or unvisited graph components, respectively. This is valid if we assume that the output of the imperative program is stored in variable r[0] at the end of execution. At the beginning of program execution the inputs are held in variables r[0] to r[3]. In the corresponding graph representation the inputs denote variable sink nodes, besides the sink nodes that hold constants.

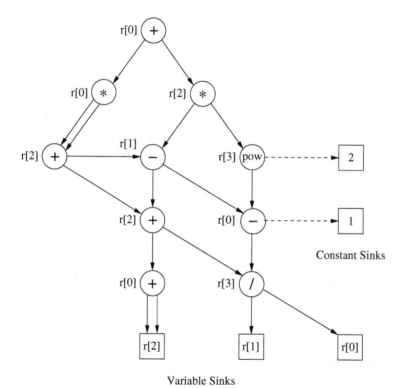

Variable Sinks

**Fig. 7.2.** Graph equivalent of the effective linear program code in Example 1. Operator nodes are labeled with the destination registers of the corresponding instructions (see Algorithm 3). Output register r[0] marks the start node. Outgoing edges are not labeled here but printed in the right order.

## 7.5   Linear Genetic Operators

Traditionally, recombination is applied in genetic programming for varying the lengths of programs. In this section we introduce an efficient variant of linear GP that works exclusively with mutations. We will demonstrate that

linear programs are manipulated more precisely by using mutations only. This will also be motivated by means of the functional interpretation discussed in Section 7.4.

The linear crossover operator exchanges two arbitrarily long subsequences of instructions between two individuals. If the operation cannot be executed because one offspring would exceed the maximum length, crossover is performed with equally long subsequences. Macro-mutations denote deletions or insertions of single full instructions here. Both crossover and macro-mutations operate on instruction level and control program growth (macro-variations). Micro-mutations by comparison, randomly replace instruction components that comprise single operators, registers, or constants.

### 7.5.1    Variation Step Sizes

In genetic programming changing a small program component may lead to almost arbitrary changes in program behavior. Variation operators that induce smaller step sizes with a high probability allow a more precise approximation to locally optimal solutions. This relies on the assumption that, on average, smaller structural variations of genetic programs result in smaller semantic variations (see also Section 7.6).

In tree programs, crossover and mutation points can be expected to be the more influential the closer they are to the root. Instead, in a linear program, each position of an instruction may have a more similar influence on program semantics. Recall that the linear graph representation is restricted in width through the number of registers provided (see Section 7.4).

Moreover, in tree-based GP, crossover only affects a single point in data flow, which is the root of the exchanged subtree. With linear crossover the contents of many effective registers may change, i.e., several points in data flow are modified simultaneously. The reason lies in the rather narrow graph structure of linear genetic programs. This graph is disrupted easily when applying linear crossover on the imperative structure. As a result, crossover step sizes may become quite large, on average. However, it has to be considered that, to a certain degree, the influence of linear crossover on the effective code is reduced implicitly by a high rate of structural intron code that emerges with this operator.

A more explicit way of reducing the effect of linear crossover is to limit the length of the exchanged instruction segments. We experienced that rather small maximum segment lengths produce the best results [5].

The above reasons suggest the use of macro-mutations instead of crossover for macro-variations in linear GP. Mutations guarantee a sufficient variation strength and freedom of variation in linear GP for the following reasons: First, in linear GP already single micro-mutations that exchange a register index in an instruction may change the data flow within a linear program. This is true because several instructions that precede the mutated instruction may become effective or non-effective respectively (see Section 7.6). On

the graph level a single edge is redirected from an effective subgraph to a former noncontiguous component. In tree-based GP micro-mutations usually manipulate the contents of single nodes but do not change edges of the tree structure.

Second, the linear program/graph structure can be manipulated with a high degree of freedom. In tree programs, by comparison, it is rather difficult to delete or insert a group of nodes at an arbitrary position. Complete subtrees might be removed along with the operation to satisfy the constraints of the tree structure [10]. This is why the tree structure is less suitable to be varied by smaller macro-mutations since modification of upper program parts necessarily involve bigger parts of code. In linear GP, depending substructures do not get lost when deleting or inserting an instruction but remain within the linear representation as structural non-effective code, i.e., unvisited graph components.

### 7.5.2 Effective Mutations

According to the definition of effective code in Section 7.2, let an effective variation denote a genetic operation that modifies the effective code of a genetic program. Note that even if the effective code is altered, the predictions of the program for a considered set of fitness cases might be the same. An effective variation is merely meant to bring about a structural change of the effective program code. There is no change of program semantics (fitness) guaranteed which is due to semantic intron code. In general, decreasing the number of non-effective variations is expected to reduce the rate of neutral variations. A genetic operation is neutral if it does not change the fitness of a program. Note that by definition, the term non-effective is meant to be related to the code level only.

We consider two different approaches to effective mutations. One variant (effmut2) uses macro- and micro-mutations to operate on effective instructions exclusively but leaves the non-effective code untouched. This is motivated by the assumption that mutations of non-effective instructions may be more likely invariant in terms of a fitness change than mutations of effective code. The other variant (effmut) allows single non-effective instructions to be deleted. In order to guarantee that the effective code is altered, an effective insertion may directly follow such intron deletions. A third approach might be to delete all emerging non-effective instructions directly after the variation.[2]

The deletion of an instruction in general is not complicated. If an instruction is inserted, its destination register is chosen in such a way that the instruction is effective at the corresponding program position. By doing so, insertions of non-effective instructions (introns) are avoided explicitly. Like the effective code, effective registers can be identified efficiently in linear runtime $O(n)$ [5].

---

[2] This has been found too restrictive for program growth because of substantial loss of genetic material.

### 7.5.3   Prediction Performance

In the next two sections we will compare the performance of linear genetic operators and their influence on solution complexity using two regression problems. The first problem is represented by the two-dimensional Mexican hat function that is to be approximated:

$$f_{mexicanhat}(x,y) = \left(1 - \frac{x^2}{4} - \frac{y^2}{4}\right) \times e^{\left(-\frac{x^2}{8} - \frac{y^2}{8}\right)} \tag{7.3}$$

The second test problem, named two points, computed the square root of the scalar product of two three-dimensional vectors $p$ and $q$:

$$f_{twopoints}(p_1, p_2, p_3, q_1, q_2, q_3) = \sqrt{p_1 \times q_1 + p_2 \times q_2 + p_3 \times q_3} \tag{7.4}$$

Tables 7.1 and 7.2 compare fitness and generalization performance for different forms of macro-variation: crossover (cross), free macro-mutations (mut) and two variants of effective macro-mutations (see above). In the same run micro-mutations are applied together with one of the four macro operators with a probability of 25 percent. Only one genetic operator, however, is applied at a time to vary a certain individual program.

First, we can see that prediction errors reduce together with the variation step sizes when using macro-mutations instead of crossover. Interestingly, an additional significant improvement can be observed with effective mutations. One explanation is that effective mutations, by definition, reduce the number of neutral variations significantly compared to mutations that are not restricted to the effective code. In this way, evolution is allowed to progress faster within the same period of time, i.e., number of generations.

**Table 7.1.** Mexican hat problem: Average prediction error over 60 runs.

| Variation | Training Error | Validation Error | Test Error | Neutral Variations |
|-----------|---------------|------------------|------------|---------------------|
| cross     | 15.46         | 17.68            | 17.73      | 200                 |
| mut       | 5.77          | 9.94             | 10.25      | 408                 |
| effmut    | 1.35          | 1.92             | 1.69       | 75                  |
| effmut2   | 1.59          | 2.25             | 2.05       | 88                  |

In the effmut2 experiments effective mutations are restricted to effective instructions. Although non-effective code is not touched by this variant, it maintains a much lower intron rate than standard macro-mutations (mut) as can be seen from Figures 7.3–7.6. Instead, the other variant of effective mutations (effmut) allows deletions of non-effective instructions (see Section 7.5.2), which reduces the rate of non-effective code further. The intron code may

**Table 7.2.** Two points problem: Average prediction error over 60 runs.

| Variation | Training | Validation | Test | Neutral |
|-----------|----------|------------|------|---------|
|           | Error    | Error      | Error | Variations |
| cross     | 192.1    | 276.5      | 274.7 | 218 |
| mut       | 95.5     | 165.6      | 195.2 | 419 |
| effmut    | 66.3     | 92.7       | 95.5 | 66 |
| effmut2   | 76.5     | 114.0      | 123.9 | 56 |

function as a protection mechanism that compensates (reduces) the influence of the user-defined deletion rate (33 percent here) on the effective code. This could explain the slightly better prediction performance compared to effmut2 solutions. By applying insertions two times more frequently than deletions, a sufficient code growth is guaranteed, especially with effective macro-mutations.

In general, when using macro-mutations, worse performance may also result from negative effects, in particular, larger mutation step sizes, by reactivation of intron code. Note that non-effective code hardly ever emerges with the effmut variant.

### 7.5.4  Control of Code Growth

Figures 7.4 and 7.6 document that effective lengths are around 50 percent smaller if solutions are developed with effective mutations (effmut) than with normal macro-mutations (mut). Most probably it is more difficult for evolution to preserve shorter effective code in the presence of many non-effective instructions. In general, a reduction of effective program size always means an acceleration of processing time too because only effective code is executed in our linear GP system (see Section 7.3).

For unlimited standard crossover the absolute length approaches the maximum size limit of 200 instructions within the first 300 generations in Figures 7.3 and 7.5. With free mutations this effect is extenuated by reaching the maximum later. Obviously, program growth is slower by using mutations only. Nevertheless, both variation forms induce a fast increase of the absolute code length. This effect is referred to as bloat and is observed in tree-based GP too [26]. If we compare the development of absolute lengths with the effective lengths in Figures 7.6 and 7.4 it becomes obvious that the bloat effect is mostly related to a growth of the structural non-effective code in linear GP. Because non-effective code does not influence fitness directly, there is no selection pressure on this part of a program. Moreover, the rate of non-effective code works as an implicit control of crossover step sizes. If crossover is used this protection function forms an additional drive for code growth.

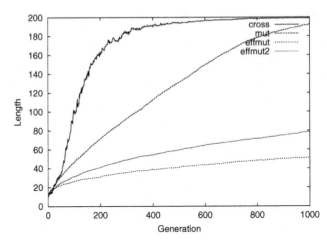

**Fig. 7.3.** Mexican hat problem: Development of absolute program length for different forms of variation. Average figures over 60 runs.

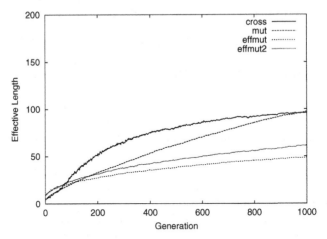

**Fig. 7.4.** Mexican hat problem: Development of effective program length for different forms of variation. Average figures over 60 runs.

Effective mutations reduce the intron code and thus, the bloat effect significantly, even without allowing deletions of non-effective instructions explicitly (effmut2). In this way, effective mutations incorporate an implicit control of code growth in linear genetic programming. In contrast to crossover, macromutations do not require non-effective code for inducing efficient solutions, neither for controlling variation strength nor for preserving code diversity.

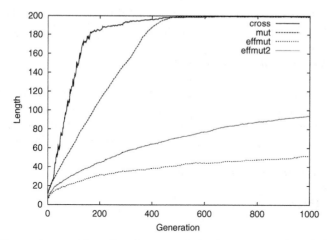

**Fig. 7.5.** Two points problem: Development of absolute program length for different forms of variation. Average figures over 60 runs.

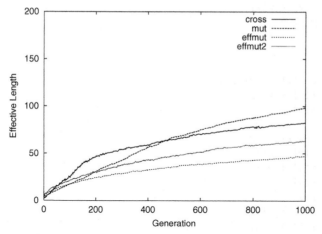

**Fig. 7.6.** Two points problem: Development of effective program length for different forms of variation. Average figures over 60 runs.

## 7.6   Control of Variation Step Size

An implicit control of structural variation distance may be realized by imposing respective restrictions on the variation operators. This has been demonstrated above by using macro-mutations instead of recombination in order to vary program length. Moreover, by concentrating variations on the effective code, structural variations become more closely related to semantic variations. Recall that only the degree of variation on the effective code decides the difference in fitness. Unfortunately, a variation operator, even if it is op-

erating on the effective code exclusively, can only guarantee for the absolute program structure that a certain maximum variation step size is not exceeded. Variation steps on the effective code instead may still be much bigger, even though this will happen with a lower probability.

The following example represents the result of applying a micro-mutation to the program example from Section 7.2. In the instruction on line 9 the operand register r[3] has been exchanged by register r[2]. As a consequence, five preceding, formerly non-effective, instructions become effective now, which corresponds to an effective distance of five (see below).

**Example 2**: (*Mutated linear program*)

```
void gp(r)
  double r[5];
{
    ...

    r[3] = r[1] - 3;
!   r[1] = r[2] * r[1];
!   r[3] = r[2] / r[1];
!   r[0] = r[1] - 1;
!   r[1] = r[2] * r[0];
!   r[1] = r[0] * r[1];
!   r[0] = r[2] + r[2];
!   r[2] = pow(r[1], r[0]);
!   r[2] = r[0] + r[2];     <- Effective mutation point
!   r[0] = r[3] - 1;
!   r[1] = r[2] - r[0];
!   r[3] = pow(r[0], 2);
!   r[2] = r[2] + r[1];
    r[0] = r[1] + 9;
    r[0] = r[1] / r[3];
!   r[0] = r[2] * r[2];
!   r[2] = r[1] * r[3];
!   r[0] = r[0] + r[2];
}
```

In order to avoid such bigger variation steps on program structure, we propose an explicit control that repeats a variation until the structural distance between parent and offspring falls below a maximum threshold [7, 8].[3] In particular, we want to find out whether a further reduction of effective mutation step size may still improve results. This requires mutation distances to be measured explicitly on the effective code. In the following section we will identify substructures of linear genetic programs that are sufficient to be distinguished by a structural distance metric.

Using an explicit control of semantic distance (fitness distance) between parent and offspring instead would require an additional fitness calculation

---
[3] Reference [7] is an extended version of [8].

for each repeated variation and can become computationally expensive, especially if a larger number of fitness cases is involved. A structural distance has to be recalculated after each iteration while its computational costs do not directly depend on the number of fitness cases. It is also difficult to find appropriate maximum thresholds for fitness distance because they are not problem-specific. Finally, it is not sensible to restrict fitness improvements at all.

## 7.6.1  Structural Program Distance

The string edit distance [12] measures the distance between two arbitrarily long character strings by counting the number of basic operations, including insertion and exchange of single elements, that are necessary to transform one string into an other. We apply this distance metric to determine the structural distance between the effective part of parent and offspring (effective variation distance) because a difference in effective code may be more directly related to a difference in program behavior. It is important to realize that the effective distance cannot be part of the absolute distance. Actually, two programs may have a small absolute distance while their effective distance is comparatively large, as has just been demonstrated at the example program above. On the other hand, two equally effective programs might differ significantly in their non-effective code.

For an efficient distance calculation we concentrate on representative substructures of linear programs and regard the sequence of operators from the effective instructions. For instance, the order of effective operators of the genetic program from Section 7.2 is

$$(+, *, *, +, pow, -, -, +, +, /)$$

when starting with the last instruction. The distance of effective operator symbols has been found sufficiently precise to differentiate between program structures, provided that the used operator set is not too small. This is due to the fact that in most cases the modification of an effective instruction changes the effectivity status of (at least) one instruction (see Section 7.6.3). Note that in contrast to the effective distance the absolute operator sequence would not be altered by the exchange of single registers.

Effective mutations work closely with the effective distance metric. As defined in Section 7.5, macro-mutations operate on full instruction level, while micro-mutations vary smaller components within instructions. In order to guarantee a sufficient growth of programs, however, the higher number of variations is performed on macro level, i.e., comprises macro-mutations. Since, in this way, the average step size is not further reducible from operator side, measuring the distance between full effective programs does not necessarily promise a higher precision. This is another reason why operator sequences represent a sufficient basis for distance calculation between linear

genetic programs. Besides, a registration of absolutely every structural difference should not be necessary if we take into account that the correlation between semantic and structural distance is probabilistic (see Section 7.6.3).

## 7.6.2   Prediction Performance

As a first benchmark problem we test the above notions with the iris data set that contains popular real-world data from the UCI Machine Learning Repository [3]. Fitness is the classification error, i.e., the number of wrongly classified inputs. The second test problem is a parity function of dimension eight (even-8-parity). This function outputs 1 if the number of set input bits is even, otherwise the output is 0.

**Table 7.3.** Restriction of effective mutation distance. Average error over 200 runs. Statistical standard error in parenthesis. Percentage difference from baseline results.

| Variation | Maximum | iris | | even-8-parity | |
|---|---|---|---|---|---|
| | Distance | *mean (std)* | $\Delta$ (%) | *mean (std)* | $\Delta$ (%) |
| effmut | — | 0.90 (0.06) | 0 | 16 (1.2) | 0 |
| | 10 | 0.72 (0.06) | 20 | 13 (1.2) | 19 |
| | 5 | 0.74 (0.06) | 18 | 12 (1.2) | 25 |
| | 2 | 0.68 (0.05) | 24 | 11 (1.1) | 31 |
| | 1 | 0.54 (0.05) | 40 | 9 (0.9) | 44 |

**Table 7.4.** Average number of iterated mutations until a maximum distance is met.

| Variation | Maximum | Iterations | |
|---|---|---|---|
| | Distance | iris | even-8-parity |
| effmut | — | 1.00 | 1.00 |
| | 10 | 1.02 | 1.02 |
| | 5 | 1.05 | 1.05 |
| | 2 | 1.12 | 1.12 |
| | 1 | 1.18 | 1.20 |

Table 7.3 compares average prediction errors for different maximum mutation distances. The maximum possible distance equals the maximum program length (200 instructions) and imposes no restrictions. For both problems, the best results are obtained with the smallest effective distance (one). This is all

the more interesting if we consider that a restriction of variation distance always implies a restriction in variation freedom too. More specifically, certain modifications might not be executed at certain program positions because too many other instructions would be affected.

As we can learn from Table 7.4, the average number of iterations until a maximum effective distance is met increases only slightly if the maximum threshold is lowered. On average, only $\approx 1.2$ iterations are necessary with the smallest threshold, and the maximum number of iterations (10 here) has hardly ever been exceeded. Both aspects emphasize that freedom of variation is restricted only slightly here.

### 7.6.3   Distance Distribution and Correlation

The results in Table 7.4 correspond to the distribution of effective mutation distances in Figure 7.7, where only about 20 percent of all measured step sizes are larger than one. Obviously, larger disruptions of effective code, as demonstrated with the example program in Section 7.6, occur less likely. Effective programs emerge to be quite robust against larger mutation steps because their survival probability is higher in this way. In particular, this is achieved by the effectivity of an instruction that may depend on more than one succeeding instruction in program. Because the rate of non-effective instructions that emerge with effective mutations is comparatively low, most effective mutation distances result from deactivation of effective instructions.

**Fig. 7.7.** Distribution of effective mutation distances for iris problem (similar to even-8-parity). Average figure over 100 runs.

Controlling variation step size on effective code level necessarily requires a distance metric that measures causal connections between program changes

and fitness changes sufficiently precisely. Even if already small modifications of the program structure may result in almost arbitrary changes in program behavior, smaller variations of the genotype should lead to smaller variations of the phenotype for a higher probability. In [13] this has been demonstrated by applying the edit distance metric to program trees. Figure 7.8 clearly shows a positive correlation between the mutation distance on code level and the mutation distances on fitness level for our code-selective metric on linear genomes. The corresponding distribution of mutation distances in Figure 7.7 confirms this to be true for the vast majority of occurring distances.

**Fig. 7.8.** Correlation between program distance and fitness distance for iris problem (similar to even-8-parity). Average figure over 100 runs.

The first observation, that small mutation steps occur more frequently than large mutation steps, fulfills guideline M 2 of a metric-based evolutionary algorithm (MBEA) (see Section 6.4.2). The second observation (causality) forms a necessary precondition for the success of evolutionary algorithms in general.

## 7.7  Control of Structural Diversity

The effective distance metric between programs from Section 7.6.1 is applied here for an explicit control of genotype diversity, which is the average structural distance between individuals in the population [7]. Therefore, we introduce the two-level tournament selection shown in Figure 7.9. On the first level, individuals are selected by fitness (fitness selection). On the second level, the two individuals with maximum distance are chosen among three fitter individuals (diversity selection). More precisely, we select for the

effective edit distance minus the distance of effective lengths.[4] By doing so, the growth of effective length is not rewarded directly during selection.

While an absolute measure, such as fitness may be compared between two individuals, selection by a relative measure, such as distance or diversity necessarily requires a minimum of three individuals. In general, two from $n$ individuals are selected with the greatest sum of distances to the $n-1$ other individuals. Selection pressure on the first level depends on the size of fitness tournaments. Pressure of diversity selection on the second level is controlled by the number of these tournaments. Additionally, we use a selection rate parameter in order to tune the selection pressure on the second level more precisely.

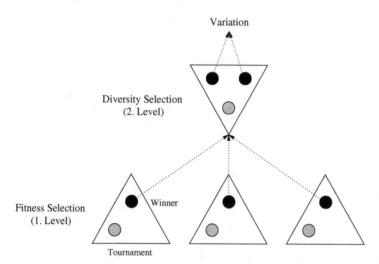

**Fig. 7.9.** Two-level tournament selection.

The two-level tournament selection constitutes a multiobjective selection method that finds individuals that are fitter and more diverse in relation to others while selection for fitness keeps the higher priority. Selecting individuals only by diversity for a certain probability instead does not necessarily result in more different search directions among better solutions in the population. Dittrich et al. [11] report on a spontaneous formation of groups when selecting the most distant of three individuals from a population of real numbers.

The number of fitness calculations and the processing time, respectively, do not increase with the number of tournaments if the fitness of individuals is saved and is updated only after variation. Only diversity selection itself becomes more computationally expensive the more individuals participate in

---

[4] This is possible because both measures, edit distance and length distance, operate on instruction level here.

it. $n$ individuals require $\binom{n}{2}$ distance calculations. Hence, program distance should be efficiently computable.

A fitness landscape on the search space of programs is defined by a structural distance metric between programs and a fitness function that reflects the quality of program semantics. The application of a genetic operator corresponds to performing one step on this landscape. The active control of structural diversity increases the average distance of individuals. Graphically, the individuals spread more widely over the fitness landscape. Thus, there is a lower probability that the evolutionary process gets stuck in a local minimum and more different search directions may be explored in parallel.

Note that the two-level selection process can also be used for an explicit control of code growth (complexity selection). Therefore, in linear GP a selection pressure may be imposed specifically on the smallest effective or noneffective program length.

### 7.7.1   Prediction Performance

For the two test problems introduced in Section 7.6, Table 7.5 summarizes average error rates obtained with and without selecting for structural diversity. Different selection pressures have been tested. For the minimum number of fitness tournaments (three) we used selection probabilities 50 percent and 100 percent. Higher selection pressures are induced by increasing the number of tournaments (up to four here). For each problem and variation operator, the performance increases continuously with the influence of diversity selection. The highest selection pressure that has been tested results in about a 2 to 3 times better prediction error on average, than without increasing diversity actively.

Again, as already demonstrated in Section 7.5 for regression tasks, linear GP performs significantly better by using effective macro-mutations instead of crossover. Obviously, the linear program representation is more suitable for being developed by mutations only, especially if those concentrate on effective instructions. Nonetheless, the experiments with linear crossover show that diversity selection does not depend on a special type of variation. Moreover, the application of this technique is demonstrated with a population-dependent operator.

### 7.7.2   Development of Diversity

Figures 7.10–7.13 illustrate the development of structural diversity during runs for different selection pressures and different variation operators. Obviously, the higher the selection pressure is adjusted, the higher is the diversity. Even without applying diversity selection, the average effective program distance does not drop towards the end of runs. While the diversity of effective code increases with crossover until a certain level and stays rather constant then, diversity with effective mutations increases more linearly.

**Table 7.5.** Second-level selection for structural diversity with different probabilities and different numbers of fitness tournaments (#T). Average error over 200 runs. Statistical standard error in parenthesis. Percentage difference from baseline results.

| Variation | Selection | | iris | | even-8-parity | |
| --- | --- | --- | --- | --- | --- | --- |
| | % | #T | *mean (std)* | $\Delta$ (%) | *mean (std)* | $\Delta$ (%) |
| cross | 0 | 2 | 2.11 (0.10) | 0 | 58 (3.4) | 0 |
| | 50 | 3 | 1.42 (0.08) | 33 | 35 (2.4) | 40 |
| | 100 | 3 | 1.17 (0.07) | 44 | 27 (2.2) | 53 |
| | 100 | 4 | 1.09 (0.07) | 48 | 19 (1.8) | 67 |
| effmut | 0 | 2 | 0.84 (0.06) | 0 | 15 (1.2) | 0 |
| | 50 | 3 | 0.63 (0.05) | 25 | 12 (1.0) | 20 |
| | 100 | 3 | 0.60 (0.05) | 29 | 10 (1.1) | 33 |
| | 100 | 4 | 0.33 (0.04) | 61 | 7 (0.8) | 53 |

Two major reasons can be found to explain this behavior: First, genetic programming is working with a variable length representation that grows continuously during a run. In linear GP this is especially true for the effective program length, which may still grow even if the absolute length has reached the maximum. The longer effective programs become the bigger effective distances are possible. Actually, the growth of effective code is restricted earlier with crossover because of a much higher proportion of non-effective

**Fig. 7.10.** Iris problem: Diversity with crossover and different selection pressures. Selection pressure controlled by selection probability and number of fitness tournaments (T). Average figures over 100 runs.

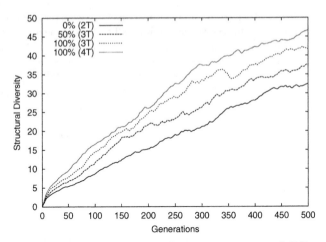

**Fig. 7.11.** Iris problem: Diversity with effective mutations and different selection pressures. Selection pressure controlled by selection probability and number of fitness tournaments (T). Average figures over 100 runs.

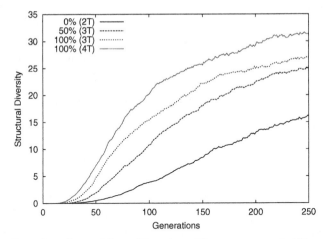

**Fig. 7.12.** Even-8-parity problem: Diversity with crossover and different selection pressures. Selection pressure controlled by selection probability and number of fitness tournaments (T). Average figures over 100 runs.

code that emerges with this operator (approximately 50 to 60 percent here). It is important to note that the average effective program length is hardly influenced by the distance selection in comparison to the average program distance.

Second, both forms of variation, linear crossover and effective mutation, maintain program diversity over a run implicitly in terms of the structural distance metric. This is true even without the explicit selection for diversity.

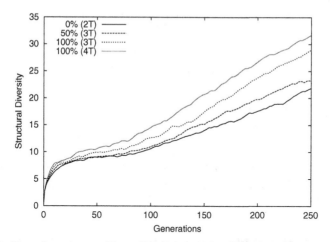

**Fig. 7.13.** Even-8-parity problem: Diversity with effective mutations and different selection pressures. Selection pressure controlled by selection probability and number of fitness tournaments (T). Average figures over 100 runs.

For linear crossover the reason might be in its high variation strength. Additionally, the high amount of non-effective code contributes to a preservation of effective code diversity with this operator. When using mutations exclusively instead, a high degree of innovation is introduced continuously into the population. This leads to a higher diversity of effective code than it occurs with crossover (see Figures 7.10–7.13) in consideration of the fact that the average effective length is about the same here for crossover and effective mutations in the final generation.

The stronger one selects for diversity, however, the more diversity gains ground in crossover runs. Obviously, a stronger influence on population diversity can be observed with crossover than with effective mutations. In comparison to mutation the success of recombination depends more strongly on the composition of the genetic material in the population. The more different two recombined solutions are, the higher is the expected innovativity of their offspring.

In the previous sections we have concentrated on improving the linear GP approach. In particular, the development of methods has been based on the distinction between used and unused parts of the linear representation. In tree programs, by comparison, redundant substructures may emerge only from program semantics, not from program structure.

In the following sections we will apply genetic programming for solving different problems from machining technology. Even if, in principle, linear GP may be used as well as tree-based GP in this problem domain, we have decided on the latter approach here. An introduction to this standard GP

variant is given along with a description of the problem-specific configurations that have been chosen here.

## 7.8  Genetic Programming in Machining Technology

Modeling the chip-building process in cutting has been in the center of interest for a long time. The chip geometry, the material movement and the thermal processes that take place at the center of the cutting zone are decisive for high quality machining, reduction of machining times, and tool wear. Most existing approaches incorporate finite element or molecule dynamic methods as well as analytical techniques based on cutting force models [1, 14, 23, 25, 29, 37] (see Figure 7.14). CI methods are used for this problem

**Fig. 7.14.** Finite element (left), molecular dynamics (center) and analytical (right) methods.

as well [20]. These approaches focus on neural networks. Neural networks can model the general behavior of correlations, but the models cannot be used aside from neural networks for further usage or investigations.

Another CI method is symbolic regression via genetic programming. Symbolic regression is a widely used method for reconstructing mathematical correlations. Here a new graphical representation of the individuals is also presented. This new three-dimensional representation allows the user to recognize certain possibilities to improve the setup of the process parameters. Furthermore, this new representation allows the visualization of the generated three-dimensional objects with nearly every CAD program for further use.

The approach of symbolic regression using genetic programming is quite similar to that of a human observer [29] watching a certain process and trying to describe it in physical terms. Furthermore, the system may produce some unexpected but valid results, which do not correspond to any known properties of the process. If such phenomena occur, new knowledge about the process has been generated. This knowledge may be helpful in supporting the development of new simulation tools, which are able to increase the productivity

of the process as well as its reliability. In chip modeling, these models can be described in a two dimensional way by relations of this type:

$$f(t) = \begin{pmatrix} f_x(t) \\ f_y(t) \end{pmatrix} \tag{7.5}$$

To succeed in this objective, some problems have to be solved. The first task is to obtain the trajectories of the particles in the crystalline structure of the metal workpiece, shown in Sections 7.9 and 7.10. The second is to find a suitable model describing the physical correlations that lead to these trajectories. For this task the method of symbolic regression via genetic programming is used [2, 16]. This is shown in Section 7.11.

## 7.9  Experimental Setup

The primary goal of the experiments was to gather some exemplary data from the processes. In order to show the power of the developed algorithms, several different processes and methods of image processing have been tried. They all have a similar principle setup in common. The first step consists of taking pictures of the process without disturbing the cutting process or influencing the chip formation. This step is done on two different experimental setups to gather a wide range of data. The second step is to transform these pictures into a digital form.

### 7.9.1  Extracting Particle Trajectories from a Turning Process

The first experimental setup was used on the turning of aluminum alloys. This setup was similar to that described by Warnecke [29]. In contrast to his setup, no microscope is used. In this experiment a high speed camera Weinberger Speedcam+, which takes up to 4500 pictures per second, was used. A schematic view of this setup is shown in Figure 7.15. A Grob BZS-600 turning machine was used for this process to cut the aluminum rings. The whole chip-building process was filmed with the high-speed camera. About two seconds of every run were recorded. The experiment was carried out with various process parameters. Two different alloys were used as workpiece material. The workpiece was a ring-shaped profile, which was rotating while the tool was fixed. The cutting speed $v_c[\frac{m}{min}]$, the feed $f[mm]$, and the back engagement of the cutting edge in $a_p[mm]$ were varied.

### 7.9.2  Extracting Particle Trajectories from a Drilling Process

The second setup to receive data from a cutting process was a modified drilling process. A short, rigid, and torsion-resistant drill was used. Therefore the workpieces could be formed as cylinders with a slightly reduced diameter

**Fig. 7.15.** Experimental setup for turning of aluminum alloys.

**Table 7.6.** Parameters used for the turning process.

| Indexable Insert (Seco) | workpiece | $v_c[\frac{m}{min}]$ | $f[mm]$ | $a_p[mm]$ |
|---|---|---|---|---|
| DNMG 150608-MF2 TP200 | 9SMn28k | 265 | 0.20 | 4 |
| DNMG 150608-MF2 TP200 | 9SMn28k | 265 | 0.50 | 4 |
| DNMG 150608-MF2 TP200 | 9SMn28k | 380 | 0.20 | 4 |
| DNMG 150608-MF2 TP200 | 9SMn28k | 380 | 0.40 | 4 |
| DNMG 150608L-95 HX | AlCuMgPb | 500 | 0.05 | 2 |
| DNMG 150608L-95 HX | AlCuMgPb | 500 | 0.45 | 2 |
| DNMG 150608L-95 HX | AlCuMgPb | 50 | 0.05 | 4 |
| DNMG 150608L-95 HX | AlCuMgPb | 50 | 0.40 | 4 |
| DNMG 150608-MF2 TP200 | C60 | 50 | 0.08 | 2 |

compared to the diameter of the drill. Thus, the high-speed camera could be positioned at the side of this cylinder to enable the filming of the chip flow at the cutting edge of the workpiece. Figure 7.16 shows a schematic view of this setup as well as a photo of the machine (NBH70 from Hüller-Hille). Due to its rigid design, this machine allows very low turning speeds without loosing precision. In this setup only the feed was varied. One difficulty was to determine optimal parameters for the camera setup to obtain the best possible images of the process. For this purpose the parameters of the camera were also varied (Table 7.6). The parameters are the frequency of the image recording and the intensity of the illumination by the stroboscopic lights. The pictures that were taken in this drilling experiment were edited afterwards, using image processing software to increase contrast and thereby the visibility of the chip flow. Two different types of chips occur in this setup because of the two cutting edges of the used drill. Figure 7.17 shows 6 frames extracted from the processed movie. All pictures show the work of the inner cutting edge. Low cutting speed at the inner cutting edge leads to a shear chip formation.

**Fig. 7.16.** The used machine (left) and a schematic view of the setup (right).

**Table 7.7.** Parameters used for the drilling process.

| $v_c[\frac{m}{min}]$ | $f[mm]$ | $f[\frac{1}{s}]$ |
|---|---|---|
| 50 | 0.40 | 1000 |
| 50 | 0.40 | 2000 |
| 50 | 0.15 | 4000 |
| 100 | 0.15 | 1000 |
| 150 | 0.15 | 1000 |
| 150 | 0.15 | 2000 |
| 150 | 0.15 | 4500 |

It can be seen how the chip is sheared and pushed together. The effect of the chip breaker on the chip can be seen.

## 7.10 Gathering Data

After gathering the image data from the different cutting processes significant trajectories of particles in the material need to be extracted. This was done in two steps. First, a segmentation of the particles in the filmed material was carried out. Afterwards, the trajectories of the particle flow were tracked. This can be done either manually or automatically, depending on the type of the material. In filmed sequences, an automatic tracking method is useful to reduce the manual effort.

### 7.10.1 Segmentation

The task of determining which pixels of an image belong to a single element of the filmed object is called segmentation. Here, single particles of a metallic

**Fig. 7.17.** Shear chip formation at the inner edge of the drill filmed at $4500f/sec$.

workpiece have to be located to simplify the tracing of the flow in the cutting process. The problem is to find sets of points in the image that belong to a single crystalline or another solid part in the flow. This can be any part of the chip. All particles in the flow which follow significant paths are suitable. Therefore, the main difficulty is to find the same arbitrarily chosen but significant particle in every frame of the filmed data.

### 7.10.2 Automatic Extraction

Automatic extraction of trajectories from filmed material requires a certain quality of the film sequences to enable a software to trace the particles. There-

fore, the software used for this tracking needs to be able to recognize the same particle in different frames of the filmed material. Traced particles need to be on the filmed surface during the whole process of tracking. Otherwise the particles need to be segmented on every frame of the filmed material.

In oder to extract the particle trajectories, the program WINAnalyze from Mikromak was used. Provided that the images had a certain quality, the software was able to follow a segmented particle automatically. A screenshot of the software performing a full automatic segmentation is shown in Figure 7.18.

**Fig. 7.18.** The software WINAnalyze v.13 (Mikromak).

The program generates ASCII trace files, which can analyze the trajectories of the particles by the GP system. The positions of the particles can now be interpreted as the values of a two-dimensional relation.

In addition to the fully automatic method, manual segmentation was used. This method was used when, due to optical or material deformation effects, it was not possible to locate one single particle in every frame automatically. After this step, the software can carry out the automatic extraction of the

trajectories. This semi-automatic method was used on the data extracted from the turning process as described in Section 7.9.1.

### 7.10.3  Manual Extraction

As a second method, a manual tracing method was used to extract the particle trajectories. This was used in cases where the extraction of the particle dynamics could not be obtained automatically due to optical or material deformation effects. These data from the drilling process (see Section 7.9.2) had to be prepared with this fully manual method, due to the strong deformation of the chip volumes and the low image resolution resulting from the high frame rate of the movies (up to 4500 images per second), see Figure 7.19.

**Fig. 7.19.** The manually analyzed chip volumes generated in the drilling process.

## 7.11  Tree-Based GP for Symbolic Regression

Symbolic regression via genetic programming was used to model the physical correlations that underlie the processes shown in Section 7.9. The implementation of the genetic programming kernel contains several aspects from evolution strategies [24] and genetic programming [16] as well.

### 7.11.1  GP System

The evolution starts with a randomly generated set of functions (initial population). Due to the fact that the algorithm has to evolve parametric functions, two symbolic representations have to be generated in parallel. The genetic operations are applied to this first generation of functions and a succeeding generation is produced (see Figure 7.20). The fitness of an individual is measured directly by evaluating the evolved formulae and comparing the geometric shape of these parametric functions with the given point set, i.e., the trajectory of the chip. Technically the formulae are represented as instances of a tree-based data structure, which is implemented in C++ using

**Fig. 7.20.** General GP scheme.

the Standard Template Library [28]. Hence, the fitness function can be evaluated in a fast and efficient way. It is also possible to use an interpreter, which may be more flexible in some cases (e.g., for debugging tasks), but slows down the evaluation. The structure of the program follows a basic GP scheme [15, 16, 19]. The probability of choosing a constant during mutation, the maximum size of an individual, or the probability of changing a function or a terminal, and the number of changes is defined in a parameter file. The data structure and the genetic operators will be described next.

### 7.11.2 Data Structure

Analogous to the infix notation of mathematical functions, a tree-based representation forms the genotype. A sample individual, consisting of two functions $f_x(t)$ and $f_y(t)$, is shown in Figure 7.21. The primary goal of obtaining the mathematical representations of two-dimensional curves leads to the phenotypic representation of an individual as graphs of a parametric function. This difference between genotypic and phenotypic representation and the need for genetic operators in which small changes in the genotype results in small changes in the phenotype yields two problems. First, a set of genetic operators that performs these small changes on the genotype must be used. Second, a fitness function which correctly evaluates the fitness of one individual has to be defined.

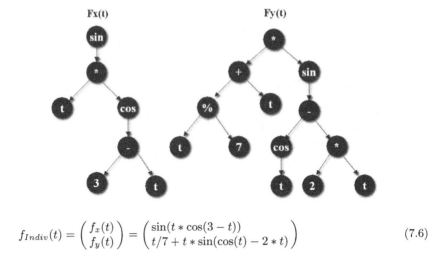

$$f_{Indiv}(t) = \begin{pmatrix} f_x(t) \\ f_y(t) \end{pmatrix} = \begin{pmatrix} \sin(t * \cos(3 - t)) \\ t/7 + t * \sin(\cos(t) - 2 * t) \end{pmatrix} \tag{7.6}$$

**Fig. 7.21.** Sample structure of a single individual and the corresponding functions.

### 7.11.3   Genetic Operators

The genetic operators are known as mutation and recombination. Mutation takes place on a single individual and changes one or more positions of the individual, here, a single node of the tree. In contrast to classical GP [2] inner nodes as well as terminal nodes can be mutated in real-value steps. In Figure 7.22 mutation is illustrated by a single mutational tree insertion of a terminal node with value 8. To allow insertion of new nodes without erasing older nodes the randomly (out of all possible operators) chosen operator + was inserted together with the new node. The second genetic operator is the recombination between two arbitrary individuals (see Figure 7.23). The genetic recombination operator selects an arbitrary node in the first individual and replaces it by an arbitrarily selected node from the second individual. These two genetic operators allow a change in the genotype, yielding small changes in the phenotype of the individual. The amount of change in the phenotype can be reduced and increased by setting the amount of mutation (e.g., number of nodes to be chosen).

A closer look at these operators is given in [31, 32, 34, 36].

### 7.11.4   Deterministic Correction of the Individuals

In order to improve the efficiency of the naive approach, some variations have been implemented [31]. This leads to a faster adaption of the curves without adding any extra knowledge to the algorithm.

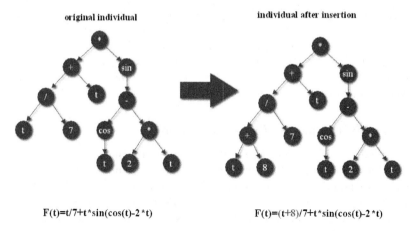

original individual

individual after insertion

F(t)=t/7+t*sin(cos(t)-2*t)

F(t)=(t+8)/7+t*sin(cos(t)-2*t)

**Fig. 7.22.** Mutational insertion on a individual.

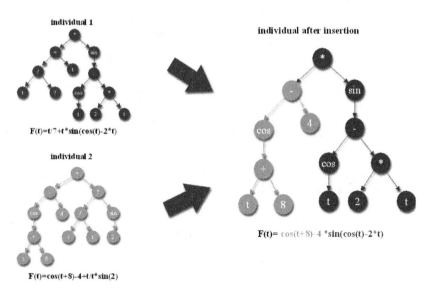

individual 1

F(t)=t/7+t*sin(cos(t)-2*t)

individual 2

F(t)=cos(t+8)-4+t/t*sin(2)

individual after insertion

F(t)= cos(t+8)-4 *sin(cos(t)-2*t)

**Fig. 7.23.** Recombination of individual 1 with individual 2.

Due to the fact that many solutions of the GP algorithm show shapes that are similar to the correct solution but differ only in orientation and size, a deterministic adaptation step is inserted between the mutation and the fitness calculation of the individual.

First, a translation vector is determined by placing the starting point of the individual at the position of the first point of the trajectory. The

translation is performed by adding to each point of the genetically generated function. In a second step, the individual will be scaled to the size of the trajectory curve. The scaling values are determined by setting the maxima of the individual to the same value as the maxima of the trajectory points. This leads to a faster adaption of the curves without adding any extra knowledge to the algorithm.

### 7.11.5   Multiobjective Fitness Function

Here (Figure 7.24) a multiobjective fitness function that compares the function values of the individual with corresponding points of the trajectory is used. If all points of the trajectory lie on the function plot this individual will represent a perfect solution to the problem. The fitness values will be represented by a weighted point to value distance scheme (7.7), and they will be summed up by a root mean square sum.

$$F_{dist} = \sum_{t=0}^{t_{max}} \left( |\overrightarrow{a_t}| \right) \tag{7.7}$$

**Fig. 7.24.** The used point to value distance scheme.

### 7.11.6   Weighting the Size of the Individuals

Individuals in genetic programming can grow very rapidly [16, 18]. This is referred as code bloat and can be observed in linear genetic programming or in tree-based genetic programming. Quadratic [17] and exponential growth are documented [22]. To stop individuals from growing and thus, from slowing down the evaluation, an extra weighting function is inserted into the fitness function. Both fitness values, the weighted point to value distance scheme (7.7), and the weighted number of nodes, are combined:

$$F_{indiv} = F_{dist} * \delta_w - F_{node} * \nu_w \tag{7.8}$$

**Fig. 7.25.** Individual extruded in a space-time plot.

$$F_{indiv} = \sum_{t=0}^{t_{max}} \left(|\vec{a_t}|\right) * \delta_w - (((\alpha) - (\alpha_e))^4) * \nu_w \qquad (7.9)$$

**Table 7.8.** The parameters used in equation 7.9.

| parameter | description |
|---|---|
| $\delta_w$ | distance weight $[0-1]$ |
| $\nu_w$ | number of nodes weight $[0-1]$ |
| $\alpha$ | number of nodes of an individual $[> 0]$ |
| $\alpha_e$ | estimated optimal number of nodes $[> 0]$ |

The estimated number of nodes is added to the function to force the solution into the estimated global optimum. To verify this procedure a set of tests have been carried out. These tests have been performed to determine the sufficient weighting of a punishment of a node count. The results of these tests are shown in Section 7.13.3.

## 7.12   Graphical Representation

Due to the fact that choosing the right parameters for setting up a fast and safe reconstruction of the estimated functions is difficult, a new graphical representation of a complete GP run was developed. In order to allow an overview of the whole chronology of the reconstruction process [33], a third dimension was added to the two dimensions (function values of $f_x(t)$ and $f_y(t)$) of the current best individual of the run (see Figure 7.25). This third

dimension is the index in the order of the arrival of the best individuals in the reconstruction. This is achieved by dumping all best individuals in this current run, ordered by the time of their appearance. Therefore, a standard CAD/CAM file format was used. This stereo lithography (STL) format is well defined and can be used in various CAD/CAM systems or computer graphic systems. This opens various possibilities for rendering and later usage of the generated objects. The data is generated by building triangles that have two of their vertices in two consecutive points of one individual and one vertex in another neighboring individual. That is, one triangle consists of the points $p[in][pn]$ and $p[in][pn+1]$ in the individual indicated by $in$ and a third point $p[in-1][pn]$ in a second individual indicated by $in-1$. It is obvious that, in order to get a closed surface, the following triangle has two points in the individual indicated by $in-1$, all three points for this individual are $p[in-1][pn]$, $p[in-1][pn+1]$ and $p[in][pn+1]$. This scheme is shown in Figure 7.26. In Figure 7.25 a sample rendering of such a plot is shown. The

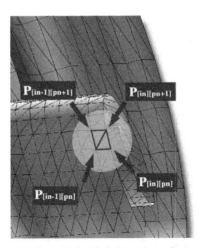

**Fig. 7.26.** The scheme for the triangulation at two triangles.

rendering was carried out using a third party software called 3D Exploration from Righthemisphere, Inc. The timeline goes from left (back) side of the figure to final stage at the right hand-side (front). In this final stage the individual has reached a fitness of 0.98. One thing that can be seen from this picture is that only a small number of steps is needed to reconstruct the general shape of the individual and that after this phase the algorithm behaves more like an evolution strategy [24] for reaching the final shape.

## 7.13 Results of the GP Kernel

### 7.13.1 Test Function

In this case a known function (7.10) was used in order to illustrate the operation of the symbolic regression algorithm

$$\boldsymbol{f}(t) = \begin{pmatrix} f_x(t) \\ f_y(t) \end{pmatrix} = \begin{pmatrix} \sin(\frac{\pi}{180} * t) * \frac{t}{10} \\ \cos(\frac{\pi}{180} * t) * \frac{t}{10} \end{pmatrix}. \tag{7.10}$$

The shape of the graph of $\boldsymbol{f}(t)$ resembles a chip. The values of the parameter $t$ range from 0 to 360. For simulation purposes this function was sampled yielding discrete points, which represent points that could have been sampled from a film sequence. The final stage of the reconstruction, the plot and the graph of the best individual, is shown in Figure 7.27. In spite of weighting

**Fig. 7.27.** The final stage of the reconstruction of the test function.

the length of the individual in the fitness calculation, the relation function (7.11) of the best individual after this reconstruction contains more nodes than the original function (7.10).

$$\boldsymbol{f}(t) = \begin{pmatrix} \frac{\sin(\frac{t}{57.2969}) * \frac{t}{27.8692}}{\sin(25.4997)} \\ \cos(\frac{t*\sin(149.902)}{44.6694}) * 12.2718 * \sin(40.5023) * \frac{t}{40.738} \end{pmatrix} \tag{7.11}$$

After the resulting function has been simplified by hand (7.12) it can be seen that the real structure of the searched relation function is found by the algorithm. This basic structure can be used for further extrapolations. The reduced form is:

$$\begin{pmatrix} f_x(t) \\ f_y(t) \end{pmatrix} = \begin{pmatrix} \sin(0.0175 * t) * t * 0.0920 \\ \cos(0.0178 * t) * t * 0.1000 \end{pmatrix} \tag{7.12}$$

Another way to analyze the behavior of the population is to store a snapshot of the whole population into one figure. All individuals are sorted in order of their fitness values to enable a closer look at the progression of the solution within the whole population. Figure 7.28 shows such a plot of the reconstruction of the test function at particular times of the reconstruction process. Table 7.9 shows the parameters for this test run. It can be seen that due to the application of a $(\mu + \lambda)$-selection strategy taken from evolution strategies [24] in the regression process, the best individual propagates very fast through the whole population. The use of a type of a parallel multipopulation strategy may lead to a higher genetic diversity (see Section 7.14).

**Table 7.9.** Parameter table for diversity test.

| value | parameter |
| --- | --- |
| 200 | Number of parents per generation. |
| 300 | Number of children per generation. |
| 0.001 | Weight of the node count fitness to overall fitness. |
| 1 | Weight of the distance fitness to overall fitness. |
| 10 | Estimated best node count in this reconstruction. |

### 7.13.2   Real-world Data

After the verification of the GP scheme using a known test function, the algorithm is applied to data extracted from a filmed sequence of the turning process, as described in Section 7.15. To get information about the physical correlations that lead to the shape of the chip in the turning process, the positions of the chip are manually segmented (see Figure 7.29). One difference in the preceding test function is the lack of knowledge of the exact number of nodes needed to obtain a good model of the physical process. Therefore, the number of necessary nodes has to be estimated, and as a consequence, the weighting factor of the number of nodes in the fitness calculation had to be reduced. The parameters for the estimated best node count are increased to 20 nodes, and the average initial size of the individuals is increased to 20 nodes. Therefore, the resulting relation functions (7.13) are not as short and handy as the results from the test function. Figure 7.30 shows that the algorithm has a tendency to smooth the trajectories. The approximations can be improved locally by increasing the number of sampling points. This way the reconstructions are forced to follow also small bendings of the particle flow.

Fitness 0.101527          Fitness 0.150081          Fitness 0.203431

Fitness 0.282540          Fitness 0.306753          Fitness 0.359802

Fitness 0.461218          Fitness 0.627615          Fitness 0.669400

Fitness 0.898750          Fitness 0.920857          Fitness 0.957200

**Fig. 7.28.** The population at different generations of the reconstruction. Each picture shows one complete population at discrete values of their overall fitness. The individuals are sorted according to their fitness from lower fitness at the left to greater fitness at the right side of the 3D shapes. During the starting phase where low fitness values appear a strong diversity of the individuals in the population can be observed. With increasing fitness values the population looses its diversity. At this stage the main reconstruction work is performed by variation of the similar individuals.

$$f(t) = \begin{pmatrix} \cos\left(\dfrac{-0.8995}{\sin\left(\frac{\cos(102.448)}{t+40.5886}\right)}\right) * t + 3 * t * \sin\left(\frac{0.447}{t}\over 0.0163\right) + 0.3679 * t \\ 0.700 + t+\sin(t)+\frac{0.020}{t} - 1655.335*t \\ t + 0.624 + 0.432 \end{pmatrix} \quad (7.13)$$

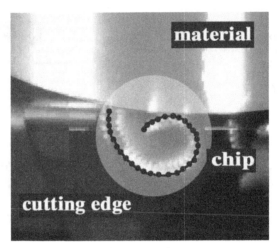

**Fig. 7.29.** Manually traced particle flow (picture size 30 mm by 30 mm).

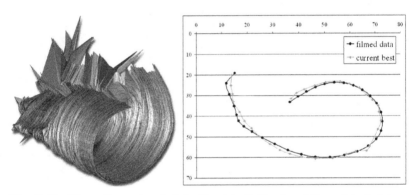

**Fig. 7.30.** Three dimensional plot of best individual and the original data.

### 7.13.3   The Dependency of the Number of Nodes

To evaluate and to verify the strategy of weighting the lengths of the individuals (see Section 7.11.6), several tests were performed. For the tests, two different target functions to the parameters were used. About 400 single reconstruction runs were carried out per series.

Two parameters were varied to examine their influence on the fitness evaluation. The value for the estimated best number of nodes per individual was varied from 0 to 20. To examine the influence of the weighting of the number of nodes, the weighting factor in every run was estimated by the result of $1/d$ where $d$ was increased from 0 to 20 over the whole test run.

**Table 7.10.** Parameter table for both test series.

| value | parameter |
| --- | --- |
| 30 | Number of parents per generation. |
| 60 | Number of children per generation. |
| 1 | Weight of the distance fitness to overall fitness. |
| 20 | Estimated best nodecount in this reconstruction. |
| 5000 | Maximum number of generations. |
| 20 | Average size of new individuals. |
| 0.9 | Probability of inserting a node. |
| 0.9 | Probability of deleting a node. |
| 0.9 | Probability of mutation a inner node. |
| 0.6 | Probability of inserting a value. |
| 0.5 | Probability of inserting a function. |
| 10 | Amount of mutation to a value. |
| -100 | Maximum value of a new value node. |
| 100 | Minimum value of a new value node. |

In Figure 7.31 the first two plots at the top are taken from one test series that was done using the known test function. The two plots at the bottom are taken from a second run with the real-world relation extracted in the experiment described in Section 7.9.1.

## 7.14   Parallelization

To overcome the problem of low diversity in a population, a distributed parallel approach was implemented into the symbolic regression approach. The task here was to prove the possibility of speed improvement that could be achieved by multiple computer systems or processors [27].

The general idea of the parallelization is based on the experiences of [30] and [35]. For this reason a strong similarity between the algorithms used in both applications [31, 34] was very useful for the implementation. As was done in [35], a blackboard communication model has been applied. Each instance of the population posts a certain number $n$ of individuals to the blackboard every $g$ generations. The parameters $n$ and $g$ are specific to each population. So the migration rate can be chosen according to the speed of the computer that carries the population. The implementation is based on file sharing mechanisms. This allows a decentralized and flexible parallelization.

### 7.14.1   Tests

Due to the splitting of the population, the problem of low diversity was reduced. First tests show noticeably higher convergence rates. As it can be

**Fig. 7.31.** Fitness of the test function (top) and the real-world data (bottom), distance fitness $F_{dist}$ (left), overall fitness $F_{indiv}$ (right).

seen in Figure 7.32, convergence increases, and the overall computing time decreases. A similar behavior can be seen in [35]. The parameters for all runs are equal, as shown in Table 7.11. Migration frequency, i.e., the number of generations after which a certain number of individuals migrate, should not be higher than 1/1000 to stop good individuals from spreading around all populations. This is similar to conclusions Cantu-Paz has published in [9]. It is presumed that this behavior again may result from the usage of the $(\mu + \lambda)$ strategy (see Figure 7.28).

Further investigations will be carried out to examine the behavior of different parameterizations in every different population. So the influence of different node weighting parameters may be interesting. Additionally, some examinations will be made regarding the behavior of a comma strategy to stop the decreasing diversity.

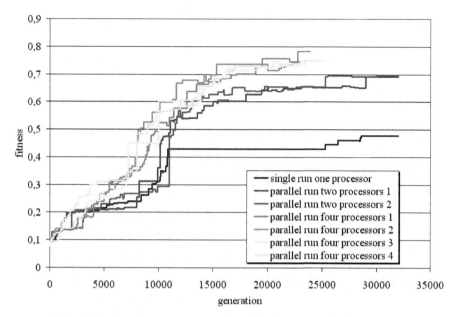

**Fig. 7.32.** Convergence tests on the distributed implementation.

**Table 7.11.** Parameter table for all parallel test series.

| value | usage of this parameter |
|---|---|
| 200 | Number of parents per generation. |
| 200 | Number of childrens per generation. |
| 0.000001 | Weight of the nodecount fitness to overall fitness. |
| 1 | Weight of the distance fitness to overall fitness. |
| 10 | Estimated best nodecount in this reconstruction. |
| 100000 | Maximum number of generations. |
| 10 | Average size of new individuals. |
| 0.9 | Probability of inserting a node. |
| 0.9 | Probability of deleting a node. |
| 0.9 | Probability of mutation a inner node. |
| 0.6 | Probability of inserting a value. |
| 0.5 | Probability of inserting a function. |
| 5 | Amount of mutation to a value. |

## 7.15   Conclusion

This chapter reported on genetic programming including advanced methodic aspects and applications to machining-technology problems. Development of methods concentrated on the linear program representation. In particular, the identification of effective and unused parts of code was addressed. On the

basis of this information program evolution could be accelerated, not only in computing time but also on the level of generations. Efficient mutation operators were developed for linear programs that increased speed of fitness convergence significantly. Even though the mutation-based LGP variant provides a sufficient preservation of code diversity already implicitly, increasing the level of diversity explicitly by a multiobjective selection mechanism further improved prediction performance. In general, mutation-based variation was found to be much more successful than crossover-based variation in linear GP. The main reason is that mutations induce smaller step sizes on the linear program structure that allow a more continuous approximation to solutions of higher quality. Additionally, a code-selective distance metric was applied in order to reduce mutation step sizes more precisely on the effective code.

It has been shown that the use of symbolic regression is suitable for constructing and developing descriptive models for physical relations. The process of reconstruction can be automatized. Experiments have proved that the reconstruction of correlations that take place in the chip building process in metal cutting is possible. There are also results showing that the quality of the found relations allows the reconstruction of other, not only physical correlations. The correlations may be taken from other scientific fields where process models still do not exist or should be renewed. This may be in applied social studies or applied economics. Another possible field is the reconstruction of signals, such as audio signals, or the task of filtering disturbances from given signals of low quality. The quality of the models found is good enough to use them for extrapolation or for further analyse.

A new method for analyzing the behavior of reconstruction processes of mathematical functions by symbolic regression via genetic programming has been developed. This graphical representation of the reconstruction process can be used in multiple ways to optimize or visualize the whole reconstruction process. Furthermore, it can be used to investigate specific points in time in the reconstruction. Additional considerations can be made to integrate this system into standard symbolic regression libraries.

The symbolic regression application was used for a distributed parallel approach to speed up the regression and to increase the convergence probability. First investigations show the efficiency of a blackboard parallelization method. This allows a high grade of parallelization without getting too much overhead from communication. The application is portable and flexible, and can be used on various types of systems. Further examinations need to be done on the optimal parameterization of the migration rate and the parameters of the population.

In order to develop a physical model of the turning process, additional considerations have to be made. For example, adding parameters, such as the

cutting speed or the tool angle to the reconstruction process. Furthermore, restrictions have to be developed to constrain the length of the reconstructed individuals.

# References

1. D. Bähre, M. Müller, and G. Warnecke. Basic characteristics on cutting effects in correlation to dynamic effects. In *Technical Papers of the 25th North American Manufacturing Research Conference*, pages 21–26. Lincoln, VT, May 20–23, 1997.
2. W. Banzhaf, P. Nordin, R. E. Keller, and F. D. Francone. *Genetic Programming – An Introduction. On the Automatic Evolution of Computer Programs and its Application.* dpunkt/Morgan Kaufmann, Heidelberg/San Francisco, CA, 1998.
3. C. L. Blake and C. J. Merz. UCI Repository of Machine Learning Databases [http://www.ics.uci.edu/~mlearn/MLRepository.html].
4. M. Brameier and W. Banzhaf. A comparison of linear genetic programming and neural networks in medical data mining. *IEEE Transactions on Evolutionary Computation*, 5(1):17–26, 2001.
5. M. Brameier and W. Banzhaf. Effective linear program induction. Technical Report CI-108/01, Collaborative Research Center 531, University of Dortmund, 2001.
6. M. Brameier and W. Banzhaf. Evolving teams of predictors with linear genetic programming. *Genetic Programming and Evolvable Machines*, 2(4):381–407, 2001.
7. M. Brameier and W. Banzhaf. Explicit control of diversity and effective variation distance in linear genetic programming. Technical Report CI-123/01, Collaborative Research Center 531, University of Dortmund, 2001.
8. M. Brameier and W. Banzhaf. Explicit control of diversity and effective variation distance in linear genetic programming. In *Proceedings of the Fifth European Conference on Genetic Programming (EuroGP-2002)*, Kinsale, Ireland, April 3–5, 2002. (Accepted).
9. E. Cantu-Paz. Migration policies and takeover times in genetic algorithms. In W. Banzhaf, J. Daida, A. E. Eiben, M. H. Garzon, V. Honavar, M. Jakiela, and R. E. Smith, editors, *Proceedings of the Genetic and Evolutionary Computation Conference (GECCO'99)*, volume 1, page 775. Morgan Kaufmann, San Francisco, CA, July 13–17, 1999.
10. K. Chellapilla. Evolving computer programs without subtree crossover. *IEEE Transactions on Evolutionary Computation*, 1(3):209–216, 1997.
11. P. Dittrich, F. Liljeros, A. Soulier, and W. Banzhaf. Spontaneous group formation in the seceder model. *Physical Review Letters*, 84:3205–3208, 2000.
12. D. Gusfield. *Algorithms on Strings, Trees and Sequences*. Cambridge University Press, New York, 1997.
13. C. Igel and K. Chellapilla. Investigating the influence of depth and degree of genotypic change on fitness in genetic programming. In W. Banzhaf, J. Daida, A. E. Eiben, M. H. Garzon, V. Honavar, M. Jakiela, and R. E. Smith, editors, *Proceedings of the Genetic and Evolutionary Computation Conference (GECCO'99)*, volume 1, pages 1061–1068. Morgan Kaufmann, San Francisco, CA, July 13–17, 1999.

14. T. Inamura. Brittle/ductile phenomena observed in computer simulations of machining defect-free monocrystalline silicon. *Annals of the CIRP*, 46:31–34, 1997.

15. R. E. Keller, J. Mehnen, W. Banzhaf, and K. Weinert. *Surface Reconstruction from 3D Point Data with a Genetic Programming/Evolution Strategy Hybrid*, chapter 2, pages 41–65. Advances in Genetic Programming 3. MIT Press, Cambridge, MA, 1999.

16. J. R. Koza. *Genetic Programming*. MIT Press, Cambridge, MA, 1992.

17. W. B. Langdon. Quadratic bloat in genetic programming. In D. Whitley, D. Goldberg, E. Cantu-Paz, L. Spector, I. Parmee, and H.-G. Beyer, editors, *Proceedings of the Genetic and Evolutionary Computation Conference (GECCO-2000)*, pages 451–458. Morgan Kaufmann, San Francisco, CA, July 10–12, 2000.

18. W. B. Langdon, T. Soule, R. Poli, and James A. Foster. The evolution of size and shape. In L. Spector, W. B. Langdon, U.-M. O'Reilly, and P. J. Angeline, editors, *Advances in Genetic Programming 3*, chapter 8, pages 163–190. MIT Press, Cambridge, MA, 1999.

19. J. Mehnen. Evolutionäre Flächenrekonstruktion. PhD thesis, University of Dortmund, 2000.

20. M. Müller. Prozeßidentifikation beim Drehen mit Hilfe künstlicher neuronaler Netze. *FBK - Produktionstechnische Berichte*, 22, 1996.

21. P. Nordin. A compiling genetic programming system that directly manipulates the machine-code. In K.E. Kinnear, editor, *Advances in Genetic Programming*, pages 311–331. MIT Press, Cambridge, MA, 1994.

22. P. Nordin and W. Banzhaf. Complexity compression and evolution. In L. Eshelman, editor, *Genetic Algorithms: Proceedings of the Sixth International Conference (ICGA95)*, pages 310–317. Morgan Kaufmann, San Francisco, CA, 1995.

23. H. Schulz and K. Bimschas. Optimisation of precision machining by simulation of the cutting process. *Annals of the CIRP*, 1993.

24. H.-P. Schwefel. *Evolution and Optimum Seeking*. Wiley, New York, 1995.

25. S. Shimada, N. Ikawa, H. Tanaka, and J. Uchikoshi. Structure of micromachined surface simulated by molecular dynamics analysis. *Annals of the CIRP*, 1994.

26. T. Soule, J. A. Foster, and J. Dickinson. Code growth in genetic programming. In J. R. Koza, D. E. Goldberg, D. B. Fogel, and R. L. Riolo, editors, *Proceedings of the Genetic Programming Conference (GP'96)*, pages 215–223. MIT Press, Cambridge, MA, 1996.

27. J. Sprave. Ein einheitliches Modell für Populationsstrukturen in evolutionären Algorithmen. PhD thesis, University of Dortmund, 1999.

28. B. Stroustrup. *The C++ Programming Language*. Addison-Wesley, Reading, MA, 1998.

29. G. Warnecke. *Spanbildung bei metallischen Werkstoffen*. Technischer Verlag Resch, Gräfelfing, 1974.

30. K. Weinert, J. Mehnen, and G. Rudolph. Dynamic neighborhood structures in parallel evolution strategies. Technical Report CI-114/01, Collaborative Research Center 531, University of Dortmund, 2001.

31. K. Weinert and M. Stautner. Reconstruction of particle flow mechanisms with symbolic regression via genetic programming. In L. Spector, E. D. Goodman, A. Wu, W. B. Langdon, H. M. Voigt, M. Gen, S. Sen, M. Dorigo, S. Pezeshk,

M. H. Garzon, and E. Burke, editors, *Proceedings of the Genetic and Evolutionary Computation Conference (GECCO-2001)*, pages 1439–1443. Morgan Kaufmann, San Francisco, CA, July 7–11, 2001.

32. K. Weinert and M. Stautner. Reconstruction of physical correlations using symbolic regressions. Technical Report CI-116/01, Collaborative Research Center 531, University of Dortmund, 2001.

33. K. Weinert and M. Stautner. A new view on symbolic regression. In *Proceedings of the Fifth European Conference on Genetic Programming (EuroGP-2002)*. Kinsale, Ireland, April 3–5, 2002. (Accepted).

34. K. Weinert, T. Surmann, and J. Mehnen. Evolutionary surface reconstruction using CSG-NURBS-hybrids. In L. Spector, E. D. Goodman, A. Wu, W. B. Langdon, H. M. Voigt, M. Gen, S. Sen, M. Dorigo, S. Pezeshk, M. H. Garzon, and E. Burke, editors, *Proceedings of the Genetic and Evolutionary Computation Conference (GECCO-2001)*, page 1456. Morgan Kaufmann, San Francisco, CA, July 7–11, 2001.

35. K. Weinert, T. Surmann, and J. Mehnen. Parallel surface reconstruction. In *Proceedings of the Fifth European Conference on Genetic Programming (EuroGP-2002)*. Kinsale, Ireland, April 3–5, 2002. (Accepted).

36. K. Weinert and A. Zabel. Modelling chip-building in orthogonal cutting by using a cellular automata/genetic programming approach. In C. Fye, editor, *Proceedings of the Second ICSC Symposium on Engineering of Intelligent Systems – EIS 2000*. ICSC International Computer Science Conference, University of Paisley, Scotland, June 27–30, 2000.

37. J. Q. Xie, A. E. Bayoumi, and H. M. Zbib. Fea modeling and simulation of shear localized chip formation in metal cutting. *International Journal of Machine Tools and Manufacture*, 38:1057–1087, 1998.

# Part III

# Machine Learning and Optimization

# 8 Novel Learning Tasks, Optimization, and Their Application

Guido Daniel[1], Jan Dienstuhl[2], Sebastian Engell[1], Sven Felske[4],
Karl Goser[2], Ralf Klinkenberg[3], Katharina Morik[3], Oliver Ritthoff[3], and
Henner Schmidt-Traub[4]

[1] University of Dortmund, Department of Chemical Engineering, Process Control,
44221 Dortmund, Germany
{g.daniel, s.engell}@ct.uni-dortmund.de
[2] University of Dortmund, Faculty of Electrical Engineering & Information
Technology, Microelectronics, 44221 Dortmund, Germany
{dienstuhl, goser}@luzi.e-technik.uni-dortmund.de
[3] University of Dortmund, Department of Computer Science, Informatik VIII,
44221 Dortmund, Germany
{klinkenberg, morik, ritthoff}@Ls8.cs.uni-dortmund.de
[4] University of Dortmund, Department of Chemical Engineering, Plant Design,
44221 Dortmund, Germany
{felske, schmtr}@ct.uni-dortmund.de

**Summary.** This chapter describes methods for learning and optimizing solutions
to engineering problems. Where standard learning tasks do not fit the requirements
of the applications, they are varied. In particular, the use of prior knowledge and
unlabeled examples may ease a learning task. In contrast, learning target concepts
that change over time and constructing new features are challenging learning tasks.
The changes of the learning tasks that ease as well as those that challenge learning
cover a broad range of application problems. Solutions to the new, nonstandard
learning tasks are developed, integrating numerical optimization strategies, evolu-
tionary algorithms, support vector machines, neural networks, and fuzzy controllers.
A workbench for learning methods allows their systematic evaluation with respect
to the engineering problems.

The chapter is structured as follows: First, extensions of standard learning tasks
are defined, and methods for their solution are proposed and evaluated. Second, a
problem in chemical engineering is presented in Section 8.2, namely the determi-
nation of suitable models for batch chromatography. This process is a standard
separation method in the pharmaceutical industry. The critical model parameters
of physical and grey-box models of the process are determined from measured con-
centration profiles. Mathematical parameter estimation and neural networks are
used for this purpose. Third, from the field of electrical engineering, circuit de-
sign, device modeling, and quality assessment tasks are covered in Section 8.3.
An evolutionary algorithm for the optimization of electronic circuit components is
presented. Furthermore the construction of simulation models for electronic device
development and circuit control is investigated. Different algorithms and hardware
implementations of neural networks for microelectronics and process control are
examined.

## 8.1    Novel Learning Tasks in Machine Learning

### 8.1.1    Introduction

According to Tom Mitchell's well-known definition, "a computer program is said to learn from experience $E$ with respect to some tasks $T$ and performance measure $P$, if its performance at tasks in $T$ as measured by $P$ improves with experience $E$ [82]." This definition shows clearly how closely optimization and machine learning are related. Some classes of learning problems have been well-posed and investigated. The problem that has been understood best is the one of concept learning (classification): Given training examples $X$ in a representation language $L_E$, which are classified according to an unknown Boolean function $c(x)$ and a space of hypotheses $L_H$, learn a Boolean-valued function $h \in L_H$, such that $h(x) = c(x)$ for all $x$.

The quality of the learning result is evaluated with respect to a test set of examples and a quality measure, most often accuracy ($h(x) = c(x)$) and coverage (all $x$ are handled by $h(x)$). If the function to be learned is real-valued, the learning task is called regression. If the examples are not labeled but have to be grouped together by the learning algorithm, the task is to learn from observations (unsupervised learning).

Learnability plays a central role in the field of computational intelligence (CI). Not only artificial neural networks, but also evolutionary algorithms show learning capabilities, since the underlying optimization process can been seen as a search for the optimal solution (hypothesis) in the space of all possible solutions (hypothesis space) given by the evolutionary operators. The complexity of concept learning has been shown for many pairs of $L_E$ and $L_H$, in the paradigm of probably approximately correct (PAC) learning [127] as well as in inductive logic programming (ILP) [61, 62, 92]. Ros [106] has studied a class of genetic algorithms that only use a crossover operator and renounce a mutation operator for learning Boolean functions in the context of PAC learnability. Ros showed that specific classes of Boolean conjunctions and disjunctions (including k-CNF and k-DNF) are PAC-learnable in polynomial time and thus can be considered as PAC learning methods. Applications of evolutionary algorithms for learning are referred to in [32, 33, 46].

In both paradigms, PAC and ILP, the assumptions of the concept learning task have been varied. For instance, the learning algorithm can be allowed to ask questions to an oracle, which eases the learning task. A variance, which has been shown to be harder than pure concept learning, includes background knowledge. Where deviations from the standard learning tasks in learnability research are driven with the aim of determining the exact border between polynomially learnable and not learnable concepts, here we introduce nonstandard learning tasks driven by real-world problems.

In practice we often meet modified learning tasks that deviate from the standard scenario of classification and regression. In other words, given an application problem, the assumptions made when using a standard learning

task are often not appropriate. The consequence of this inappropriateness is either that the learning performance in terms of the predictive error (accuracy) decreases, or some hypotheses simply cannot be learned. The overall goal of this project was to study nonstandard learning tasks on the basis of given application problems.

A problem that occurs frequently when confronting the standard learning tasks with real-world problems is the change of concepts over time. In reality we cannot always assume a concept to be given once for all. This means that the distribution of examples regarding the labels is not constant but changes over time. The corresponding nonstandard learning task is called learning drifting concepts [68, 131].

It is not necessary for real-world problems to be more difficult than the standard problems. Sometimes, we are given more knowledge than the standard learning tasks assume. Vapnik has raised the question, why we should induce a hypothesis on the basis of examples first and then apply our learning result to a test set if the given test set is all of our problem [130]. This means that sometimes we know in advance the position of the test examples that we have to classify by our learning result. If this is so, why not ease the learning task by directly approaching the test set? The nonstandard learning task of learning from labeled as well as unlabeled examples is called transduction.

Another piece of information the user of machine learning methods may well have is the importance and cost of particular examples. Where the standard tasks consider all examples of equal worth and achievability, in medicine, for instance, some critical observations can be made only rarely, but are considered very important. If a patient's state can be measured by a particular catheter and this catheter is only seldomly applied, we want to strengthen the influence of these measurements on the learning result. As opposed to background knowledge (i.e., a theory), prior knowledge about examples can ease learning.

In cases where the example language $L_E$ is not expressive enough, the correct hypothesis cannot be found by the learner. If, for instance, a learning algorithm can only separate data linearly, but the positive examples are separated by a curve (e.g., a polynomial) from the negative examples, the target concept cannot be determined. A mapping from the original feature space to another one may shape the examples, such that the learner can recognize the target concept. The transformation of $L_E$ in order to ease learning is called feature generation.

A contrary situation occurs if the example language contains too many irrelevant or redundant features. This often degrades the learning performance. Intuitively, we can imagine that the true concept is hidden in the noise of irrelevant descriptions. A solution to this problem is feature selection, i.e., the process of removing redundant and accordingly, irrelevant features from the original feature set.

## Overview

In the following section we give a short introduction into the principles of support vector machines (SVM) (Section 8.1.2) since, due to its positive theoretical and practical properties, this learning method was intensively used in our experiments. Subsequently, we present in detail the results achieved from the investigations of nonstandard learning tasks. The first learning task that we considered is learning of drifting concepts (Section 8.1.3), followed by the topic of transduction (Section 8.1.4). The next section deals with the task of learning with prior knowledge (Section 8.1.5). The nonstandard learning tasks feature generation and selection are the topic of Section 8.1.6. To study these nonstandard learning tasks, especially feature generation and selection as well as learning with prior knowledge, a flexible learning environment is needed. An elaborate description of an environment that fulfills the necessary requirements is given in Section 8.1.7. Finally, Section 8.1.8 summarizes the most important results of our work and gives a short outlook to open research topics in the field of nonstandard learning tasks.

### 8.1.2    Support Vector Machines

#### Support Vector Machines and Structural Risk Minimization

Support vector machines are based on the principle of structural risk minimization (SRM) [129, 130]. The idea of structural risk minimization is to find the hypothesis in the hypothesis space for which one can probabilistically guarantee the smallest true error. The smallest true error of a hypothesis in the context of classification problems is the probability of making an error on an unseen, randomly drawn test example. The following bound combines the true error of a hypothesis $h$ with its error on the training examples and the complexity of $h$ [130]. Let $\delta$ be the maximal probability that this bound is false, i.e., that the true error is larger than expected:

$$P(error(h)) \leq train\_error(h) + 2\sqrt{\frac{d \cdot (ln\frac{2n}{d} + 1) - ln\frac{\delta}{4}}{n}} \tag{8.1}$$

The bound holds with a probability of $1 - \delta$. $n$ is the number of training examples and $d$ is the Vapnik-Chervonenkis (VC) dimension (VCdim) [130]. The VC dimension is a measure for the complexity of the hypothesis space from which the examples are drawn. The upper error bound reflects the well known trade-off between the complexity of the hypothesis space and the training error. A hypothesis space that is too small (low VCdim) will probably not contain a good hypothesis and thus lead to a high training error and a high true error. On the other hand, a hypothesis space that is too large will probably contain hypotheses with a low training error, but the second addend on the right hand side of the inequality will become large. This situation is

generally known as overfitting, i.e., low training error but poor generalization performance on unseen examples. Thus the choice of the hypothesis space is crucial. The structural risk minimization principle takes the complexity of the hypothesis space into account by structuring the hypotheses spaces. This structure subdivides the hypothesis space into subsets of increasing VCdim. Let $d_i$ be the VCdim of the hypothesis space $H_i$:

$$H_1 \subset H_2 \subset H_3 \subset ... \subset H_i \subset ... \quad \text{and} \quad \forall i : d_i \leq d_{i+1} \qquad (8.2)$$

Only those hypotheses $h$ in the hypothesis space $H_i$ are chosen that minimize the above error bound. The remarkable point is that the dimension of the attribute space does not appear in this context. Thus an SVM potentially shows good generalization results in very high dimensional spaces. Support vector machines are very universal learners for classification, regression, and density estimation. By the use of so-called Mercer kernels, SVM can be parameterized so that they generalize polynomial classifiers, radial basis networks, and two-layered sigmoidal neural feed-forward networks [12]. Beyond the remarkable learning-theoretical properties, SVM are also interesting from a complexity-theoretical point of view. The training phase of an SVM can be ascribed to a semi-definite quadratic optimization problem, for which a global optimum can be efficiently computed.

**Optimization Problems Solved by a Support Vector Machine**

In their basic form, SVM learn linear decision rules,

$$h(\boldsymbol{x}) = sign\{\boldsymbol{w} \cdot \boldsymbol{x} + b\} = \begin{cases} +1, & \text{if } \boldsymbol{w} \cdot \boldsymbol{x} + b > 0 \\ -1, & \text{else} \end{cases} , \qquad (8.3)$$

described by a weight vector $\boldsymbol{w}$ and a threshold $b$. The idea of structural risk minimization is to find a hypothesis $h$ for which one can guarantee the lowest probability of error. For SVM, [130] shows that this goal can be translated into finding the hyperplane with maximum soft-margin.[1] Computing this hyperplane is equivalent to solving the following optimization problem:

**Optimization Problem 1 (SVM (primal))**

$$\textit{Minimize:} \quad V(\boldsymbol{w}, b, \boldsymbol{\xi}) = \frac{1}{2}\boldsymbol{w} \cdot \boldsymbol{w} + C \sum_{i=1}^{n} \xi_i \qquad (8.4)$$

$$\textit{subject to:} \quad \forall_{i=1}^{n} : y_i[\boldsymbol{w} \cdot \boldsymbol{x}_i + b] \geq 1 - \xi_i \qquad (8.5)$$

$$\forall_{i=1}^{n} : \xi_i \geq 0 \qquad (8.6)$$

In this optimization problem, the Euclidean length $||\boldsymbol{w}||$ of the weight vector is inversely proportional to the soft-margin of the decision rule. The constraints

---

[1] See [22] for an introduction to SVM.

(8.5) require that all training examples are classified correctly up to some slack $\xi_i$. If a training example lies on the "wrong" side of the hyperplane, the corresponding $\xi_i$ is greater or equal to 1. Therefore $\sum_{i=1}^{n} \xi_i$ is an upper bound on the number of training errors. The factor $C$ in (8.4) is a parameter that allows trading-off training error vs. model complexity.

For computational reasons, it is useful to solve the Wolfe dual [31] of optimization problem 1 instead of solving optimization problem 1 directly [130].

**Optimization Problem 2 (SVM (dual))**

$$\text{Minimize:} \quad W(\boldsymbol{\alpha}) = -\sum_{i=1}^{n} \alpha_i + \frac{1}{2}\sum_{i=1}^{n}\sum_{j=1}^{n} y_i y_j \alpha_i \alpha_j (\boldsymbol{x}_i \cdot \boldsymbol{x}_j) \tag{8.7}$$

$$\text{subject to:} \quad \sum_{i=1}^{n} y_i \alpha_i = 0 \tag{8.8}$$

$$\forall_{i=1}^{n} : 0 \le \alpha_i \le C \tag{8.9}$$

Joachims has succeeded in developing a fast algorithm for computing the solution to this optimization problem [47, 50]. We use this algorithm implemented in $SVM^{light}$ for the classification and transduction[2] experiments described in this chapter.[3]

Support vectors are those training examples $\boldsymbol{x}_i$ with $\alpha_i > 0$ at the solution. From the solution of optimization problem 2, the decision rule can be computed as

$$\boldsymbol{w} \cdot \boldsymbol{x} = \sum_{i=1}^{n} \alpha_i y_i (\boldsymbol{x}_i \cdot \boldsymbol{x}) \text{ and } b = y_{usv} - \boldsymbol{w} \cdot \boldsymbol{x}_{usv} . \tag{8.10}$$

The training example $(\boldsymbol{x}_{usv}, y_{usv})$ for calculating $b$ must be a support vector with $\alpha_{usv} < C$. Finally, the training losses $\xi_i$ can be computed as

$$\xi_i = max(1 - y_i [\boldsymbol{w} \cdot \boldsymbol{x}_i + b], 0).$$

For solving optimization problem 2 as well as applying the learned decision rule, it is sufficient to be able to calculate inner products between feature vectors. Exploiting this property, Boser et al. introduced the use of kernels $\mathcal{K}(\boldsymbol{x}_1, \boldsymbol{x}_2)$ for learning nonlinear decision rules [12]. Depending on the type of kernel function, SVM learn polynomial classifiers, radial basis function (RBF) classifiers, or two layer sigmoid neural nets. Such kernels calculate an inner product in some feature space and replace the inner product in the formulas above.

---

[2] The learning task of transduction and the corresponding optimization problems are described in Section 8.1.4.

[3] $SVM^{light}$ is available at http://www-ai.informatik.uni-dortmund.de/svm_light

### 8.1.3   Learning Drifting Concepts

Most machine learning approaches assume that the distribution underlying the training examples and new unseen test examples is static, and that a model, once learned on the training data, can be applied to any new test example and does not need any adjustments later on. For many real-world machine learning tasks, however, where data is collected over an extended period of time, its underlying distribution is likely to change over time. A typical example is information filtering, i.e., the adaptive classification of text documents with respect to a particular user interest. Information filtering techniques are used, for example, to build personalized news filters, which learn about the news-reading preferences of a user or to filter e-mail. Both the interest of the user and the document content change over time. A filtering system should be able to adapt to such concept changes.

This section describes a new method proposed in [68] to recognize and handle concept changes with support vector machines. The method maintains a window on the training data. The key idea is to automatically adjust the window size so that the estimated generalization error is minimized. The new approach is both theoretically well founded as well as effective and efficient in practice. Since it does not require complicated parameterization, it is simpler to use and is more robust than comparable heuristics. Experiments with simulated concept drift scenarios based on real-world text data compare the new method with other window management approaches. We show that it can effectively select an appropriate window size in a robust way.

After reviewing other work on adaptation to changing concepts, this section explains the new window adjustment approach and evaluates it in three simulated concept drift scenarios on real-world text data. The experiments show that the approach effectively selects an appropriate window size and results in a low predictive error rate.

### Concept Drift

The problem of concept drift for the pattern recognition problem can be formalized in the following framework. Each example $z = (x, y)$ consists of a feature vector $x \in \mathbb{R}^N$ and a label $y \in \{-1, +1\}$ indicating its classification. Data arrives over time in batches. Without loss of generality these batches are assumed to be of equal size, each containing $m$ examples:

$$z_{(1,1)}, \ldots, z_{(1,m)}, z_{(2,1)}, \ldots, z_{(2,m)}, \cdots, z_{(t,1)}, \ldots, z_{(t,m)}, z_{(t+1,1)}, \ldots, z_{(t+1,m)} \, ,$$

$z_{(i,j)}$ denotes the $j$th example of batch $i$. For each batch $i$, the data is independently identically distributed with respect to a distribution $\Pr_i(x, y)$. Depending on the amount and type of concept drift, the example distribution $\Pr_i(x, y)$ and $\Pr_{i+1}(x, y)$ between batches will differ. The goal of the learner $\mathcal{L}$ is to sequentially predict the labels of the next batch. For example, after batch $t$ the learner can use any subset of the training examples from batches

1 to $t$ to predict the labels of batch $t + 1$. The learner aims to minimize the cumulated number of prediction errors.

In machine learning, changing concepts are often handled by time windows of fixed or adaptive size on the training data [69, 81, 131] or by weighting data or parts of the hypothesis according to their age and/or utility for the classification task [122]. The latter approach of weighting examples has already been used for information filtering in the incremental relevance feedback approaches of [2] and [7]. Here the earlier approach maintaining a window of adaptive size is explored.

For windows of fixed size, the choice of a "good" window size is a compromise between fast adaptability (small window) and good generalization in phases without concept change (large window). The basic idea of adaptive window management is to adjust the window size to the current extent of concept drift.

The task of learning drifting or time-varying concepts has also been studied in computational learning theory. Learning a changing concept is infeasible if no restrictions are imposed on the type of admissible concept changes[4], but drifting concepts are provably efficiently learnable (at least for certain concept classes) if the rate or the extent of drift is limited in particular ways.

Helmbold and Long [42] assume a possibly permanent but slow concept drift and define the extent of drift as the probability that two subsequent concepts disagree on a randomly drawn example. Their results include an upper bound for the extend of drift maximally tolerable by any learner and algorithms that can learn concepts that do not drift more than a certain constant extent of drift. Furthermore, they show that it is sufficient for a learner to see a fixed number of the most recent examples. Hence a window of a certain minimal fixed size allows to learn concepts for which the extent of drift is appropriately limited.

While Helmbold and Long restrict the extend of drift, [73] determine a maximal rate of drift that is acceptable by any learner, i.e., a maximally acceptable frequency of concept changes, which implies a lower bound for the size of a fixed window for a time-varying concept to be learnable, which is similar to the lower bound of Helmbold and Long.

In practice, however, it usually cannot be guaranteed that the application at hand obeys these restrictions, e.g., a reader of electronic news may change interests almost arbitrarily often and radically. Furthermore the large time window sizes, for which the theoretical results hold, would be impractical. Hence more application-oriented approaches rely on far smaller windows of fixed size or on window adjustment heuristics that allow far smaller window sizes and usually perform better than fixed and/or larger windows [69, 131]. While these heuristics are intuitive and work well in their particular appli-

---

[4] E.g., a function randomly jumping between the values one and zero cannot be predicted by any learner with more than 50 percent accuracy.

cation domain, they usually require tuning their parameters, are often not transferable to other domains, and lack a proper theoretical foundation.

## Window Adjustment by Optimizing Performance

Our approach to handling drift in the distribution of examples uses a window on the training data and support vector machines (Section 8.1.2, [130]) as its core learning algorithm. This window should include only those examples which are sufficiently close to the current target concept. Assuming the amount of drift increases with time, the window includes the last $n$ training examples. Previous approaches, including one developed by us [69], used similar windowing strategies. Their shortcomings are that they either fix the window size [81] or involve complicated heuristics [69, 131]. A fixed window size makes strong assumptions about how quickly the concept changes. While heuristics can adapt to different speed and amount of drift, they involve many parameters that are difficult to tune. Here, we present an approach to selecting an appropriate window size that does not involve complicated parameterization. This is the first approach that is neither restricted to static concepts nor requires an extensive parameterization and that effectively and efficiently learns drifting concepts. They key idea is to select the window size so that the estimated generalization error on new examples is minimized. To get an estimate of the generalization error we use a special form of $\xi\alpha$-estimates [49]. $\xi\alpha$-estimates are a particularly efficient method for estimating the performance of a SVM.

**$\xi\alpha$-Estimators:** $\xi\alpha$-estimators are based on the idea of leave-one-out estimation [77]. The leave-one-out estimator of the error rate proceeds as follows. From the training sample $S = ((\boldsymbol{x}_1, y_1), \cdots, (\boldsymbol{x}_n, y_n))$ the first example $(\boldsymbol{x}_1, y_1)$ is removed. The resulting sample $S^{\backslash 1} = ((\boldsymbol{x}_2, y_2), \cdots, (\boldsymbol{x}_n, y_n))$ is used for training, leading to a classification rule $h_{\mathcal{L}}^{\backslash 1}$. This classification rule is tested on the held-out example $(\boldsymbol{x}_1, y_1)$. If the example is classified incorrectly it is said to produce a leave-one-out error. This process is repeated for all training examples. The number of leave-one-out errors divided by $n$ is the leave-one-out estimate of the generalization error.

While the leave-one-out estimate is usually very accurate, it is very expensive to compute. With a training sample of size $n$, one must run the learner $n$ times. $\xi\alpha$-estimators overcome this problem using an upper bound on the number of leave-one-out errors instead of calculating them brute force. They owe their name to the two arguments they are computed from. $\boldsymbol{\xi}$ is the vector of training losses at the solution of the primal SVM training problem. $\boldsymbol{\alpha}$ is the solution of the dual SVM training problem. Based on these two vectors – both are available after training the SVM at no extra cost – the $\xi\alpha$-estimators are defined using the following two counts: With $R_\Delta^2$ being the maximum difference of any two elements of the Hessian (i.e.,

$R_\Delta^2 \geq \max_{\boldsymbol{x},\boldsymbol{x}'}(\mathcal{K}(\boldsymbol{x},\boldsymbol{x}) - \mathcal{K}(\boldsymbol{x},\boldsymbol{x}')))$,

$$d = |\{i : (\alpha_i R_\Delta^2 + \xi_i) \geq 1\}| \qquad (8.11)$$

counts the number of training examples, for which the quantity $\alpha_i R_\Delta^2 + \xi_i$ exceeds one. Since the document vectors are normalized to unit length in the experiments described in the following subsection, here $R_\Delta^2 = 1$. It is proven in [49] that $d$ is an approximate upper bound on the number of leave-one-out errors in the training set. With $n$ as the total number of training examples, the $\xi\alpha$-estimator of the error rate is

$$Err_{\xi\alpha}^n(h_\mathcal{L}) = \frac{|\{i : (\alpha_i R_\Delta^2 + \xi_i) \geq 1\}|}{n}. \qquad (8.12)$$

The theoretical properties of this $\xi\alpha$-estimator are discussed in [49]. It can be shown that the estimator is pessimistically biased, overestimating the true error rate on average. Experiments show that the bias is acceptably small for text classification problems and that the variance of the $\xi\alpha$-estimator is essentially as low as that of a holdout estimate using twice as much data. It is also possible to design similar estimators for precision and recall, as well as combined measures like $F1$ [49].

**Window Adjustment Algorithm:** A window adjustment algorithm has to solve the following trade-off. A large window provides the learner with much training data, allowing it to generalize well, given that the concept did not change. On the other hand, a large window can contain old data that is no longer relevant (or even confusing) for the current target concept. Finding the right size means trading-off the quality against the number of training examples.

To answer this question the window adjustment algorithm proposed in the following uses $\xi\alpha$-estimates in a particular way. At batch $t$, it essentially tries various window sizes, training a SVM for each resulting training set.

$$z_{(t,1)}, ..., z_{(t,m)} \qquad (8.13)$$

$$z_{(t-1,1)}, ..., z_{(t-1,m)}, z_{(t,1)}, ..., z_{(t,m)} \qquad (8.14)$$

$$z_{(t-2,1)}, ..., z_{(t-2,m)}, z_{(t-1,1)}, ..., z_{(t-1,m)}, z_{(t,1)}, ..., z_{(t,m)} \qquad (8.15)$$

$$\vdots$$

For each window size, it computes a $\xi\alpha$-estimate based on the result of training. In contrast to the previous section, the $\xi\alpha$-estimator used here considers only the last batch, that is, the $m$ most recent training examples $z_{(t,1)}, ..., z_{(t,m)}$:

$$Err_{\xi\alpha}^m(h_\mathcal{L}) = \frac{|\{i : 1 \leq i \leq m \wedge (\alpha_{(t,i)} R_\Delta^2 + \xi_{(t,i)}) \geq 1\}|}{m}. \qquad (8.16)$$

This reflects the assumption that the most recent examples are most similar to the new examples in batch $t + 1$. The window size minimizing the $\xi\alpha$-estimate of the error rate is selected by the following algorithm:

- input: $S$ training sample containing $t$ batches of $m$ examples each
- for $h \in \{0, ..., t-1\}$
  - train SVM on examples $z_{(t-h,1)}, ..., z_{(t,m)}$
  - compute $\xi\alpha$-estimate on examples $z_{(t,1)}, ..., z_{(t,m)}$
- output: window size which minimizes $\xi\alpha$-estimate

## Experiments

**Experimental Setup:** Each of the following four data management approaches is evaluated in combination with the SVM. In the full memory approach, the learner generates its classification model from all previously seen examples, i.e., it cannot "forget" old examples. With no memory, the learner always induces its hypothesis only from the most recent batch. This corresponds to using a window of the fixed size of one batch. A window of the fixed size of three batches is also used. Finally, for a window of adaptive size, the window adjustment algorithm proposed in the previous section adapts the window size to the current concept drift situation.

The experiments are performed in an information filtering domain, a typical application area for learning drifting concept. Text documents are represented as attribute-value vectors (bag of words model), where each distinct word corresponds to a feature whose value is the "ltc"-TF/IDF-weight [118] of that word in that document. Words occurring less than three times in the training data or occurring in a given list of stop words are not considered. Each document feature vector is normalized to unit length to abstract from different document lengths.

The performance of a classifier is measured by the metrics prediction error, recall, and precision. Recall is the probability, that the classifier recognizes a relevant document as relevant. Precision is the probability, that a document classified as relevant actually is relevant. All reported results are estimates averaged over ten runs.

**Table 8.1.** Probability of being relevant for a document of the specified category at the specified time step (batch) in the concept change scenarios A, B, and C

| Scenario | Category | Relevance probability for documents at a specified time step | | | | | | | | | | | |
|---|---|---|---|---|---|---|---|---|---|---|---|---|---|
| | | 0 | 1 | 2 | ... | 7 | 8 | 9 | 10 | 11 | 12 | ... | 19 |
| A | 1 | 1.0 | 1.0 | 1.0 | ... | 1.0 | 1.0 | **1.0** | **0.0** | 0.0 | 0.0 | ... | 0.0 |
| | 3 | 0.0 | 0.0 | 0.0 | ... | 0.0 | 0.0 | **0.0** | **1.0** | 1.0 | 1.0 | ... | 1.0 |
| B | 1 | 1.0 | 1.0 | 1.0 | ... | 1.0 | 0.8 | 0.6 | 0.4 | 0.2 | 0.0 | ... | 0.0 |
| | 3 | 0.0 | 0.0 | 0.0 | ... | 0.0 | 0.2 | 0.4 | 0.6 | 0.8 | 1.0 | ... | 1.0 |
| C | 1 | 1.0 | 1.0 | 1.0 | ... | 1.0 | **1.0** | **0.0** | 0.0 | **1.0** | 1.0 | ... | 1.0 |
| | 3 | 0.0 | 0.0 | 0.0 | ... | 0.0 | **0.0** | **1.0** | 1.0 | **0.0** | 0.0 | ... | 0.0 |

The experiments use a subset of 2608 documents of the data set of the Text REtrieval Conference (TREC) consisting of English business-news texts. Each text is assigned to one or several categories. The categories considered here are 1 (Antitrust Cases Pending), 3 (Joint Ventures), 4 (Debt Reschedul- ing), 5 (Dumping Charges), and 6 (Third World Debt Relief). For the experi- ments, three concept change scenarios are simulated. The texts are randomly split into 20 batches of equal size containing 130 documents each.[5] The texts of each category are distributed as equally as possible over the 20 batches.

Table 8.1 describes the relevance of the categories in the three concept change scenarios A, B, and C. For each time step (batch), the probability of being relevant (interesting to the user) is specified for documents of categories 1 and 3, respectively. Documents of the classes 4, 5, and 6 are never relevant in any of these scenarios. In the first scenario (scenario A), first documents of category 1 are considered relevant for the user interest and all other doc- uments irrelevant. This changes abruptly (concept shift) in batch 10, where documents of category 3 are relevant and all others irrelevant. In the second scenario (scenario B), again, first documents of category 1 are considered rel- evant for the user interest and all other documents irrelevant. This changes slowly (concept drift) from batch 8 to batch 12, where documents of cat- egory 3 are relevant and all others irrelevant. The third scenario (scenario C) simulates an abrupt concept shift in the user interest from category 1 to category 3 in batch 9 and back to category 1 in batch 11.

**Results:** Figure 8.1 compares the prediction error rates of the adaptive window size algorithm with the non-adaptive methods. The graphs on the left show the prediction error on the following batch. In all three scenarios, the full memory strategy and the adaptive window size algorithm essentially coincide as long as there is no concept drift. During this stable phase, both show lower prediction error than the fixed size and the no memory approach. At the point of concept drift, the performance of all methods deteriorates. While the performance of no memory and adaptive size recovers quickly after the concept drift, the error rate of the full memory approach remains high, especially in scenarios A and B. Like before the concept drift, the no memory and the fixed size strategies exhibit higher error rates than the adaptive window algorithm in the stable phase after the concept drift. This shows that the no memory, the fixed size, and the full memory approaches all perform suboptimally in some situation. Only the adaptive window size algorithm can achieve a relatively low error rate over all phases in all scenarios. This is also reflected in the average error rates over all batches given in Table 8.2. The adaptive window size algorithm achieves a low average error rate on all three scenarios. Similarly, precision and recall are consistently high.

---

[5] Hence, in each trial, out of the 2608 documents, eight randomly selected texts are not considered.

**Fig. 8.1.** Comparison of the prediction error rates for scenario A (top), B (middle), and C (bottom). The x-axis denotes the batch number and the y-axis the average prediction error. The results are averaged over 10 runs

**Table 8.2.** Error, recall, and precision of all window management approaches for all scenarios averaged over 10 trials with 20 batches each (standard sample error in parentheses)

|  | Full Memory | No Memory | Fixed Size | Adaptive Size |
|---|---|---|---|---|
| **Scenario A:** | | | | |
| Error: | 20.36% (4.21%) | 7.30% (1.97%) | 7.96% (2.80%) | 5.32% (2.29%) |
| Recall: | 51.69% (8.37%) | 74.42% (4.61%) | 77.64% (6.07%) | 85.35% (4.93%) |
| Precision: | 64.67% (8.38%) | 91.29% (5.10%) | 87.73% (5.93%) | 91.61% (5.11%) |
| **Scenario B:** | | | | |
| Error: | 20.25% (3.56%) | 9.08% (1.57%) | 8.44% (2.00%) | 7.56% (1.89%) |
| Recall: | 49.35% (7.01%) | 67.22% (5.04%) | 73.85% (5.51%) | 76.70% (5.42%) |
| Precision: | 65.09% (6.80%) | 88.86% (3.67%) | 87.19% (4.18%) | 88.48% (3.89%) |
| **Scenario C:** | | | | |
| Error: | 7.74% (3.05%) | 8.97% (2.84%) | 10.17% (3.30%) | 7.07% (3.16%) |
| Recall: | 76.54% (6.26%) | 63.68% (5.27%) | 68.18% (7.05%) | 78.17% (6.34%) |
| Precision: | 83.15% (6.69%) | 87.67% (7.06%) | 79.00% (8.09%) | 87.38% (6.99%) |

A more detailed discussion of the results along with additional figures illustrating the performance of the different approaches in the three scenarios and the automatically chosen window sizes can be found in [68].

Other work on machine learning under time effects at the Artificial Intelligence Unit of the University of Dortmund included a study on the use of machine learning techniques for the analysis of time series data, in particular the analysis of telecommunication networks and the detection of overload situations as well as the load prediction [79]. Aside from changing load pattern scenarios in the data, a major challenge in this application domain was the large amount of time dependent data.

In [83], we provide a survey of representations, learning tasks, and data preprocessing techniques involved in machine learning from different types of time-dependent data for different types of learning approaches. [72] also contains a collection of several approaches to handling temporal data in machine learning.

## Summary and Conclusions

In this section, we proposed a new method for handling concept drift with support vector machines. The method directly implements the goal of discarding irrelevant data with the aim of minimizing generalization error. Exploiting the special properties of SVM, we adapted $\xi\alpha$-estimates to the window size selection problem. Unlike for the conventional heuristic approaches, this gives the new method a clear and simple theoretical motivation. Furthermore, the new method is easier to use in practical applications, since it involves less parameters than complicated heuristics. Experiments in an information fil-

tering domain show that the new algorithm achieves a low error rate and selects appropriate window sizes over very different concept drift scenarios.

An open question is how sensitive the algorithm is to the size of individual batches. Since in the current version of the algorithm, the batch size determines the estimation window, the variance of the window size is likely to increase with smaller batches. It might be necessary to select the estimation window size independent of the batch size. A shortcoming of most existing algorithms handling concept drift (an exception is [74]) is that they can detect concept drift only after labeled data is available. That is, after the learning algorithm starts making mistakes. While this appears unavoidable for concept drift with respect to $\Pr(y|\boldsymbol{x})$, it might be possible to detect concept drift in $\Pr(\boldsymbol{x})$ earlier by using transductive support vector machines. The approach to learning drifting concepts proposed here has been extended to make use of unlabeled examples using the principle of transduction (see Section 8.1.4 and [65, 67]).

### 8.1.4  Transduction: Learning from Labeled and Unlabeled Data

### Learning from Labeled and Unlabeled Data

The idea of transduction was introduced by Vapnik (see Chapter 10 in [129]). Often it is not necessary to learn a general classification rule for arbitrary test examples. Instead, the easier problem of only classifying a specific already known test set with as few errors as possible has to be solved. First experiments Thorsten Joachims conducted in a text classification domain showed that the minimization of the probability of an error on a specific test set leads to better results than learning a general classification rule [48, 50].

Obviously the unlabeled test examples provide additional information for a transductive learner not only using the labeled training examples, information about the distribution of the test data that the learner can leverage on. This section investigates when and why transduction works better than the standard induction approach. After reviewing transductive support vector machines and providing a positive and a negative case study, we discuss properties of a domain that enable transduction to be helpful.

### Transductive Support Vector Machines (TSVM)

Inductive support vector machines have already been described in Section 8.1.2 along with the primal and dual optimization problems 1 and 2 they solve. The setting of transductive inference was introduced by Vapnik (see Chapter 10 in [129]). The description here follows the presentation in [48]. For a learning task $P(\boldsymbol{x}, y) = P(y|\boldsymbol{x})P(\boldsymbol{x})$ the learner $\mathcal{L}$ is given a hypothesis space $H$ of functions $h : X \longrightarrow \{-1, 1\}$ and an i.i.d. sample $S_{train}$ of $n$ training examples

$$(\boldsymbol{x}_1, y_1), (\boldsymbol{x}_2, y_2), ..., (\boldsymbol{x}_n, y_n). \tag{8.17}$$

Each training example consists of a document vector $x \in X$ and a binary label $y \in \{-1, +1\}$. In contrast to the inductive setting, the learner is also given an i.i.d. sample $S_{test}$ of $k$ test examples

$$x_1^*, x_2^*, ..., x_k^* \tag{8.18}$$

from the same distribution. The transductive learner $\mathcal{L}$ aims to select a function $h_{\mathcal{L}} = \mathcal{L}(S_{train}, S_{test})$ from $H$ using $S_{train}$ and $S_{test}$ so that the expected number of erroneous predictions

$$R(\mathcal{L}) = \int \frac{1}{k} \sum_{i=1}^{k} \Theta(h_{\mathcal{L}}(x_i^*), y_i^*) dP(x_1, y_1) \cdots dP(x_k^*, y_k^*)$$

on the test examples is minimized. $\Theta(a, b)$ is zero if $a = b$, otherwise it is one. As shown in [130], for linearly separable problems this leads to the following optimization problem:

**Optimization Problem 3 (Transductive SVM (lin. sep. case))**

*Minimize over* $(y_1^*, ..., y_k^*, w, b)$:

$$\frac{1}{2}||w||^2$$

*subject to:* $\forall_{i=1}^n : y_i[w \cdot x_i + b] \geq 1$

$\forall_{j=1}^k : y_j^*[w \cdot x_j^* + b] \geq 1$

Solving this problem means finding a labeling $y_1^*, ..., y_k^*$ of the test data and a hyperplane $< w, b >$, so that this hyperplane separates both training and test data with maximum margin. To be able to handle non-separable data, we can introduce slack variables $\xi_i$ similarly to the way we do with inductive SVM.

**Optimization Problem 4 (Transductive SVM (non-sep. case))**

*Minimize over* $(y_1^*, ..., y_k^*, w, b, \xi_1, ..., \xi_n, \xi_1^*, ..., \xi_k^*)$:

$$\frac{1}{2}||w||^2 + C \sum_{i=0}^{n} \xi_i + C^* \sum_{j=0}^{k} \xi_j^*$$

*subject to:* $\forall_{i=1}^n : y_i[w \cdot x_i + b] \geq 1 - \xi_i$

$\forall_{j=1}^k : y_j^*[w \cdot x_j^* + b] \geq 1 - \xi_j^*$

$\forall_{i=1}^n : \xi_i \geq 0$

$\forall_{j=1}^k : \xi_j^* \geq 0$

$C$ and $C^*$ are parameters set by the user. They allow trading off margin size against misclassifying training examples or excluding test examples. Joachims developed an algorithm to solve this optimization problem efficiently [48, 50]. In this chapter, $SVM^{light}$ is also used for computing the solution of this optimization problem (see also Section 8.1.2).

## Using Transduction in Text Classification Tasks and for Learning Drifting Concepts from Labeled and Unlabeled Data

Empirical results show that unlabeled data can help to significantly improve the performance of text classifiers [48, 75, 84]. Thorsten Joachims showed that especially in case of few labeled examples and many unlabeled examples, the use of unlabeled data can improve the classification performance significantly [48]. The transductive use of unlabeled data also lets the performance drop more gracefully if the number of labeled examples is reduced. As pointed out in [48, 50], it is well-known in information retrieval that words in natural language occur in strong co-occurrence patterns [128]. While some words are likely to occur in one document, others are not. This type of information is independent of the document labels and can be exploited, if unlabeled data is used.[6] In addition to the TSVM described above, there are other semi-supervised methods for exploiting unlabeled data (see [48] or [65] for a short overview).

In Section 8.1.3 information filtering was named as a typical application, where the target concept in a classification task, in this case the user interest, may change over time, which makes the learning task more difficult because the system should adapt to such changes. Another difficulty in this example domain is the fact that users often give little feedback and expect a filtering system to achieve a good performance, even if only a few labeled training examples are provided.

This section proposes an extension of the method for recognizing and handling concept changes with support vector machines (Section 8.1.3 and [68]). The extended method uses unlabeled data to reduce the need for labeled data [65, 67]. Its basic idea is to first use the algorithm described in [68] to find the right window size on the labeled training data, $win_{labeled}$, using $\xi\alpha$-estimates for an inductive SVM. Then an almost identical algorithm is used to determine a good window size on the unlabeled data, $win_{unlabeled}$, on the same stream of documents using $\xi\alpha$-estimates for a transductive SVM to estimate the prediction error on the test set, leaving the window size $win_{labeled}$ unchanged.

Why are separate window sizes $win_{labeled}$ and $win_{unlabeled}$ maintained for labeled and unlabeled data respectively? The probability $P(y|x)$, which describes the user interest, i.e., the drifting concept, and which is captured by the labeled data, may change at a rate other than the probability $P(x)$, which describes the distribution of documents identically underlying both the labeled and unlabeled examples independent of their labels. Hence it is

---

[6] For example the words *pepper* and *salt* as well as the words *atom* and *physics* have a high co-occurrence, while e.g., *pepper* and *atom* only rarely co-occur. This is for example helpful, if cooking recipes have to be separated from texts about atom physics, and the training documents contain only some of these words, while the test documents contain some of the others along with the corresponding co-occurrences.

262    Daniel et al.

sensible to use separate windows to obtain the best information from both probability distributions.

The window adjustment algorithm for the labeled data using an inductive SVM and the training data only has already been described above in Section 8.1.3. The modified window adjustment algorithm for the unlabeled data [65] using a transductive SVM is explained below using the same notation as Section 8.1.3.

According to the transductive setting, the test set, i.e., the examples in the batch $t+1$ are used as unlabeled examples in the optimization for learning a TSVM. For text classification tasks with stable, i.e., non-drifting concepts, the performance improvement of a TSVM as compared to an inductive SVM is maximal for small sets of labeled training examples and large sets of unlabeled examples used in addition [48]. Therefore, here not only the relatively small test set is considered useful unlabeled data, but also the unlabeled data in the current time window of size $win_{labeled}$, i.e., in the currently used part of the training set, and all training examples outside this time window, which are all treated as unlabeled examples, are considered potentially useful.

The batch size is usually imposed by the application, e.g., by the number of news texts arriving per day, and hence it is the same for the this part of the algorithm and the part described in the previous section. While only the labeled examples were used by the algorithm described in the previous section, both labeled and unlabeled examples are used by the algorithm described in this section. Since the earlier only uses the labeled examples, its virtual batch size $m$ seems to be smaller than the real batch size $m'$ of the latter containing all examples.

The algorithm for finding the window for the unlabeled data (and the final hypothesis) can be summarized as follows:

- input: $S_{train}$ training sample containing $t$ batches of $m'$
  examples each and $S_{test}$ test sample
- for $h \in \{0, ..., t-1\}$
  - train TSVM on examples $z_{(t-h,1)}, ..., z_{(t,m')}$, considering all
    training examples outside the window of size $win_{labeled}$
    as unlabeled, and on the test examples $z_{(t+1,1)}, ..., z_{(t+1,m')}$
  - compute $\xi\alpha$-estimate on examples $z_{(t+1,1)}, ..., z_{(t+1,m')}$
- output: window size which minimizes $\xi\alpha$-estimate $(win_{unlabeled})$

Klinkenberg [67] describes encouraging experimental results with this approach.

## Comparison of Inductive and Transductive Learning in a Semiconductor Rare Fault Detection Domain

In order to evaluate the use of transduction in a completely different application domain, inductive and transductive support vector machines (SVM

and TSVM, respectively) were applied to a classification task from electrical engineering, namely a semiconductor rare fault detection task. This task has previously been used as a benchmark for computational intelligence methods within the DFG Collaborative Research Center on Computational Intelligence (SFB 531) (see especially pages 230–231 in [36]). Section 8.3 describes further applications of computational intelligence methods to tasks in electrical engineering.

The semiconductor rare fault detection data consist of the bond resistance values measured in stress tests with 34 high performance transistors until these transistors failed. Given the bond resistance values of such a transistor of the last four time steps, the task is to predict whether or not the transistor will fail within the next seven time steps. The 34 time series of measurements are split into 29 series for training and 5 time series for testing, which result in 1157 training and 204 test examples when a time window of size four time steps is used on these time series as described above.

In order to evaluate whether transduction can help to improve the performance of a support vector machine in this domain, we trained SVM and TSVM with linear, RBF, and polynomial kernel on the training data and compared their performance on the test set. To estimate the potential for transduction on this data set, we allowed the TSVM to use the complete test set (without its labels of course) in the training phase.[7] Since the goal of these experiments was not to find an optimal parameter setting for SVM on this data set but to estimate the potential of using transduction to improve the performance of SVM in this domain, no extensive testing of a large number of parameter combinations was performed. However, only a few arbitrarily picked combinations of the parameters $C \in \{1, 100, 1000, 100000\}$ (capacity constant) and $\epsilon \in \{0.001, 0.1, 1\}$ (tolerable error) were tested. In case of the RBF kernel, values in $\{0.01, 0.1, 1, 10, 100\}$ were tried as factor $g$ in the exponent of the exponential radial basis function. In case of the polynomial kernel, the polynom degrees 1 to 7 were tested. The Table 8.3 shows the best results obtained by SVM and TSVM with these kernels using these few parameter variations. The best results for the linear kernel were obtained with $C = 1000$ and $\epsilon = 0.1$. The best results for the RBF kernel were achieved with $C = 1000$, $\epsilon = 1$, and $g = 1$. The best polynomial kernel used the same parameter values for $C$ and $\epsilon$ as the RBF kernel and polynom degree 5. The experiments were performed using the SVM implementation $SVM^{light}$ [47, 50][8] developed at the Artificial Intelligence Unit of University

---

[7] In a more realistic setting, TSVM should only be able to see some part of the test set during training, i.e., previous examples in an incremental setting, or only the next measurement at each time step instead of all future measurement of each time series. Hence, if transduction leads to a significant improvement in this simplified setting, this indicates a potential for the use of transduction. However, the amount of the improvement should be expected to be smaller in a more realistic setting.

[8] $SVM^{light}$ is available at http://www-ai.informatik.uni-dortmund.de/svm_light

**Table 8.3.** Error of several previously used computational intelligence methods (left) and of inductive and transductive support vector machines, SVM and TSVM, respectively, with linear and RBF kernel, respectively (right), on the semiconductor rare fault detection classification task

| Error | Feedforward RBF Neural Network | Recurrent RBF Neural Network | Genetic Program- ming | Fuzzy- ROSA Method | SVM, linear kernel | SVM, RBF kernel | SVM, poly- nom kernel |
|---|---|---|---|---|---|---|---|
| Induction: | 12.9 % | 11.6 % | 10-15 % | 8.00 % | 10.80 % | 9.20 % | 9.20 % |
| Transduction: | — | — | — | — | 9.60 % | 10.00 % | 10.00 % |

of Dortmund. While for the linear kernel, the SVM seems to benefit from transduction, it does not for the other two kernels, i.e., RBF and polynomial kernel.

Table 8.3 also provides the best results previously achieved by other computational intelligence methods, namely feed-forward RBF neural networks, recurrent RBF neural network, genetic programming (GP),[9] and Fuzzy-ROSA Method (FRM).[10] Obviously the performance of both, SVM as well as TSVM, compares well to these previous results. The differences in performance cannot be considered significant, since the data set is used, exactly as for the previously tested methods, on a fixed single split of the data set into training and test data only. In contrast to the positive results for transduction in the text classification and information filtering domain, there seems to be not much potential for transduction on this failure detection data set.

### Conclusions: When Does Transduction Work Well?

While transduction leads to performance improvements in text classification tasks, where unlabeled data can be used to lower error rates or to reduce the need for labeled data, it does not seem to be very helpful in the second case study of semiconductor failure detection. So the questions arises, what the difference between these two domains is, and when transduction can be expected to be helpful, i.e., which properties a domain should have to benefit from transduction.

There are two intuitive reasons why transduction works well in text classification domains. First, as explained above, texts of similar or identical topics show strong word co-occurrence patterns in a sparse feature space. These co-occurrence patterns often indicate a topic membership before the

---

[9] Results obtained in the DFG Collaborative Research Center on Computational Intelligence (SFB 531) in the period 1997–1999 (see [36] especially pages 230–231).

[10] Result obtained by Daniel Schauten in the DFG Collaborative Research Center on Computational Intelligence (SFB 531) in 2001.

topic label of a document is known. Hence unlabeled examples can be used to better estimate the boundaries of a topic than is possible from fewer labeled examples alone. Second, as shown in [48, 67], a learner particularly benefits from transduction if many unlabeled examples and only few labeled examples are available. In most text classification domains, there are huge numbers of unlabeled documents available, which usually by far exceed the number of available labeled documents. So in text classification tasks, transduction can leverage from these two aspects.

The second domain investigated here, the rare fault detection task, exhibits neither of these two properties. First, the number of test examples is rather small compared to the number of training examples. Second, the feature space is not sparse, but very dense, and it is not clear whether there are any strong co-occurrence patterns within the two classes "will fail (in seven time steps)" and "will not fail." Summarizing this discussion, good prerequisites for transduction to work well seem to be strong co-occurrence patterns between examples of the same class and a large number of test examples compared to the number of training examples.

### 8.1.5    Learning with Prior Knowledge

In machine learning one can distinguish two kinds of knowledge, namely background knowledge and prior knowledge. In the case of learning with prior knowledge, the underlying learning task can be eased by introducing additional, domain-specific knowledge. This knowledge restricts the hypothesis space and thus allows faster and/or better learning. Background knowledge extends the examples and thus the search space and therefore makes the learning task more difficult.

One classical setting for the learning task learning with background knowledge is the following: The learned hypothesis $H$ has to be consistent not only with the given example set $E$, but also with the given background knowledge $T$, and any positive but no negative example should be deducible from the union of hypothesis $h$ and background knowledge $T$. This inclusion of background knowledge complicates the given learning task [92], and restrictions of the representation formalism are necessary to make learning possible in polynomial time [62]. The inclusion of background knowledge makes learning more difficult because on the one hand, more complex hypotheses can be built. On the other hand, the integration of new knowledge becomes harder the more knowledge is given, since the number of possible inconsistencies increases.

In contrast to this kind of knowledge that enlarges the hypothesis space, the use of prior knowledge permits the scaling down of the hypothesis space and/or the control of the search in the search space. By restricting the hypothesis space adequately, one can accelerate learning in terms of running time of the algorithm and improve the learning performance, e.g., in terms of classification/regression error on the test examples.

In the following section we present an approach to SVM, which makes use of prior knowledge about the importance of particular examples in an example set and adapts the optimization problem solved by SVM appropriately.

## Using Additional User Knowledge in Support Vector Machines

Many applications of machine learning involve the learning of classifiers. Given a set of labeled training examples, the task is to learn a classifier for predicting the labels of previously unseen examples. By providing the labels of the training examples, the users already specify a lot of their knowledge about the classification problem at hand. In some cases, however, the users may not be satisfied with the result provided by the learning method. Hence the users may want to specificy additional knowledge about the problem or constraints on the desired solution and they may want the learner to provide a classifier that better fits their needs.

The goal of this research is to allow the users to specify additional knowledge about the classification problem and to incorporate this knowledge into the learning process. In [66] support vector machines [130] (Section 8.1.2) were chosen as learning methods for classifiers, and weights for individual training examples provided by the user are considered in the training of SVM.

User knowledge about a domain can often be expressed by weights for the examples provided to the learner. Consider for example a medical domain, where it is essential to classify critical cases correctly, e.g., because a person's life may depend on it, while incorrect classifications of non-critical cases may be more acceptable. Similarly, one can use weights to enforce the proper classification of typical cases of a disease and to put less emphasis on non-typical cases, i.e., cases where even the expert is not sure or where the appearance of a disease is not clearly recognizable. Weights could also be used to express the confidence one has in the correctness of the classification of the examples, e.g., to express the reliability a certain example source.

Klinkenberg [66] extends the formulation of the primal SVM optimization problem (see optimization problem 1 in Section 8.1.2) for weighted examples and derives the corresponding dual SVM optimization problem. For each training example $x_i \in \mathbb{R}^N$, the user provides a label $y_i \in \{-1, +1\}$, indicating its classification and a weight $\beta_i \in [0, 1]$ (or alternatively $\beta_i \in \mathbb{R}_0^+$) indicating the importance of the correctness of the prediction of the class label of this example $(i = 1, ..., n)$.

## Optimization Problem 5 (SVM (primal with weights))

$$\textit{Minimize: } V(\boldsymbol{w}, b, \boldsymbol{\xi}) = \frac{1}{2}\boldsymbol{w} \cdot \boldsymbol{w} + C \cdot \sum_{i=1}^{n} \beta_i \cdot \xi_i \qquad (8.19)$$

$$\textit{subject to: } \forall_{i=1}^{n} : y_i[\boldsymbol{w} \cdot \boldsymbol{x}_i + b] \geq 1 - \xi_i \qquad (8.20)$$

$$\forall_{i=1}^{n} : \xi_i \geq 0 \qquad (8.21)$$

The derivation of the corresponding dual optimization problem is provided in [66].

### 8.1.6 Feature Selection and Generation

The nonstandard learning tasks that are considered in this section are feature selection and especially feature generation. Feature selection simply means to eliminate irrelevant and redundant features[11] from the original feature set to improve the performance of a consecutive learning method. Several definitions of feature relevance are stated in [11]. An extensive overview of different feature selection methods is given in [28].

Most machine learning methods for classification and regression are designed to find the most suitable, i.e., relevant, features in a given feature set. Thus, feature selection is already implicitly done. Nevertheless we often need an additional preprocessing step prior to the application of the actual learning method. One reason is that the prediction error of many learning algorithms, including decision tree learners such as C4.5 [94], increases when we add irrelevant and redundant features respectively [51]. Another problem that particularly affects the computation time is the lacking scalability of many learning methods since their applicability significantly decreases on large-scale data sets [91]. In the case of learning with support vector machines, feature selection is not implicitly done. The experimental results described later in this section show that the elimination of irrelevant features increases the learning performance of SVM.

Feature generation can be seen as the process of creating features by applying feature constructors, taking the original features as input and producing a modified feature set containing additional constructed features. Conceptionally, one can consider feature generation as the modification of a given search space, which is created by the enlarged feature set. The main motivation for the use of feature generation is to create hypotheses that were not representable in the original representation language [78, 133].

### Feature Generation Using Type-Restricted Generators

We propose a general framework for the generation of features [95], subdividing the generation process into the following main steps: Given a set of features $X = x_1, .., x_n$ and feature generators $F = f_1, .. f_k$, we first choose a particular feature generator $f_j \in F$ for the generation process. Then, by checking the types of all features in $X$, the compatible feature subsets $X_c = X_{c_1}, .., X_{c_l} \subseteq X$ are determined with regard to the type restrictions

---

[11] Note: In Chapter 3 the term variable is used instead of the term feature because of the slightly different terminologies in the fields of electrical engineering and control there and machine learning here.

of the generator at hand.[12] Finally, the feature generator $f_j$ is applied to
the set of compatible features (or a subset thereof) and the resulting fea-
tures $w_i = f_j(X_{c_i})$ are added to the original feature set $X$. Any $x_i \in X_{c_i}$ is
not limited to original features but can also contain compound features that
have already been created by a generator. In contrast to the feature gen-
eration concept presented in Section 4.5.4, this allows a recursive and thus
arbitrarily complex feature generation.[13]

An obvious advantage of using type-restricted constructors is the reduc-
tion of the set of constructible features to a well-formed subset and thus the
restriction of the search space extension to useful subspaces. Feature construc-
tors may include general functions, such as basic mathematical operations,
as well as domain-specific generators. The generator concept presented here
is used as part of the following approach that incorporates feature generation
and selection in a common framework.

**Combined Approaches to Feature Generation and Selection**

Feature generation and selection are not two totally independent issues. They
can be viewed as two sides of the representation problem, i.e., the problem
of finding an adequate representation language for the learning task at hand.
In cases where this language is not sufficient to describe the problem, feature
generation helps to enrich the language by constructing compound features.
In cases where the representation language contains more features than nec-
essary, subset selection helps to simplify the language. Since some of the con-
structed features are redundant and irrelevant accordingly, feature selection
can remove these features (two exemplary approaches that combine feature
selection and generation are described in [10, 76]).

In this section we introduce a wrapper-based approach, using a modi-
fied genetic algorithm that incrementally selects and generates new features
as well as an induction algorithm as learning and evaluation method [95].
The general idea is to combine the positive search properties of conventional
genetic algorithms (GA) with the option to adapt the search space incre-
mentally. Genetic algorithms have proved to work well on feature selection
problems, where the search space produced by the initial feature set already
contains the searched hypothesis [6, 93, 126, 134]. In cases where this premise
is not fulfilled, we have to create new features to adequately extend the search
space. This can be done by applying the given feature generators to feature
subsets with compatible feature types.

The combined approach is an adapted version of the wrapper-approach
presented in [70]. According to our approach, the feature selection and gen-
eration by means of a modified genetic algorithm exists as a wrapper around

---

[12] The concept of meta-knowledge in terms of feature types is described in more
detail in Section 8.1.7.
[13] In Sections 4.5.3 and 4.5.4, features are also referred to as input variables.

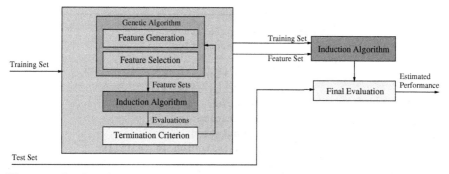

**Fig. 8.2.** Combined wrapper-based feature selection and generation approach

the induction algorithm.[14] The genetic algorithm conducts the search for a good feature subset using the induction algorithm for the evaluation of the current feature subsets. The induction algorithm is run on a data set, which is usually partitioned into internal training and hold-out sets, with different sets of features removed from and added to the data accordingly. The process of generating feature sets, using the modified genetic algorithm, and evaluating these sets is repeated until a given termination criterion is fulfilled. The resulting feature set is chosen as the final set on which to run the induction algorithm. The final evaluation of the resulting classifier is done using an independent test set that was not used during the search (see Figure 8.2).

Conceptionally, the key idea in using the incremental feature generator in the context of a genetic algorithm is that any feature can be generated with a probability $P > 0$ in a finite number of iterations, i.e., generations, given a particular generator and the original features. Thus, more formally, given a feature set $\boldsymbol{f_t}$ at time-point $t$, a generator $G$ with $G(\boldsymbol{f_t}) = \boldsymbol{f_{t+1}}$ and $f$ being the searched feature, the following proposition is assumed:

$$f \in \lim_{t \to \infty} G(\boldsymbol{f_t})$$

The basis for our combined approach to feature generation and selection is a canonical genetic algorithm, as described in [35, 44, 80]. It consists of an $n$-tuple of binary strings $b_i$ of length $l$, where the bits of each string are considered to be genes of an individual chromosome and where the $n$-tuple is said to be a population. Following the terminology of biologic evolution the operations performed on the population are called mutation, crossover, and selection. Each individual represents a feasible solution of a given problem and its objective function value $\phi(b_i)$ is said to be its fitness, which has to be maximized.

Now, in our approach, a chromosome is interpreted as a binary representation of an underlying feature set, i.e., any gene is associated with a particular

---

[14] The induction algorithm we chose was a support vector machine, although any other induction algorithm could have been used.

```
create initial population
evaluate initial population
repeat
      perform selection
      perform modified crossover
      perform mutation
      perform generation
      evaluate population
until termination criterion is fulfilled
```

**Fig. 8.3.** The main algorithm of the combined approach

feature and a gene value 1 means that this feature is selected for the current feature set, whereas 0 means that it is deselected. The task is to search for a good representation by selecting and generating features. We therefore partially adapted the standard genetic algorithm operators. Figure 8.3 shows the corresponding operator sequence. Here, crossover recombines different feature sets and mutation selects and accordingly deselects single features on a particular chromosome, i.e., feature set. In addition to the set of standard GA operators for feature selection, we applied the feature generator presented.[15] To ensure that the best individual of a generation remains in the population, an elitist selection strategy was used. The whole evolutionary cycle of selection, recombination, and evaluation proceeds until a given termination criterion is fulfilled. This criterion may be a maximum number of generations, a fixed time limit, the achievement of a required fitness value, or convergence of the fitness values.

To evaluate the framework presented, we compared it to an approach using a genetic algorithm for feature selection. We applied both approaches to an artificial data set whose searched target function $f$ was given by

$$f(\boldsymbol{a}) = a_1 + \ldots + a_5,$$

i.e., the sum of five given features containing three original features $a_1$ to $a_3$ and two constructed features $a_4 := a_1 * a_2$ and $a_5 := a_2/a_3$. The performance evaluation was done using two nested cross validations. The inner cross validation was used to find a good feature set, while the outer cross validation was used to evaluate the performance of this feature set on an extra validation set. The wrapper parameters were set to 200 for the number of generations, 30 for the population size, 0.5 for the crossover probability, 0.2 for the mutation, i.e., feature selection probability, and 0.5 for the feature generation probability. The induction algorithm we chose was a regression SVM [116, 120] using a dot kernel.

---

[15] The application of the feature generator leads to a variable length chromosome, i.e., feature set.

**Table 8.4.** Comparing the absolute prediction error of the genetic algorithm for feature selection with the combined approach on an artificial data set

| Given feature set | Genetic algorithm | Combined approach |
|---|---|---|
| Original features | 0.0587 | 0.0584 |
| Original features + one random feature | 0.0546 | 0.0468 |
| Original features + two random features | 0.0590 | 0.0547 |
| Original features + three random features | 0.0591 | 0.0583 |
| Original features - one generated feature | 1.4569 | 1.1114 |
| Original features - two generated features | 1.4669 | 0.6986 |

Table 8.4 shows that the combined approach clearly outperforms the genetic algorithm in the case of missing relevant features, especially where no generated feature is given. In cases of different amounts of noise, i.e., irrelevant features and even in the optimal case where the original feature set was given, this approach turned out to be superior to the genetic algorithm. Furthermore, in the case where no generated features from the original feature set were given in advance, the new approach exactly reconstructed these missing features.

An open question in this context is how the feature generation process can be controlled to improve the prediction performance and accelerate the search process. Therefore, criteria are needed to guide, for example, the choice of a specific generator and feature subset in a particular situation. Furthermore, a comparison of our framework with different approaches in symbolic regression, especially those using GP concepts, is planned.[16]

## Experimental Evaluations

In this section, a comparison of the appropriateness of different chains of feature generation and selection methods for an application problem from chromatography is presented. Chromatography is used in chemical industry to separate temperature sensitive substances (see Section 8.2). A mixture of components is injected for a certain amount of time into a column filled with porous particles. Due to the different adsorption strengths of the substances on the porous particles, the components have various velocities in the column and reach its end at different times, where they can be separated.[17]

The learning task considered here is to predict the four characteristic coefficients of a two-component mixture, given the corresponding chromatogram

---

[16] An approach using tree-based GP for constructing descriptive models for physical relations is presented in Section 7.11

[17] For a more detailed description of this application domain see Section 8.2.2.

**Table 8.5.** Average absolute error for the target values Henry and Langmuir on a two-substance mixture, tested on different operator chains (average relative error in parentheses)

| Applied operator chain | Henry constant substance 1 | Langmuir constant substance 1 | Henry constant substance 2 | Langmuir constant substance 2 |
|---|---|---|---|---|
| Original data | 2.2045 (0.6285) | 21.9081 (0.6497) | 2.4719 (0.7046) | 22.3291 (0.6883) |
| Feature generation | 0.5416 (0.1551) | 17.6388 (0.4695) | 0.6146 (0.1671) | 17.7408 (0.4645) |
| Feature gen. & FS | 0.1542 (0.0336) | 11.1394 (0.3271) | 0.2291 (0.0349) | 19.0193 (0.5407) |
| Feature gen. & BE | 0.0872 (0.0215) | 0.1079 (0.0028) | 0.0923 (0.0238) | 0.1045 (0.0028) |
| Feature gen. & GA | 0.0790 (0.0193) | 0.1105 (0.0028) | 0.0918 (0.0232) | 0.1059 (0.0029) |
| Combined approach | 0.0730 (0.0187) | 0.1054 (0.0027) | 0.0774 (0.0199) | 0.1045 (0.0028) |

time series. Two constants, namely the Henry and Langmuir coefficients respectively, determine the chromatogram of a substance. The data set contains 200 examples with 5000 attributes each with measurements at 2500 equidistant time points of the chromatogram time series for each of the two components. Since the learning task is a regression problem, the learning method used throughout the following experiments is a regression SVM [120, 116]. For the support vector machine we chose a radial basis kernel with $\gamma = 1$, $C = 1000$, and $\epsilon = 0.1$.[18] The learning performance is evaluated using absolute and relative error, comparing the predicted and the real values of the Henry and Langmuir constants.[19] Based on the structure of the overall learning task, we systematically compare the performance of different learning chains, comprising feature generation and selection. Table 8.5 shows the experimental results.

In the first experiment we simply used the original (time series) features to learn and evaluate an SVM model without any preprocessing steps. The corresponding learning chain (comprising learner, model applier, and evaluator) was enclosed by a four-fold cross validation. Due to noise in the simulated

[18] These parameter values have been determined in preliminary tests.
[19] In Section 8.2 the target values of the regression task, i.e., the Henry and Langmuir constants, are also referred to as *parameters*, and the task is referred to as parameter estimation.

measurements of the original features, this operator chain only yielded a poor prediction performance.

The second chain additionally contained a preprocessing operator that generated numeric characteristics from the original time series data. These new features include the position (time point) and the value (concentration of the substance of the column output) of the maximum of the chromatogram and the positions and values of the two inflection points left and right of the maximum. Since the sensor readings may be noisy and perhaps slightly shifted along the time axis between different measurements, the individual feature in the time series of an example, i.e., the concentration measured at one particular point of time, is not very reliable. Accordingly, the construction of these robust features significantly improved the learning result.

Since it was not obvious which of the new and which of the original features were really helpful in solving the learning task, a feature selection step was performed subsequent to the feature generation step to obtain a feature set well-suited to the given learning task. Different feature selection wrappers, namely forward selection (FS), backward elimination (BE) [1], and a genetic algorithm were employed reducing the set of features and increasing the learning performance in terms of the prediction error. FS starts with an empty feature set and iteratively adds the single feature whose addition to the feature set improves the performance of the wrapped learning method the most, until no further improvement can be achieved by adding single features. BE starts with the complete set of all features and iteratively removes the single feature whose removal improves the performance of the wrapped learning method the most, until further removals lead to a decrease in performance. In the case of the genetic algorithm for feature selection, we used 50 generations, a population size of 30 individuals, a mutation probability of 0.05, a crossover probability of 0.5, fitness proportional selection, and one-point crossover.

Finally, we tested the combined approach on the given learning problem, using the same parameter settings as in the case of the genetic algorithm for feature selection. In this setting, feature generation and selection were not used as subsequent preprocessing steps but intertwined in a feature wrapper approach. This approach yielded an improved prediction performance compared to the feature selection wrapper using a genetic algorithm.

In this section we showed that, in cases where the original feature set is inadequate regarding the given learning task, one can significantly improve the learning performance by adding relevant features by means of feature generation and by removing irrelevant features by applying feature selection methods. Furthermore, we presented a framework that connects feature generation and selection in a combined approach using a modified genetic algorithm [95]. This approach demonstrated its usefulness in cases where the given set of features is not sufficient to find a good hypothesis in the hypoth-

esis space of the employed learner, but where such a sufficient feature set can be constructed from the given features.

### 8.1.7 Learning Environment YALE

As we have seen in the last sections, data often has to be preprocessed to be usable by a chosen machine learning method and accordingly, to achieve an acceptable level of performance in prediction. The central problem in this context is the representation of the examples by a good set of features, i.e., a set of features that allows the learning method to find a candidate hypothesis solving the learning task at hand within its hypothesis search space. Hence, finding a suitable set of features may be far more important for the overall learning success than the choice of a particular learning method. Therefore, it is often necessary to use complex chains, combining different preprocessing and learning steps, rather than using a particular single learning scheme.

This section proposes yet another learning environment (YALE) [96],[20] which allows the easy specification and execution of such learning chains for preprocessing, especially comprising feature generation and selection as well as multistrategy learning. This modular, noncommercial environment supports nested operator chains and the exchange of individual operators and thereby the systematic evaluation and comparison of different operators and operator chains for the same task or subtask.

### Existing Environments

There already exist several machine learning and data mining environments that provide a number of methods from machine learning, statistics, and pattern recognition. Two of the most popular noncommercial environments are WEKA[21] (Waikato Environment for Knowledge Analysis) [132], developed at University of Waikato, NZ, and MLC++[22] [71], first developed at Stanford University and then extended by Silicon Graphics, Inc. (SGI).

WEKA is a collection of machine learning algorithms implemented in Java. It supports a large number of learning methods for classification and regression. WEKA has some preprocessing algorithms for the manipulation of features as well as three basic feature selection schemes, namely the feature correlation based approach [41], a wrapper approach [70], and a filter approach [63]. Additionally, WEKA provides meta classifiers like bagging [13] and boosting [119].

MLC++ is a library of C++ classes for supervised machine learning. It provides a number of learning schemes similar to those used in WEKA.

---

[20] http://yale.cs.uni-dortmund.de/
[21] http://www.cs.waikato.ac.nz/ml/weka/
[22] http://www.sgi.com/tech/mlc/

Additionally, wrappers around these basic inducers, such as a discretization filter, a bagging wrapper, and a feature selection wrapper are provided.

Unfortunately, none of these learning and data mining environments is suited to handle the considered nonstandard learning tasks, because both of them neither support the composition and analysis of complex operator chains consisting of different nested preprocessing, learning, and evaluation steps nor sophisticated feature generators for the introduction of new features. MLC++ supports operator chains in a rather restrictive way. One can only build wrappers around basic inducers (learning schemes) but not around nested operator chains. The same applies to WEKA, where nesting can only be realized by numerous command line calls by creating copies of (subsets of) the data set, and manual data file management or by writing your own experiment and data management program. An additional shortcoming of WEKA is that it lacks scalability. It expects the example set to fit completely into main memory, which is not feasible for many data mining tasks. Furthermore, it is very slow on large data sets.

**Operators and Operator Chains**

Real-world learning tasks are often solved by a sequence or combination of several data preprocessing and machine learning methods. In YALE, each of these methods is regarded as an operator. A sequence of such operators is called an operator chain. An operator chain again is an operator, both in the sense of a definition as well as in the object-oriented programming sense. Operators may enclose other operators or operator chains and are then often referred to as wrappers. Typical examples of wrappers are cross validation and feature selection wrappers [70]. One central aspect is, that by enclosing other operators or operator chains, operators are arbitrarily nestable, so that even complex experimental setups can be built. For example a nested cross validation could be used to first optimize some parameters of a data preprocessing and learning chain (inner validation) and to then evaluate the performance of the whole experimental setup (outer validation).

Operators require certain objects to be present in their input and guarantee others to be in their output. During its execution, an operator may modify, remove, or add objects before passing the objects to the next operator in the operator chain. Objects that are present in the input of an operator without being required are usually ignored and passed on to the next operator. YALE verifies that each operator receives its required inputs before executing an operator chain. Among the objects passed between operators in an operator chain are attribute sets, example sets, classification and regression models, example sets with additional labels, and performance evaluation results.

Operators are easily exchangeable by other operators in order to support the comparison of different operators and operator chains for the same subtasks. The only premise is that the subsequent operators in an operator

chain, or more generally, all operators that share a common interface, have conformable input and output types.

## Data Handling

The data is described in an attribute description file, and the experiment is specified in a configuration file containing a description of the employed operator chain. Two additional types of files are needed to execute an experiment. One file type contains the examples (attribute values file) and the other contains the labels of the examples for a particular classification or regression task (label file). YALE can process data sets that can be described in a single table, i.e., in an attribute-value vector format, in which each example is described by an attribute-value vector of equal fixed length.

For almost all operators in YALE it is transparent whether the examples are read in from a text file, from main memory or (as planned as a future option) from a database, and whether only one or all examples are kept in main memory at a time. The environment only uses an internal structure to iterate over the instances in an example set and hence does not need to distinguish between the different possible data sources and ways of managing the data. The source and way of handling the data only depends on the chosen example source operator and its parameterization.

## Meta-Knowledge in YALE

While the use of meta-knowledge is standard in database and software engineering systems, many learning systems do not consider meta-knowledge. YALE supports the optional use of meta-knowledge, especially for the task of feature generation.

Attributes in YALE are described by two data structures, the `valuetype` and the `blocktype`. While the `valuetype` of an attribute (or label) specifies the data type of the individual attribute, e.g., `numerical` or `nominal`, the `blocktype` contains some meta-data about the attribute, e.g., if it is just an individual attribute or part of a more complex data structure, such as an interval boundary or a time series. For each data structure, there exists an ontology describing the hierarchical is-more-general-than-/is-superset-of-relation between the different types.[23] The information about the attribute types of an example is useful for feature generators, which can check their applicability on given attributes by verifying their attribute types. A generator extracting the maximum value of a time series can, for example, restrict its application to all value series in the attribute value vector of an example and generate the maximum for each such series separately while ignoring, for example, all individual attributes and interval attributes (see Section 8.1.6).

---

[23] For a more detailed description of the ontologies see [96].

**Summary**

The proposed machine learning environment YALE [96] allows the easy specification and automatic execution of learning chains. Operators in a chain are exchangeable by alternative operators for the same subtask. YALE thereby supports systematic comparisons and evaluations of individual methods as well as compound operator chains. The constructible operator chains are nestable and therefore, complex tasks and performance evaluations like e.g., the feature selection and generation wrapper presented in Section 8.1.6 can be carried out. To enhance the reusability and applicability to new tasks, the machine learning environment is easily expandable by additional operators. The scalability and applicability of the system is achieved by a sophisticated data handling concept. Data can be read from files, main memory, or a database, whichever seems to be most appropriate for the current task, without making changes in the learning operators when varying the data source or switching between keeping all or just one example in main memory at a time. The latter may be the preferable approach in cases of very large data sets. Furthermore, YALE allows the usage of additional meta-data, which is especially helpful in the context of feature generation tasks.

The proposed environment [96] can be used to carry out elaborate studies of different nonstandard learning tasks. The usability of this environment has been successfully tested on several feature generation and selection tasks (see Section 8.1.6).

**Outlook**

Since YALE can easily be extended by additional preprocessing and learning methods, e.g., from the field of evolutionary computation, and since nested validation chains can easily be formulated, YALE provides a good platform for a structured comparison of different learning methods in the collaborative research center. For that purpose YALE is made accessible to all members of the collaborative research center.

**8.1.8   Conclusion and Perspective**

The assumptions that are implicitly made when using standard learning tasks, such as classification or regression, often do not hold in practical applications. The consequence of this inappropriateness is that the learning performance decreases and accordingly, particular hypotheses cannot be learned. In this chapter, several nonstandard learning tasks, including learning drifting concepts, transduction, learning with prior knowledge, and feature generation and selection handling different kinds of inappropriateness have been studied.

For many real-world classification tasks, the concept to be learned may change over time and hence does not remain static as assumed in the standard

278 Daniel et al.

learning task. Existing methods usually can only handle static concepts or require extensive, often domain-specific parameterization, to be adaptable to changing concepts and to achieve a satisfiable performance. In Section 8.1.3 we described and evaluated an effective and efficient method to learn such drifting concepts without extensive parameterization [68].

In other classification tasks, unlabeled data is available and can be used transductively to improve the performance of the classifier or to reduce the number of labeled examples needed to achieve an acceptable performance (Section 8.1.4 and [65, 67]). We showed how unlabeled examples can be used to support the learning of drifting concepts and identified domain properties enabling transduction to lead to improved performance results. When a domain meets these properties, transduction can significantly reduce the need for labeled data and/or lead to lower prediction errors than a comparable inductive approach.

Sometimes the user wants to influence the learning result and provides additional knowledge about the domain. Section 8.1.5 showed a way how this may be done by extending SVM to use weighted examples [66].

In Section 8.1.6 we stated that, in cases where the original feature set is inadequate regarding the given learning task, one can significantly improve the learning performance by adding relevant features by means of feature generation and removing irrelevant features by applying feature selection methods. Furthermore, a framework was presented that applies feature generation and selection in a combined approach using a modified genetic algorithm as a feature wrapper [95]. This approach was shown to be useful in cases where the set of necessary features is not given in advance.

Section 8.1.7 proposed YALE [96], a learning environment supporting the study of nonstandard learning tasks, especially the tasks mentioned above. YALE allows the execution and evaluation of complex nested chains of data preprocessing and machine learning operators.

In addition to the nonstandard learning tasks presented here, we intend to analyze the task of repeated learning of similar tasks. The main idea of this task is to transfer knowledge between learning tasks in order to ease learning in similar tasks, i.e., to accelerate the learning process or to improve the learning performance. The transfered knowledge may, for example, include feature sets useful for particular learning tasks, which may be useful for new similar tasks, but which may also need some adaptation to the new tasks.

## 8.2 Parameter Estimation for Chromatographic Processes

### 8.2.1 Motivation

Life Science Products are of increasing importance for the chemical industry. Chromatographic separations are normally applied to high-value products

**Fig. 8.4.** Principles of batch-chromatography

with separation factors near unity (components with similar distillation characteristics) or when thermally instable substances are involved. The separation costs are often very high and frequently dominate the overall production costs. Thus, it is important to operate these separation processes at their economic optimum. In order to apply mathematical optimization, suitable models of the processes are needed. Therefore, it is essential to determine the structure and the corresponding parameters of those models. Mathematical parameter estimation and neural networks have been used for this purpose.

### 8.2.2   Principles of Batch Chromatography

Chromatography is a thermal separation process, which is used to extract one or more target components with high purity from a mixture. A fluid mixture flows through a column filled with porous particles (see Figure 8.4). The components exhibit different affinities to the porous surface of the particles. This leads to different velocities of the molecules in the column. The resulting differences of the residence time give rise to the separation mechanism of chromatography [117]. This separation principle works for low temperatures and therefore, this process is ideal for temperature-sensitive molecules, such as pharmaceutics and bio-molecules.

In batch chromatography a portion of the feed mixture of two or more components is injected into the column. The injection time $t_{inj}$, the feed concentration $C_{in}$, and the volume flow $Q$ can be chosen to determine the productivity of the process. At the column outlet, the component with the lowest adsorption affinity leaves the column first, the component with the strongest leaves last. The different components can be collected in different vessels with a high purity. The injection of the feed is repeated after a cycle time $t_{cyc}$, which is also a free operating parameter. The equilibrium of the concentration of the components in the solvent and the concentration on the surface of the particles depends on the molecules, the particle material, the solvent, and the temperature. For a constant temperature it can be described by an adsorption isotherm equation:

$$q_i = f(c_1, ..., c_n), \tag{8.22}$$

where $q_i$ denotes the concentration of the component $i$ on the surface of the particles, and $c_1, ..., c_n$ are the concentrations of the components in the fluid phase.

## Physical Modeling of the Process

The most general modeling approach for chromatographic columns, which takes into consideration all important effects, is the general rate model [39]. The non-idealities of chromatographic processes are mainly caused by the effects of axial dispersion and mass transfer resistance. In most cases the assumption of a linear driving force approach describing the effects of mass transfer resistance holds. The solid phase consists of porous, uniform, and spherical particles. Local equilibrium is established between the pores. The resulting partial differential equations for the fluid phase (8.23) and solid phase (8.24) are:

$$\frac{\partial c_i}{\partial t} = D_{ax} \frac{\partial^2 c_i}{\partial x^2} - u \frac{\partial c_i}{\partial x} - \frac{3(1-\varepsilon)k_{l,i}}{\varepsilon r_p} (c_i - c_{pi}(r_p)) \tag{8.23}$$

$$(1 - \varepsilon_p) \frac{\partial q_i}{\partial t} + \varepsilon_p \frac{\partial c_{pi}}{\partial t} = \varepsilon_p D_{p,i} \left[ \frac{1}{r^2} \frac{\partial}{\partial r} \left( r^2 \frac{\partial c_{pi}}{\partial r} \right) \right], \tag{8.24}$$

where $\epsilon$ represents the void fraction, $L$ the column length, $u$ the velocity, $c_i$ the concentration of component $i$ in the fluid phase, and $q_i$ its concentration in the solid phase. $D_{ax}$ is the axial dispersion coefficient, $c_{pi}(r_p)$ the concentration of component $i$ within the particle at the particle boundary, $r_p$ the particle radius, $k_{l,i}$ the film mass transfer coefficient, $r$ the radial coordinate of the particle, $c_{p,i}$ the concentration in the pores of the particle, $\epsilon_p$ the particle porosity, and $D_{p,i}$ the diffusion coefficient in the particle pores.

When the intra-particle diffusion is neglected, the solid-phase equation changes to (8.25), where $k_{eff,i}$ represents the effective mass transfer coefficient:

$$(1 - \varepsilon_p) \frac{\partial q_i}{\partial t} + \varepsilon_p \frac{\partial c_{pi}}{\partial t} = \frac{3k_{eff,i}}{r_p} (c_i - c_{pi}(r_p)). \tag{8.25}$$

This model is called equilibrium transport dispersive model. It is represented by (8.23) and (8.25). The most common nonlinear adsorption equilibrium relation is the competitive Langmuir isotherm :

$$q_i = \frac{H_i c_i}{1 + \sum_j L_{i,j} c_j}, \tag{8.26}$$

with $H_i$ being the Henry coefficient and $L_{i,j}$ the Langmuir coefficients.

Since the speed of the different transport phenomena often differs in several orders of magnitude, the resulting system of equations becomes very stiff. There exist different approaches for the numerical solution of the above systems of equations [40, 57]. A very efficient numerical solution has been proposed by Gu [38] for the complex general rate model. A Galerkin finite element method for the discretization of the fluid phase and orthogonal collocation for the solid phase is used. The solution has been implemented within a FORTRAN program together with a stiff ordinary differential equation solver [15].

The axial dispersion in both models can be calculated by the empirical correlation of Chung and Wen [25], and the porosities can be determined via off-line experiments [3]. The intra-particle diffusion is often assumed to be fast. Thus, the interesting parameters are the film transfer coefficients $k_{l,i}$ and the isotherm parameters $H_i$ and $L_{i,j}$.

### 8.2.3  Focus of the Project

In current industrial practice, chromatographic separations are operated, such that the required product purities are met with a considerable safety margin. The potential for the reduction of separation cost arises from the ability to operate closer to the limits. The more accurate the process model is, upon which an optimization is based, the narrower the required safety margin will be.

Especially in the case of multi-component mixtures, the determination of the model parameters by laboratory experiments can be extremely tedious and costly. The ideal situation would be to determine the important model parameters during the first cautiously performed separation experiments. From such experiments, only the concentrations profiles at the column outlet are available. This is called a chromatogram or elution profile. Then the operation parameters would be improved using the resulting, possibly only locally accurate model, e.g., following the strategy described in [30, 64]. The next chromatograms could be used for a further improvement of the model, etc.

The key step in such a procedure is the determination of the critical model parameters from measured elution profiles, most importantly, the adsorption isotherm. In the sequel, we describe several approaches to this problem using physical models and mathematical parameter estimation, grey box models, and neural nets that were trained as inverse models.

### 8.2.4  Approaches to Model Parameter Estimation

#### Parameter Adaption in Rigorous Models

The physical model is given by (8.23), (8.24) or (8.23), (8.25) and an isotherm equation with free parameters. The first decision is which model is chosen and which form of the isotherm is assumed. In the sequel, the general rate model

(8.23), (8.24) is considered, and the adsorption isotherms are formulated as asymmetric Langmuir equations, i.e., they are of the general form (8.26). Thus for a two-component mixture, 6 isotherm parameters and two mass transfer coefficients $k_{l,i}$ have to be determined from the chromatograms.

The parameter estimation problem is formulated as the standard least squares problem to find the parameter vector $p$ which minimizes

$$\min_{p} \Phi(p) = \sum_{k=1}^{N} \sum_{i=1}^{2} \left( c_{sim,i}(v, p, t_k) - c_{meas,ik} \right)^2. \qquad (8.27)$$

Here, $c_{sim,i}$ is the simulated elution profile of the two components and $c_{meas,i}$ is the measured chromatogram. The summation is over all measurements and the two components. $v$ denotes the vector of operation conditions, $t_k$ is the sampling time, and $N$ represents the total number of measurement points.

A good fit between the simulated and the measured elution profiles is necessary for a good model, but in general, it is not sufficient. Despite a good fit at the operation point under consideration, some model parameters may deviate considerably from their optimal values because the simulation is not sensitive to these parameters, and thus the extrapolation capability of the model may be limited or even very poor.

A standard tool from statistical analysis is the evaluation of the Fisher matrix. It is defined by:

$$F(p^*) = \sum_{k=1}^{N} \frac{\partial c}{\partial p}^T C(t_k)^{-1} \frac{\partial c}{\partial p}, \qquad (8.28)$$

where $p^*$ is the estimated parameter value, $\frac{\partial c}{\partial p}$ is the matrix of the partial derivatives of the data points (concentrations) with respect to the parameters, and $C(t_k)^{-1}$ is the inverse of the covariance matrix of the measurement error. The Cramer-Rao theorem [8] states that the inverse of the Fisher information matrix is a lower bound for the parameter estimation error covariance matrix. When the resulting lower bound is large, some parameters cannot be estimated with high precision. An eigenvalue and eigenvector decomposition of the Fisher matrix reveals which parameters or combinations of parameters have a large error covariance and thus, their estimates are strongly influenced by noise and numerical errors.

In the parameter estimation experiments, the "measurement data" were generated from simulations of the general rate model with nominal parameters and white noise added to the samples (2 percent relative error) and a constant offset of 0.005 of the feed concentration. The nominal parameters are shown in the first line of Table 8.7.

For the simulation of the measurement data, we considered three different experimental setups which differ with respect to the injection time (see Table 8.6). The sample time is two seconds.

**Table 8.6.** Used operating points

| Experiment | Injection time |
| --- | --- |
| 1 | 50 |
| 2 | 100 |
| 3 | 200 |

The injection time is important here because it determines whether the two components are presented simultaneously over a considerable length of the column or not. For short injection times, the interaction takes only place in the initial part of the column, thereafter, the two pulses are already separated, and almost no interaction occurs.

For parameter adaptation, initial values are needed. These must be chosen, such that the difference between the observed and the simulated chromatogram is not too large. If the elution profiles are very different, the sensitivity of the cost function to changes of the parameters is low and this causes problems in the optimization. From the retention times (the time that elapses until the two peaks are seen at the output), the Henry coefficients in the nominator of the adsorption isotherm can be estimated, assuming negligible kinetic effects, no interaction and linear isotherms [39]. We initialized the Henry coefficients in this fashion. The mass transfer coefficients were initially set to $10^{-4}$ $cm/s$ (the usual range is $10^{-3}$ to $10^{-5}$) and all interaction parameters were set to 100 which implies slight nonlinear adsorption behavior.

Tests were performed with different optimization routines from the Optimization Toolbox of MATLAB [45]. It turned out that the sequential quadratic programming (SQP) solver and the Levenberg-Marquardt algorithm gave the best results, and as the Levenberg-Marquardt algorithm exhibited superior convergence properties, it was used throughout the experiments.

First, the parameter estimation was carried out for experiment 1 (short injection time). The results are shown in Table 8.7. The approximation error for the experiments was almost exactly the noise level. But as can be seen from Table 8.7, the parameter estimates differ considerably from the true values. $p^0$ is the starting point of the optimization and $p^*$ the parameter vector at the optimum.

The computation of the Fisher matrix revealed that at the estimated values of the parameters, the interaction parameters $L_{12}$ and $L_{21}$ cannot be estimated well. This is not surprising for a short injection time. Experiment 1 can be used to compute estimates of the Henry and mass transfer coefficients, with the Langmuir parameters set to zero (see Table 8.8).

The resulting parameters for Henry and mass transfer were then chosen as initial values for an estimation using the data from experiment 2. As

**Table 8.7.** Real and estimated parameter values for operating point one

|          | $H_1$ | $H_2$ | $k_{l,1}$ | $k_{l,2}$ | $L_{11}$ | $L_{12}$ | $L_{21}$ | $L_{22}$ |
|----------|-------|-------|-----------|-----------|----------|----------|----------|----------|
| $p^{real}$ | 23.26 | 8.78 | 3.70 | 5.00 | 333.1 | 94.5 | 211.4 | 43.5 |
| $p^0$ | 8.52 | 4.08 | 1 | 1 | 100 | 100 | 100 | 100 |
| $p^*$ | 22.92 | 8.40 | 3.45 | 5.40 | 323.00 | 16.64 | 54.13 | 38.95 |

**Table 8.8.** Estimated parameter values for operating point one (linear adsorption)

|          | $H_1$ | $H_2$ | $k_{l,1}$ | $k_{l,2}$ |
|----------|-------|-------|-----------|-----------|
| $p^0$ | 8.52 | 4.08 | 1 | 1 |
| $p^*$ | 18.85 | 7.79 | 2.75 | 5.5 |

Table 8.9 shows, the parameter estimates are considerably better now, but still the parameter $L_{21}$ cannot be estimated well.

**Table 8.9.** Real and estimated parameter values for operating point two

|          | $H_1$ | $H_2$ | $k_{l,1}$ | $k_{l,2}$ | $L_{11}$ | $L_{12}$ | $L_{21}$ | $L_{22}$ |
|----------|-------|-------|-----------|-----------|----------|----------|----------|----------|
| $p^{real}$ | 23.26 | 8.78 | 3.70 | 5.00 | 333.1 | 94.5 | 211.4 | 43.5 |
| $p^0$ | 18.85 | 7.79 | 2.75 | 5.50 | 100 | 100 | 100 | 100 |
| $p^*$ | 23.32 | 8.51 | 3.79 | 5.49 | 333.2 | 96.5 | 141.7 | 39.9 |

Figure 8.5 shows the extrapolation capability of the model from Table 8.9 for different operating points. The maximum height and width of the curves increases with rising injection times. It is obvious, that the models are quite accurate for injection times up to 200 seconds, but for 300 seconds there is a considerable error. The Fisher matrix supports this finding: the parameters $L_{12}$ and $L_{21}$ cannot both be estimated accurately from the available data.

For experiment 3, the situation is different. In the parameter estimation procedure, the Henry and mass transfer coefficients were initialized as before by the values obtained from experiment 1, and the Langmuir parameters were arbitrarily set to 100. Table 8.10 shows that the parameters now are estimated very accurately, and the extrapolation to smaller and to larger injection times is perfect, as shown by Figure 8.6.

In summary, it is possible to estimate the critical parameters of the rigorous model (8.23), (8.24) from the elution profiles, provided that the injection time is sufficiently long to lead to significant interaction of the two components in the column. The analysis of the Fisher matrix gave a good indication

**Fig. 8.5.** Comparison of measurement and estimation for injection times of 50, 100, 200 and 300 s

**Table 8.10.** Real and estimated parameter values for operating point three

|  | $H_1$ | $H_2$ | $k_{l,1}$ | $k_{l,2}$ | $L_{11}$ | $L_{12}$ | $L_{21}$ | $L_{22}$ |
|---|---|---|---|---|---|---|---|---|
| $p^{real}$ | 23.26 | 8.78 | 3.70 | 5.00 | 333.1 | 94.5 | 211.4 | 43.5 |
| $p^0$ | 18.85 | 7.79 | 2.75 | 5.50 | 100 | 100 | 100 | 100 |
| $p^*$ | 23.31 | 8.81 | 3.68 | 4.89 | 335.2 | 93.85 | 215.28 | 44.0 |

when problems with the extrapolation capacity could be expected. However, as this analysis is based on a linearization around the estimated values of the parameters, it cannot be used to determine the range of validity of the model. Of course, the crucial assumption here is that the process can indeed be described by the assumed model, in particular, that the adsorption isotherm can be represented by a Langmuir law.

## Estimation of Model Parameters from Elution Profiles Using Neural Networks

For the mathematical optimization of the parameters of a rigorous model, good initial values are helpful, if not crucial. Also the computation time can

**Fig. 8.6.** Comparison of measurement and estimation for injection time of 50, 100, 200 and 300 s

be reduced considerably if a good guess of the parameters can be obtained directly from the measured profiles. For this purpose, neural networks were generated to predict the key parameters of a simplified model from characteristic points of the elution profiles.

To train these neural networks, simulations were performed for a large number of parameter values that cover the relevant range for the processes under consideration. The sample time is one second. From the simulated profiles, characteristic values were computed. These values are the inputs to a neural net, and the known values of the parameters of the model are the samples that the neural network has to "learn." The neural nets thus are trained as inverse models.

As the complexity of the neural nets increases with a rising number of inputs, only a few characteristic values were used, namely the peak values ($P_{max}$) and the two inflection points ($Ip_1$, $Ip_2$) for each concentration profile. As variable model parameters, the Henry and the mass transfer coefficient and one Langmuir coefficient per component were chosen, i.e., a symmetric Langmuir isotherm was assumed, $L_{11} = L_{21}$, $L_{12} = L_{22}$. For the generation of the training data, the parameter space was covered by a grid and all combinations at the grid points were simulated. The distance between two successive grid points are 1 for the Henry coefficient, 7.5 for the Langmuir

and 0.25 for the mass transfer coefficient. Overall, 120 profiles were generated as training data for the single component case and 1850 profiles for two components. The test data were chosen randomly or in the center of the boxes of the grid (Figure 8.7). The network structure is an RBF net with Gaussian activation functions. For the training, the iterative procedure described in [24] was used, i.e., nodes were added until a predefined threshold on the approximation error for the training data was reached.

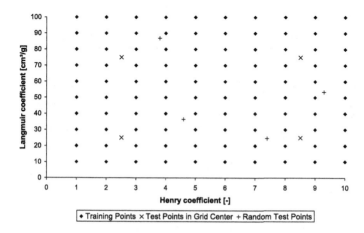

**Fig. 8.7.** Structure of the used training and test data

The approach was first tested for a single component system. The training data consisted of 4 parameter combinations, as shown in Table 8.11. The mass transfer coefficient was multiplied by 1000 for better scaling of the training data. The variance of the Gaussian functions was experimentally determined, 4 turned out to be a reasonable choice.

Two different structures of the neural networks were tested. The first one consists of three independent neural networks of the RBF type, one for each estimated parameter, the second is one integrated network with three outputs. To study the importance of the three input data points (each point is a time and a concentration), in addition to all three data points, combinations of two points were also investigated. The results are shown in Table 8.11 in the second and third columns for the network with one output, and in columns four to five for the other approach.

The results do not differ much, the best approximation was obtained by three independent nets with three input points each, at the expense of a much larger number of neurons compared with a net with three outputs. As can be expected from the small errors, the matching of the eluted profiles is very good (Figure 8.8). If the measurement of the elution profile is con-

**Table 8.11.** Training results of neural networks with one/three outputs

| Input | Neurons | Error [%] | Neurons | Error [%] | |
|---|---|---|---|---|---|
| | 65 | 0.41 | | 0.37 | $H_1$ |
| $Ip_1, P_{max}$ | 46 | 3.56 | 62 | 4.72 | $L_1$ |
| | 49 | 4.42 | | 3.69 | $k_l$ |
| | 47 | 0.29 | | 0.21 | $H_1$ |
| $Ip_1, Ip_2$ | 72 | 3.49 | 65 | 4.00 | $L_1$ |
| | 60 | 6.47 | | 8.14 | $k_l$ |
| | 52 | 0.59 | | 0.52 | $H_1$ |
| $P_{max}, Ip_2$ | 62 | 3.89 | 60 | 4.41 | $L_1$ |
| | 57 | 4.35 | | 4.40 | $k_l$ |
| | 96 | 0.19 | | 0.26 | H1 |
| $Ip_1, P_{max}, Ip_2$ | 89 | 2.25 | 118 | 3.86 | $L_1$ |
| | 57 | 3.12 | | 3.44 | $k_l$ |

**Fig. 8.8.** Comparison of the simulated profiles with parameters from neural network

taminated by noise, the determination of the inflection points is subject to considerable errors and hence the parameter estimation shows large errors. A possible remedy is to average over a number of profiles first. Figure 8.9 shows

the situation where 11 elution profiles have been averaged to determine the inflection points.

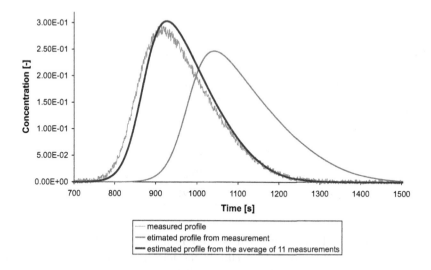

**Fig. 8.9.** Impact of noise to the quality of the determined concentration profiles

**Table 8.12.** Training area of the neural networks

| Parameter | Minima | Maxima |
|---|---|---|
| $H_1$ | 1 | 2 |
| $H_2$ | 2(3) | 6 |
| $L_{11}$ and $L_{22}$ | 0.02 | 0.4 |
| $k_{l,i}$ | 2.5e-4 | 5e-4 |

For the separation of two components with nonlinear isotherms, in the simplest case 5 parameters have to be estimated. Table 8.12 shows the distribution of the training data in this case. A symmetric isotherm was assumed, and also the mass transfer coefficients had to be equal for both components. As inputs to the networks there are now 3 points for each elution profile, 12 values all together. Again, independent networks for each parameter and one integrated network for all 5 parameters were trained and compared on test data. The results are summarized in Table 8.13 in columns 2 and 3 for the one output network and in columns 4 and 5 for the other one. While the Henry and mass transfer coefficients can be estimated quite well, the Langmuir co-

efficients are not accurately determined. The resulting simulated respectively predicted elution profiles are compared for one test case in Figure 8.10. Similar results have been obtained with a support vector machine (see Table 8.5 in Section 8.1 or [96]).

**Table 8.13.** Training results of neural networks with one/five outputs

| Neurons | Error [%] | Neurons | Error [%] | |
|---------|-----------|---------|-----------|-------|
| 74      | 1.29      |         | 4.31      | $H_1$ |
| 116     | 0.94      |         | 1.30      | $H_2$ |
| 163     | 21.44     | 63      | 15.96     | $L_{11}$ |
| 234     | 25.6      |         | 19.22     | $L_{22}$ |
| 22      | 5.56      |         | 22.27     | $k_l$ |

A problem with this approach is the enormous amount of training data that is required even for rather simple models (remember that the model considered here has only 5 parameters, whereas 8 different parameters were estimated by fitting of the rigorous model to the observed data). This data has to be generated for each operation point of the process because the networks are dependent on the injection time, feed concentration, etc. Not surprisingly, if the parameter estimation problem is difficult due to the insensitivity of the elution profiles to the parameter values, large errors result for the neural network which approximates the inverse model, similar to the optimization of the model parameters using measured data.

**Parameter Estimation for Grey Box Models**

So far we have assumed that the adsorption isotherm can be described exactly by a function of the Langmuir type. While this formal model can be justified qualitatively by the competition of the molecules for the adsorption locations of the solid particles, some combinations of mixtures and adsorbents show qualitatively different behaviors, e.g., S-shaped isotherms. This leads to the idea to approximate the isotherms by neural networks, and to use these as isotherm equations in the rigorous model (8.23), (8.24) or (8.23), (8.25), respectively.

The obvious idea is to replace (8.26) by a neural net and to estimate the parameters of this net from experimental data, either elution profiles or measurements of the isotherms at distinct concentration values. Extensive trials showed that this is not promising. Due to the large effort required to measure adsorption isotherms, there is only a small number of measurements available, and this only covers a small range of possible combinations of concentrations.

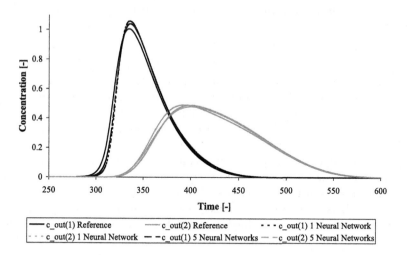

**Fig. 8.10.** Comparison of simulated two component profiles

In this situation, a structured model with few parameters is preferable, especially when a priori knowledge on the model structure is available. If a neural network with few neurons is employed, it behaves similarly to a Langmuir model. If more nodes are included, the behavior becomes more irregular, and this causes errors of the predicted elution profiles because the derivatives of the adsorption isotherm enter into the model evaluation rather than the absolute values.

A more promising approach is to approximate the slopes of the isotherms by neural networks and to train these from elution profiles, thus avoiding the effort both in time, material and manpower, to measure the adsorption isotherms in the laboratory. Neural networks are thus trained to approximate the derivatives of the isotherms with respect to the concentrations in the rigorous model (8.23), and (8.24). For a two-component separation with interaction, there are four such functions. For a physically meaningful model, these functions must satisfy:

$$\sum_{j=1}^{2} \frac{\partial q_i}{\partial c_j} \geq 0 \qquad i = 1, 2 \tag{8.29}$$

$$\left. \frac{\partial q_i}{\partial c_j} \right|_{c_i=0} = 0 \qquad i \neq j \tag{8.30}$$

Equation (8.29) means that the concentration of the adsorbed component must increase with increasing concentration in the fluid phase. Equation (8.30) implies that at very low concentrations of component $i$ regardless of

the value of the concentration of component $j$, the interactions vanish. In addition to the parameters of the neural networks for the isotherm slopes, the mass transfer coefficients must be estimated.

In order to avoid a large number of unknown parameters, the neural networks were chosen to be very simple, each having only two sigmoid neurons in the hidden layer and a linear output neuron with bias. As mentioned before, a crucial point in the parameter estimation is that the initial profile must be reasonably close to the measured one. For the neural networks, there is therefore the problem of initializing them appropriately. A randomly chosen set of initial values does not result in meaningful models. Therefore, similar to the procedure for the Langmuir isotherm parameters, the Henry coefficients and the mass transfer parameters were estimated first for small injection times. The bias terms were initialized with the Henry coefficients, and then the neural networks were trained, starting from randomly chosen weights in the input layer and output weights adapted, such that (8.29) and (8.30) hold.

Numerical experiments were performed for operation point 2 (injection time 100 seconds), and the Levenberg-Marquardt algorithm was used for the solution of the optimization problem. The results were comparable to the fitting of the Langmuir parameters, both with respect to the good approximation of the profiles at this operation point and the limited extrapolation capability. When the same procedure was applied to experiment 3 (injection time 200 seconds) models with good extrapolation quality resulted (Figure 8.11). The approximation of the profile at this operation point is within the range of the measurement error. The extrapolation is good and accurate enough for use in the optimization of the operating parameters although slightly worse than for the Langmuir isotherm. More details can be found in [27].

As the data was generated by a model with Langmuir isotherm, the estimation of the parameters of the neural network model can of course not give better results than for the Langmuir model, and also the restrictions posed by insufficient sensitivities of the elution profiles to the model parameters cannot be avoided. The advantage of the neural network model is its flexibility. A neural network isotherm model can match arbitrary behaviours and especially complex interactions in multi-component mixtures.

In order to investigate the flexibility of this model, we applied it to measured elution profiles that cannot be matched assuming a Langmuir isotherm. The data was obtained from a two-component enantiomer separation with strongly nonlinear isotherms. The isotherm models were of the same structure as before, i.e., two nonlinear neurons each and a linear output neuron with bias term. The models were fitted to data from two different experiments. In Figure 8.12, the results of the adaptation of the neural network models to this process are shown. While one experiment is matched very well, there are structural differences in the second. We presume that the reason for this

**Fig. 8.11.** Comparison of measurement and estimation for injection times of 50, 100, 200 and 300 s

results is that the flexibility of the models with only two neurons is not large enough, so the number of neurons in the hidden layer should be increased.

### 8.2.5   Conclusions and Future Work

We have explored various possibilities to use neural networks as partial models of chromatographic processes in comparison with the fitting of rigorous models. It was shown that for batch chromatography, the experimental design is crucial for the estimation of the adsorption isotherms. The components must interact sufficiently inside the column to determine the full isotherm. Experiments with small injection times are useful to determine the mass transfer parameters and the Henry coefficients. These values can then be used to initialize more complex models, which are fitted to experiments with stronger interactions. Both the estimation of the Langmuir parameters and of the weights of simple neural networks that compute the isotherm slopes are feasible and lead to comparable results. Neural networks can also be used to generate inverse models that relate characteristic points of the elution profiles and model parameters.

Keeping the goal of model-based optimization in mind, a key problem is to determine the range of validity of the models, so that the optimization

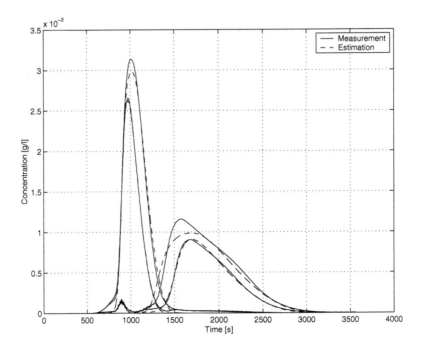

**Fig. 8.12.** Comparison of measurement and estimation for injection times of 239 and 353 s

does not employ the model for operating points where it is not reliable. This will be the focus of further research.

## 8.3 Classification and Prediction of Device Parameters with Regard to the Quality of Integrated Microelectronic Systems

In this section, various learning and optimization problems in the field of electrical engineering will be examined. Several computational intelligence methods are applied to manifold tasks, e.g., circuit design, device modeling, process control, and quality assessment. In Section 8.3.1 evolutionary algorithms for optimization of electronic circuit components are examined. When considering multiple target functions, e.g., power dissipation and signal delay, this methodology facilitates an efficient determination of Pareto-optimal design parameter sets. Arbitrary circuit concepts can be globally optimized and compared to each other. Different logic-style implementations of a full adder are investigated. The construction of simulation models for electronic device development and circuit control is investigated in Section 8.3.2. Statistical models of very complex devices, especially in nano-scale technology, can

be estimated by using evolutionary approaches. An algorithm for automated parameter determination is presented. Prospects and limitations of such an approach are studied. In Section 8.3.3 different algorithms and hardware implementations of neural networks for microelectronics and process control are presented. Finally, Section 8.3.4 is about the development and electronic realization of a complete fuzzy system. This research also deals with software development of new automatic construction tools for adaptive, self-organizing neuro-fuzzy controllers.

### 8.3.1 Circuit Analysis and Design Using Evolutionary Algorithms

The growing demand for high performance mobile electronic systems with enhanced battery lifetime requires the design of fast integrated circuits with very low power consumption. Switching speed and power optimization of digital CMOS circuits for today's submicron design flows must already be applied at the lowest abstraction levels of the design hierarchy. Considerable power savings and increases in switching speed can be achieved by addressing circuit-level design optimization. In the case of standard cell library design, basic switching elements can be implemented in various circuit techniques and transistor geometries. The estimation of the capabilities of each design choice, with regard to different objective functions, e.g., power dissipation, signal delay, chip area, etc., represents a global, high-dimensional optimization problem that cannot be sufficiently solved by applying conventional methods, such as steepest-gradient algorithms.

The evolutionary methodology provides one such global high-dimensional optimum seeking algorithm, without the necessity of starting the optimization with an already near-optimal solution in the parameter space. Thus, working designs for innovative circuit concepts can be found, even when an a priori determination of design parameters by a human expert is not simple. Furthermore, the relations between various target parameters for a given circuit topology can be extracted, so that different design concepts may be compared objectively. By observing the resulting Pareto-sets the preferable design for a specific application and given parameter constraints can be determined.

In the following section the basic idea behind circuit-level optimization will be pointed out. Some comments about classical approaches to this task will be made. Next is an introduction to the evolutionary circuit design algorithm and its capabilities. In Section 8.3.1 a full adder cell is analyzed in an exemplary fashion. Different logic styles will be analyzed with regard to high speed computation and low power dissipation. Finally, a conclusion will be drawn in Section 8.3.1.

## Circuit Optimization

A VLSI implementation of a specific binary function in a given technology is often achieved by choosing appropriate switching components from a standard cell library (SCL). This set of basic circuit elements allows an automated implementation of more complex gates for particular operation constraints. SCL generation is based on a bottom-up principle.

Transistor-level design and simulation of different circuit structures result in physical layout realizations of individual gates and registers. For the nominal optimization of these standard cells, classical linear or nonlinear programming methods are used. Well known design tools, such as DelightSpice [85], JiffyTune [26], and WiCkeD [4], are found on the nonlinear steepest-gradient algorithm. This method is based on the idea of tracking down the steepest descent in an optimization landscape. The gradient algorithm is only suitable for finding a minimum when a near optimal solution is already known because it is limited to local search. This classical approach can only optimize individual circuit implementations. The result of the optimization process is one scalar value, for example, a simulated signal delay, dissipated power, or estimated chip area. Therefore, it is not possible to extract global parameter dependencies, such as the power-delay relation. For the optimization of multiple objectives, artificial criteria have to be defined, e.g., the power-delay or energy-delay product. To enable an objectively global comparison of different design choices, the relationships among various target parameters have to be determined. The evolutionary algorithms that are introduced next can cope with this task.

## Evolutionary Circuit Design Algorithms

By applying the population-based evolutionary algorithm (EA) approach, a group of points in the parameter space is considered instead of a single point. In the case of circuit design, each individual of the population used represents a parameter vector, including transistor sizes and voltage levels, according to a specific circuit layout.

The iterative EA procedure starts with a randomly generated population $E^g$. These initial designs can also be presented by a human expert. The central mechanisms of EA, reproduction and selection, have to be realized by suitable operators [29]. The fitness rating of specific circuit designs is carried out by transient circuit simulation. Only a subset of the generated elements is selected for the next generation, so that the overall fitness of the population should increase with every iteration step, leading to improved circuit designs. The convergence of the algorithm strongly depends on the choice of the selection operator. The influence of the EA operators and parameters on the optimization strategy is further investigated in [9]. The goal of multiobjective optimization can be achieved by applying special selection mechanisms, for

example the predator-prey model or the strength pareto evolutionary algorithm (SPEA) which are further described by [136].

In a multiobjective optimization problem, the notion of an optimum is not obvious. There is no solution that is best for all the target functions, e.g., power and delay, but there exists a set of solutions that are better than other solutions in the remaining search space when considering all the objectives. This set of solutions is known as the optimal solutions of the Pareto set or nondominated solutions. All other solutions are called dominated. A solution belonging to the Pareto set is not better than another one of the same set.

EA work with a population of possible solutions. During the optimization process, a database of working solutions is build up, determining a cloud. At convergence, the cloud does not evolve any further, and its convex hull is the optimal Pareto set. For practical circuit implementation, an engineer can choose a specific design from the infinite Pareto set, which consists of all optimal design solutions.

A major advantage of applying EA to the task of circuit design is the possibility of approximating the Pareto set of objective parameters directly. Usually these parameters are power dissipation, signal delay, and chip area for a given circuit structure under specific operation constraints.

### Example: Full Adder Design

To demonstrate the basic capabilities of the presented methodology, the design of a full adder (FA) cell will be examined. The Boolean expressions for the binary adder outputs are $c_o = a_i b_i + b_i c_i + a_i c_i$ and $s_o = a_i b_i c_i + \overline{c_o}(a_i + b_i + c_i)$. The bits $a_i$ and $b_i$ are the adder inputs, $c_i$ is the carry input, $s_o$ is the sum output, and $c_o$ is the carry output. FA are the basic cells in adder arrays, e.g., carry-save adders, used in multipliers and similar components, such as dividers. In such applications, efficient full adder circuits are crucial since these building blocks are often critical and determine the operational speed, power dissipation, and chip area of the system [135]. At first, the design methodology will be applied to static logic adders. In a second step, dynamic pipelined FA will be analyzed.

Figure 8.13 illustrates the circuit arrangement for simulation of the full adder cells. The cells were designed at the transistor level in a standard $0.35\,\mu m$ CMOS process technology ($V_{tn} = 0.47\,V, V_{tp} = -0.62\,V$) and were simulated using HSPICE for a supply voltage range from $1.5\,V$ to $3.3\,V$ at $27\,°C$. Worst-case gate delays and average power dissipation, including short circuit currents, are obtained from simulation. The circuit designs from which the optimization process is started are sized by hand with the objective of balanced gate performance. Certain symmetry properties of the adder designs may be derived by taking the semantic equivalence of the input signals $a_i$ and $b_i$ into account. The related transistors are sized equally, so that the number of parameters that have to be optimized is reduced.

25

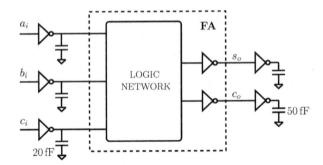

**Fig. 8.13.** Circuit arrangement for the simulation of full adders. Inverters equivalent to the adder output inverters are placed on the input and wiring capacitances of 20 fF attached in order to simulate two full adders connected in series with a fan-out of one

The comparison of static one-bit FA is carried out on a standard CMOS mirror adder (MA) and a transmission gate (TG) circuit. Some basic logic style characteristics that influence circuit performance and power dissipation are qualitatively compared in Table 8.14. In particular, the number of MOS logic networks, the number of transistors and the rail-type are given for the specific logic styles. The TG realization uses only 20 transistors instead of 28 for the MA. The basic difference of transmission gate logic compared with the CMOS logic style is that the source side of the logic transistor network is connected to some input signal instead of the power lines.

The knowledge about the circuit symmetries allows the use of a simple input pattern that will initiate a transition in each circuit path. Overlapping impulses on each input line are used, resulting in six different states [124]. For a complete analysis of power consumption, all transitions would have to be taken into account, but due to its complexity, this is not considered here. To verify the functionality of given designs, certain constraints to the output must be defined. The maximum delay time to obtain a desired output is chosen less than $2ns$ and the maximum deviation from the ideal voltage levels should be 20 percent of the supply voltage. The input signals for analyzing the static FA and the output signals along with the resulting error detection "tubes" are depicted in Figure 8.14.

The methodology presented in Section 8.3.1 is utilized for the introduced static circuit concepts. The algorithm that is used for optimization is an asynchronous SPEA with a population size of 50. In every iteration step, a set of 350 offspring is generated. In [9], the influence of the EA parameters on the optimization process is examined in detail. The output parameters that have to be minimized are the average signal propagation time $\overline{t_d}$ and the average power dissipation $\overline{p}$ of these circuits. When applying the EA to the multiobjective optimization task, a large set of solutions arises. Each of these solutions represents a working circuit design. In a simple second step,

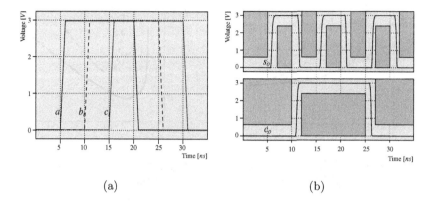

(a)                                                              (b)

**Fig. 8.14.** (a) Signal transitions for the three adder inputs $a_i$, $b_i$, and $c_i$. Six different input states can be determined. The signals raise time is chosen to be $200ps$. (b) output signals $s_o$ and $c_o$ of the 1-bit binary full adder. The sketch shows the error detection regions for logical verification as shown by [123]

the Pareto-optimal output parameter relations have to be extracted from the data sets.

One resulting characteristic is the power-delay relation for the given circuit concepts. Without a suitable vector-optimization technique like the proposed EA, the task has to be simplified by defining a scalar target function, e.g., the power-delay product. That approach ignores a significant share of the possible power saving and delay minimization. The results of the EA optimization, using the asynchronous SPEA method, are shown in Figure 8.15. The global relation between minimum average power and minimum signal delay can be determined from this plot. By looking at the voltage-power diagram in Figure 8.16 an optimal supply voltage can be found according to a particular FA logic style. The simulation results show better delay and power values for the static TG adder in the entire operation region. This is due to the efficient three-input XOR transmission gate implementation. An exemplary physical design of a static mirror adder, which is optimized for specific operating constraints is depicted in Figure 8.17.

The resulting characteristics represent only an approximation to the unknown Pareto-sets. The deviation from real parameter dependencies should become smaller the more iteration steps the optimization process takes. The results given were achieved with an approximate number of $3 \times 10^3$ iterations, which equals about $10^6$ circuit simulations when using an evolution strategy with 350 offspring. This corresponds to a simulation time in the order of days on a current small parallel computer system.

As a second example, the examination was extended to dynamic FA. The adder cells are implemented as pipeline elements with alternating eval-

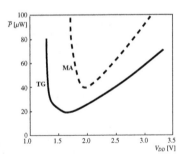

**Fig. 8.15.** Power-delay relation of the static adders

**Fig. 8.16.** Voltage-power relation of the static adders

**Fig. 8.17.** The optimization process for exemplarily selected operating constraints results in a physical design of a static mirror adder for a $0.18\mu m$ CMOS process. The three adder inputs, two outputs, and two connections for VDD and GND on the outer sides are clearly visible

**Table 8.14.** The principal logic style characteristics of the examined static and dynamic adder circuits

| LOGIC STYLE | # Rails | # MOS Networks | # FETS |
|:---:|:---:|:---:|:---:|
| CMOS | Single | $2n + 2p$ | 28 |
| TG | Dual | $n + p$ | 20 |
| C$^2$MOS | Single | $2n + 2p$ | 32 |
| CTG | Dual | $n + p$ | 24 |
| PND | Single | $n + 2$ | 22 |
| DCVS | Dual | $4n$ | 44 |

uate and hold phases. The standard circuit techniques that are considered are clocked CMOS (C$^2$MOS), clocked transmission gate logic (CTG), pn-cascode domino (PND) technique, and differential cascode voltage switch logic (DCVS). The principal logic-style characteristics of the circuits are listed

in Table 8.14. The prerequisites for the iterative optimization process can be adopted, with some slight modifications, from the static case. For example, the output signal evaluation must take the clock signals into account. Some results of the EA application are presented in Figures 8.18 and 8.19. The relationship between the parameter set of all calculated design solutions and the extracted, approximated, Pareto-optimal set is also shown.

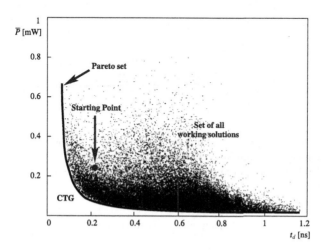

**Fig. 8.18.** EA analysis results of dynamic full adders in various logic styles. The relation between the power-delay parameter set of all calculated design solutions and the extracted, approximated, Pareto-optimal set

It can be established that the CTG design dominates the presented circuit techniques with regard to the power-delay characteristics in any operation condition. This result is generally valid for the examined circuit structures and not only for a specific design implementation. The lower limits of dissipated power and signal delay for a specific circuit technique can be obtained directly from the plotted curves. Another result of the analysis is the fact that some logic styles are more efficient than others only in specific parameter regions. For example the PND technique behaves poorly compared to the DCVS circuit, when a delay less than $0.5\,ns$ is required.

## Conclusion

The evolutionary algorithm methodology presented enables transistor-level circuit design for high speed and low power applications. The proposed methods may be used for standard cell library construction. The optimization strategy provides a high-dimensional optimum seeking algorithm, so that given circuit structures can be globally optimized with regard to multiple objectives and various parameter constraints. Independently from specific de-

**Fig. 8.19.** EA analysis results of dynamic full adders in various logic styles. Power-delay relations of the examined circuits. The applying transistor count is given in brackets. Minimum dissipated power and minimum signal delay can be determined from the plot

sign implementations, arbitrary circuit concepts can be objectively compared with each other.

The major drawback of this methodology is the massive computation effort needed to identify global parameter dependencies. This limits the complexity of circuits which the algorithm can analyze. Novel methods of circuit analysis, for example, analytical transregional models or improved power-delay calculation tools, may reduce the time required to calculate the transient circuit behavior.

The presented methodology and the implemented EA design tools have also been applied to the task of constructing threshold logic gates [23] and have been transferred to research projects at the chair of Electrical Engineering and Computer Systems, Aachen University of Technology, Germany. To evaluate the results of the examined optimization tasks, some resulting full adder designs have been manufactured in a $0.18\mu m$ CMOS process.

Further development of the EA design approach should consider the aspect of circuit robustness against statistical parameter variations of the physical circuit devices and operating or environment conditions.

### 8.3.2  Modeling of Nanoelectronic Devices Using Evolutionary Algorithms

The implementation of simulation models is an important task in electronic device development and control. Parallelizing in architecture and technology research activities increases the significance of available high-quality device models. Enhanced device complexity, especially in nano-scale technology,

shows difficulties in classical methods for parameter determination. Some of these problems can be solved by using evolutionary approaches. In the following, an algorithm for an automated parameter determination task is presented, and prospects and limitations of such an approach are studied.

## Introduction

In semiconductor device development, a precise knowledge of the technology parameters is of fundamental relevance to circuit design and fabrication control. Devices in current technologies, such as CMOS and Bipolar, are described by physical models, which show quite simple analytical characteristics such as linear correlation. Parameters in these models have to be determined by fitting to one or multiple measured input-output curves. Physically based models have the advantage that parameters are appearing as extrema, gradients, or intersections. In contrast to these analytical models, semi-empirical approaches are more complicated. They are optimized for circuit simulation purposes and achieve a better accuracy in input-output curve fitting. Parameters in such models have only indirect influence on the output characteristics. In these cases automatic parameter determination can lead to problems in the convergence behavior. Additionally some parameters can support minor effects, which increases the optimization difficulty. In the emerging nano-scale technology, models for simulation tasks show much more complexity, based on the influence of quantum and single charge effects. In these cases the classical methods for parameter determination are not suitable any more. They do not cope with the strong nonlinear dependencies of single electron transistors and resonant tunneling devices [37]. Even experts need multiple hours to fit a model for simple devices in this area. In this situation an autonomous online process control seems illusory.

## Problem

In this work an alternative parameter determination method for a resonant tunneling diode (RTD) model is presented. This quite simple device plays a key role in nano-scale technology and is a suitable object to support system design for nano-electronics [90]. The following model is proposed in [14]:

$$I_{\mathrm{RTD}}(V) = I_{\mathrm{res}}(V) + I_{\mathrm{res}}(-V) + I_{\mathrm{lkg}}(V) \tag{8.31}$$

$$
\begin{aligned}
I_{\mathrm{res}}(V) = \frac{I_P}{f}\Big(1 + \frac{2}{\pi}\arctan\Big(\frac{V_N-V}{V_W}\cdot\frac{nkT}{q(V_N-V_P)}\Big)\Big)\cdot \\
\cdot \ln\Big[1 + \exp\Big(\frac{q(V-V_P)}{nkT}\Big)\Big]
\end{aligned}
$$

$$I_{\text{lkg}}(V) = I_V \cdot \frac{\sinh\left(\frac{qV}{n_V kT}\right)}{\sinh\left(\frac{qV_V}{n_V kT}\right)}$$

$$f = 1 - \sqrt{2V_W / \pi (V_N - V_P)}$$

The current voltage dependency $I_{\text{RTD}}(V)$ is given as summation of res-
onant currents $I_{\text{res}}$ and the leakage current $I_{\text{lkg}}$. Parameter $f$ is a normal-
izing factor. The eight variable parameters in this model are: the resonant
width $V_W$, the peak voltage $V_P$, the voltage at maximum negative differen-
tial resistance $V_N$, the valley voltage $V_V$, peak current $I_P$, valley current $I_V$,
a resonant current parameter $n$, and a thermo-ionic current parameter $n_V$.
These parameters are current and voltage values except the last two, which
are numerical. These have to be adjusted to make the model curve an appro-
priate fit to the measured one. An example for the RTD DU595, inspected
by W. Prost at the University of Duisburg, shown in Figure 8.20 expresses
an essential difficulty. The size of the device is the cause for the enormous
influence of the measurement equipment on the measurement itself. As a
consequence, the descending part of the curve is hard to sense. Therefore a
global fitting criterion is not sufficient.

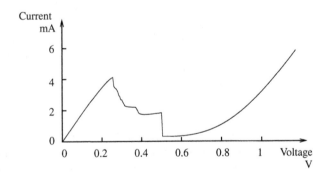

**Fig. 8.20.** IV characteristic of a resonant tunneling diode

As each parameter is in $\mathbb{R}$, the optimization problem is in $\mathbb{R}^8$. That means
a mapping $f : \mathbb{R}^8 \to \mathbb{R}$ to express a quality measure for each parameter set
is needed. Therefore a kind of distance measure between the analytical model
and the experimental curve is defined. This distance measure consists of a
weighted sum of four different criteria. For the curves rising parts, the mean
of the absolute differences, and of the gradients differences are calculated.
Additionally, the local maximum gap and the mean of negative values are

summed up. This combination is intuitively gained and cannot be derived from any information. It is the most complicated part in this automated optimization process.

Two cases have to be taken into account. First, this quality measure can only be calculated, if $I_{RTD}$ is defined over the whole input space (in our case 0 V to 1.2 V). Parameter sets performing this are called analytically correct. Consequently, the search space, defined by analytically correct items, cannot be determined a priori and may be disjointed. Second, possible parameter intervals are restricted with regard to physical limitations. This is, for example, a peak current between 1 mA and 6 mA. These predefinitions prevent the algorithm from following a senseless fitting.

## Algorithm

Parameter determination is done by utilizing evolution strategies (ES). A quality measure for each parameter set, also called fitness function in this context, is sufficient for its use. They are advantageous to genetic algorithms because of genotype and phenotype identity in this application. For a real-valued parameter space, standard methods for recombination and mutation already exist. Selection in optimization problems is easily done by taking the best individuals. In this work, a standard (15,100)-ES is used [5]. Intermediate recombination was used for the object variables and discrete for mutation variables. Mutation is realized by adding self-adjusted Gaussian noise. By prefiltering analytically incorrect parameter sets and providing an appropriate starting set, the creation of 100 analytically correct individuals in each epoch is guaranteed. Out of this set the best 15 elements, with regard to their fitness value, are taken for the next population. The fitness function for physically correct parameter sets is defined by the quality measure mentioned above. In case of physically incorrect parameter sets the distance to the area of correct objects is taken as fitness value. A more stringent handling of the physical restrictions lets the ES tend to stuck in suboptima. The complete procedure of parameter determination is divided into five steps. At first the physically correct area has to be defined by an expert. This is done by setting an allowed interval for each parameter. Within this area, 15 objects are drawn by an equally distributed random selection. The analytical correctness is checked in this process too. This gives a valid population, on which the evolution strategy works. The finally evolved parameter set or model must be controlled visually by its input-output curve to detect a failed optimization, which gets stuck in a monotonically rising curve. At last, the resulting model has to be checked against the defined restrictions. The physical boundaries may have to be respecified, respectively defined more precisely.

Simulation runs showed a sure achievement of an appropriate solution. The algorithm got a better model parameter set in every run with respect to the quality measure than a human expert did in multiple hours. One sample

run is documented in Figure 8.21. All simulations reached the expert's quality measure level within the first 200 generations.

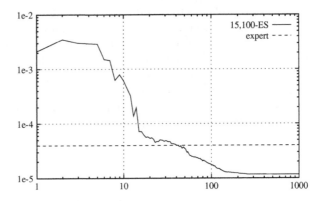

**Fig. 8.21.** Quality measure of best model in generation as solid line. The dotted line gives fitness level of the experts model

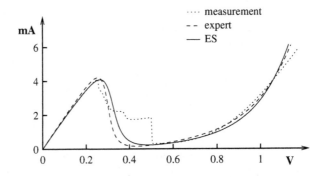

**Fig. 8.22.** Resonant tunneling diode: Measurement (dotted line) and its analytical model by an expert (dashed line) and the algorithm (solid line)

## Conclusion

It has been shown, that evolutionary algorithms are a suitable method for parameter determination for complex device modeling. The weighting of the quality measure parts is a critical aspect of the algorithm. Advances in multi-criterial optimization can lead to deterministic weight calculation. Apart from the final assessment, this is an automatic modeling process. To achieve a fully automated process, an algorithmic visual valuing is needed. This could

be done by a maximum distance measure. The state as a tool for experts is expressed by the need of a priori knowledge to line up the optimization.

### 8.3.3  Neural Networks for Microelectronics and Process Control

**SOM for Process Control**

Although in recent times the use of modern data processing procedures for analyzing processes in the power plant sector has increased, there is still much remaining to be done in the field of power plant data evaluation. In particular, the visualization and analysis of complex transient operations, such as starting up and shutting down a power plant, are at present being inadequately implemented. In the work of Otte [43, 86, 87, 88, 89], which runs in cooperation with the industry,[24] a procedure for multivariate data evaluation of transient process operations, based on the method of self-organizing feature maps, is presented. This procedure enables not only individual process states to be compared to each other, but also complex operations inside a power plant, so that divergences in the transient process phases can be detected at an early stage. On this basis, methods are presented enabling users to ascertain the fundamental causes for process divergences. The results of the work show that the data analysis procedures presented can be used for increasing a power plant's availability and for quality assurance in production processes.

**Possibilistic Neural Networks**

Possibilistic neural networks are a new approach to process modeling based on measurement data. They are derived from radial basis function networks and possibilistic reasoning. Possibilistic reasoning is based on possibility distributions of data; these are suitable for representing rare events because, in contrast to probability distributions, they do not represent the density of the data. The work of Kanstein and Glösekötter [34, 58, 60] shows radial basis function networks and related methods in statistics, clustering algorithms for training the neural network structure, how possibility distributions and possibilistic reasoning are derived from their respective probabilistic counterparts, and fuzzy logic systems and their relationship to the new model.[25] Application examples are used to discuss characteristics of the new approach. Furthermore, a hardware implementation of a computational adaptive radial basis function network is presented in this DFG project. The core circuits and the essential parts of the learning algorithm are implemented by means of functional integration in CMOS technology. The circuit also benefits an online learning algorithm. The weight storage used in the neural network uses floating gates and is handled by an intelligent programming strategy. Applications of this kind of neural hardware can primarily be found in intelligent sensor systems.

---

[24] This project was funded by the industry.
[25] This research was supported by the DFG as Project No. 379/11-12.

## Fingerprint Sensor Array Based on Cellular Neural Nets

In cooperation with the industry, Jung has constructed a capacitive fingerprint sensor array with parallel cellular neural nets,[26] implemented in CMOS technology [52, 53, 54, 55, 56]. The system acquires a binary fingerprint image and performs several image-processing steps, including thinning of the ridges of the fingerprint structure and extracting its characteristic features. Image processing is achieved by application of hexagonal local operators implemented in pixel-parallel mixed neuron-CMOS logic circuits. The massive parallelism of the cellular neural net leads to low power dissipation. Results of both simulations and measurements on a demonstrator chip agree. The approach is well suited to identification of persons, especially in small portable systems, such as smart cards or cellular phones.

## Biology Inspired Adaptive Neural Networks

Local navigation is a challenging task when a restricted cinetic system is considered. As a workhorse we employed a wheelchair that is equipped with tactile, sonar, and infrared sensors. Rossmann and Bühlmeier developed a neural network, which is composed of biologically plausible artificial neurons with adaptive Hebbian inputs [16, 17, 18, 19, 20, 21, 107, 108, 109, 110, 111, 112, 113, 114, 115]. The wheelchair learns local maneuvers according to the principles of classical and operant conditioning, which are performed in a neural hardware implementation. For this implementation we use a digital chip that evaluates the input signals at all synapses in only one clock cycle and propagates the results to the output using high-speed design techniques, such as carry-look ahead adders and parallel multipliers. The approach bases on the bio-inspired modified Hebbian learning algorithm incorporating both short- and long-term dynamics. In conjunction with time multiplexing and fast external memory, one neuron chip can emulate up to 35,000 neurons per $7.5\,\mu s$ while still preserving free configurability of the network with respect to the inter-neuronal connections. Experiments showed that after learning an obstacle avoidance task, the wheelchair, combining forward and backward movements, is able to steer autonomously through narrow doorways. In another experiment, the neural network controls a small vehicle that drives on a slot-racetrack and points out the boundary conditions. The neural hardware has to fulfill hard constraints due to the speed and the small dimensions of this setup. This research was carried out in cooperation with Prof. Manteufel from the Faculty of Computer Science at the University of Bremen and the FBN Research Institute in Dummerstorf.[27]

---

[26] This project was funded by the industry.
[27] This work was supported by the DFG as Project No. 379/12-13.

### 8.3.4  Learning in Fuzzy Systems: Cooperation with the University of Granada

Cooperation between the Microelectronics Department of the University of Dortmund and the Architecture and Computer Technology Department at Granada University began during the early 1990s with an exchange of lecturers between the two departments, as part of the Hispano-German Integrated Action programme (Conv. 1992, HA-056) entitled "Self-Organizing Map and Fuzzy Systems." Special emphasis was placed on the research activity of the groups with regard to hardware implementation and the application of architectures and intelligent systems (also termed soft computing); other concerns were the design and verification of VLSI circuits for the construction of fuzzy systems and new tools for the automatic construction of neuro-fuzzy systems. The cooperation was based on the thesis of H. Surmann about automated design of fuzzy systems, on the thesis of A. P. Ungering about design of fuzzy controllers, and on the thesis of A. Kanstein about possibilistic neural networks [59, 121, 125].

Each year between 1993 and 1995, one-month visits at the University of Dortmund were organized in order to study the different stages involved in the process of fuzzy inference and to obtain solutions to facilitate electronic implementation. As a result of this analysis, it was decided to develop and implement in VLSI new differential operators to perform fuzzy inference [101]. Due to the importance of the T-norm operators [98, 105], an initial analogue differentiable fuzzy operator prototype was built [100]. The implementation and subsequent testing of this was carried out following the specifications of the Microelectronics Institute in Dortmund and of Eurochip (the Granada group). The imprecision of the analogue option with respect to the digital method was not a serious problem, due to the fact that most control systems operate within feedback loops and do not require extreme accuracy. Finally, the electronic realization of a complete fuzzy system was carried out, using the concept of membership functions with an adaptive range. From the use of these types of function, and given that different phases of the fuzzy inference are performed by the same electronic components, a compact final implementation of the circuit was obtained [97, 99].

In the current phase of collaboration, research is centered on the software development of new tools for the automatic construction of neuro-fuzzy systems [104]. The objective of this work is to analyze the different stages involved in the process of fuzzy inference and to provide solutions to implement their implementation in adaptive, self-organizing controllers. Both the topology and the number and magnitude of rules and membership functions are constructed and adapted in an automatic way with no need for additional prior information. We have analyzed the synergy of this fuzzy algorithm using actual paradigms, namely neural networks and genetic algorithms. Ignacio Rojas has again made a stay of two months as a guest of the project (SFB531

of the German DFG). As a result of this collaboration, various projects in the field of soft computing have been carried out [102, 103].

# References

1. D. W. Aha and R. L. Bankert. A comparative evaluation of sequential feature selection algorithms. In D. Fisher and H.-J. Lenz, editors, *Learning from Data*, chapter 4, pages 199–206. Springer, New York, 1996.
2. J. Allan. Incremental relevance feedback for information filtering. In H. P. Frei, editor, *Proceedings of the Nineteenth ACM Conference on Research and Development in Information Retrieval*, pages 270–278. ACM Press, New York, 1996.
3. U. Altenhoener, M. Meurer, J. Strube, and H. Schmidt-Traub. Parameter estimation for the simulation of liquid chromatography. *Journal of Chromatography A*, 769:59–69, 1997.
4. K. Antreich, J. Eckmüller, H. Gräb, M. Pronath, F. Schenkel, R. Schwencker, and S. Zizala. WiCkeD: Analog circuit synthesis incorporating mismatch. In *IEEE Custom Integrated Circuits Conference (CICC)*, May 2000.
5. T. Bäck. *Evolutionary Algorithms in Theory and Practice: Evolution Strategies, Evolutionary Programming, Genetic Algorithms*. Oxford University Press, New York, 1996.
6. J. Bala, K. A. De Jong, J. Huang, H. Vafaie, and H. Wechsler. Hybrid learning using genetic algorithms and decision trees for pattern classification. In *Proceedings of the 14th International Joint Conference on Artificial Intelligence (IJCAI-95)*, volume 1, pages 719–724. Morgan Kaufmann, 1995.
7. M. Balabanovic. An adaptive web page recommendation service. In W. L. Johnson, editor, *Proceedings of the First International Conference on Autonomous Agents*, pages 378–385. ACM Press, New York, 1997.
8. Y. Bard. *Nonlinear Parameter Estimation*. Academic Press, New York, 1974.
9. T. Beielstein, J. Dienstuhl, C. Feist, and M. Pompl. Circuit design using evolutionary algorithms. *IEEE World Congress on Computational Intelligence (CEC-2002), Hawai*, May 2002. (Accepted).
10. E. Bloedorn and R. S. Michalski. Data-driven constructive induction. *IEEE Intelligent Systems*, 13(2):30–37, 1998. Special issue on Feature Transformation and Subset Selection.
11. A. L. Blum and P. Langley. Selection of relevant features and examples in machine learning. *Artificial Intelligence*, 97:245–271, 1997.
12. B. E. Boser, I. M. Guyon, and V. N. Vapnik. A traininig algorithm for optimal margin classifiers. In D. Haussler, editor, *Proceedings of the 5th Annual ACM Workshop on Computational Learning Theory (COLT)*, pages 144–152. ACM Press, Pittsburgh, PA, 1992.
13. L. Breiman. Bagging predictors. *Machine Learning*, 13(2):30–37, 1996.
14. T. P. E. Broekaert et al. A monolithic 4-bit 2-gsps resonant tunneling analog-to-digital converter. *IEEE Journal of Solid-State Circuits*, 33(9):1342–1347, September 1998.
15. P. Brown, G. Byrne, and A. Hindmarsh. Vode: A variable-coefficient ODE solver. *SIAM Journal on Scientific and Statistical Computing*, 10(5):1038–1051, 1989.

16. A. Bühlmeier and G. Manteuffel. Operant conditioning in robots. In O. Omid-var and P. P. van der Smagt, editors, *Neural Systems for Robotics*, pages 195–225. Academic Press, San Diego, CA, 1997.
17. A. Bühlmeier, G. Manteuffel, M. Rossmann, and K. Goser. Robot learning in analog neural hardware. In *Proceedings of the Sixth International Conference on Microelectronics for Neural Networks and Fuzzy Systems*, volume 1112 of LNCS, pages 311–316. Springer, Berlin, 1996.
18. A. Bühlmeier, G. Manteuffel, M. Rossmann, and K. Goser. Application of a local learning rule in a wheelchair robot. In *Third International Conference on Neural Networks and their Applications*, pages 177–182. Marseille, March 1997.
19. A. Bühlmeier, M. Rossmann, K. Goser, and G. Manteuffel. Adaptive local navigation. In *Proceedings of the 26th Göttingen Neurobiology Conference: New Neuroethology on the Move*, volume 1, 1998.
20. A. Bühlmeier, M. Rossmann, G. Manteuffel, and K. Goser. Multistage STM in a multilayer hebbian learning architecture for local navigation. In *Proceedings of the 8th International Conference on Artificial Neural Networks*, volume 2, pages 1139–1144. Skövde, Sweden, September 1998.
21. A. Bühlmeier, P. Steiner, M. Rossmann, K. Goser, and G. Manteuffel. Heb-bian multilayer network in a wheelchair robot. In *Proceedings of the 7th International Conference on Artificial Neural Networks, Lausanne, Switzerland*, volume 1327 of LNCS, pages 727–732. Springer, Berlin, 1997.
22. C. Burges. A tutorial on support vector machines for pattern recognition. *Data Mining and Knowledge Discovery*, 2(2):121–167, 1998.
23. C. Burwick, M. Thomas, J. Dienstuhl, and K. F. Goser. Threshold-gates in arithmetic circuits. *The $8^{th}$ IEEE International Conference on Electronics, Circuits and Systems*, II:909–912, September 2001.
24. S. Chen, C. F. N. Cowan, and P. M. Grant. Orthogonal least squares learning algorithm for radial basis function networks. *IEEE Transactions on Neural Networks*, 2(2):302–309, 1991.
25. S. Chung and C. Wen. Longitudinal diffusion of liquid flowing through fixed and fluidized beds. *AIChE Journal*, 14:857–866, 1968.
26. A. R. Conn, P. K. Coulmann, R.A. Haring, G. L. Morill, C. Visweswariah, and C. W. Wu. Jiffy tune: Circuit optimization using time-domain sensitiv-ities. *IEEE Transactions on Computer-Aided Design of Integrated Circuits and Systems*, 17(12):1292–1309, 1998.
27. G. Daniel. Parameter estimation for grey box models of batch chromatogra-phy. Technical report, Process Control Laboratory, University of Dortmund, December 2001.
28. M. Dash and H. Liu. Feature selection for classification. *International Journal of Intelligent Data Analysis*, 1(3):131–156, 1997.
29. J. Dienstuhl, C. Burwick, M. Thomas, and K. F. Goser. Evolutionary circuit design algorithms for low-power CMOS. *Kleinheubacher Berichte*, September 2001.
30. G. Dünnebier, S. Engell, A. Epping, F. Hanisch, A. Juppke, K.-U. Klatt, and H. Schmidt-Traub. Model-based control of batch-chromatography. *AIChE Journal*, 47(11):2493–2502, 2001.
31. R. Fletcher. *Practical Methods of Optimization*. Wiley, New York, 2nd edition, 1987.

32. A. Giordana and F. Neri. Search-intensive concept induction. *Evolutionary Computation*, 3(4):375–416, 1995.

33. A. Giordana and L. Saitta. Integrating multiple learning strategies in first order logics. *Machine Learning*, 27(2):209–240, 1997.

34. P. Glösekötter, A. Kanstein, S. Jung, and K. Goser. Implementation of a RBF network based on possibilistic reasoning. In *Proceedings of the 24th EUROMICRO Conference*, pages 667–682. Västeras, Sweden, August 1998.

35. D. E. Goldberg. *Genetic Algorithms in Search, Optimization and Machine Learning*. Addison-Wesley, Reading, MA, 1989.

36. K. Goser, B. Hesse, and M. Thomas. Klassifikation und Prädiktion von Bauelemente-Parametern hinsichtlich der Qualität integrierter Elektroniksysteme. In H.-P. Schwefel et al., editor, *Sonderforschungsbereich 531–Design und Management komplexer Prozesse und Systeme mit Methoden der Computational Intelligence–Arbeits-und Ergebnisbericht 1997-1998-1999*, chapter Teilprojekt C6, pages 223–237. University of Dortmund, April 1999.

37. K. Goser, C. Pacha, A. Kanstein, and M. L. Rossmann. Aspects of systems and circuits for nanoelectronics. *Proceedings of the IEEE, Special Issue on Namometer-Scale Science & Technology*, pages 558–573, April 1997.

38. T. Gu. *Mathematical Modelling and Scale Up of Liquid Chromatography*. Springer, New York, 1995.

39. G. Guichon and S. G. Shirazi. *Fundamentals of Preparative and Nonlinear Chromatography*. Academic Press, Boston, MA, 1994.

40. B. Gustavsson and P. Olsson. Fourth order difference methods for hyperbolic ibvps. *Journal of Computational Physics*, 117:300–317, 1995.

41. M. A. Hall. Correlation-based feature selection for machine learning. PhD thesis, Department of Computer Science, University of Waikato, Hamilton, New Zealand, 1999.

42. D. P. Helmbold and P. M. Long. Tracking drifting concepts by minimizing disagreements. *Machine Learning*, 14(1):27–45, 1994.

43. B. Hesse and K. Goser. Evaluating stress test data of IGBT-modules with different neural networks. In *Fourth International Conference on Neural Networks and their Applications*, pages 163–166. Marseille, France, March 1998.

44. J. H. Holland. *Adaptation in Natural and Artificial Systems*. University of Michigan Press, Ann Arbor, MI, 1975.

45. The Mathworks Inc. *Matlab Optimization Toolbox, User's Guide*. The Mathworks Inc., 1998.

46. C. Z. Janikow. A knowledge-intensive genetic algorithm for supervised learning. *Machine Learning*, 13(4):189–228, 1993.

47. T. Joachims. Making large-scale SVM learning practical. In B. Schölkopf, C. Burges, and A. Smola, editors, *Advances in Kernel Methods – Support Vector Learning*, chapter 11, pages 169–184. MIT Press, Cambridge, MA, 1999.

48. T. Joachims. Transductive inference for text classification using support vector machines. In *Proceedings of the Sixteenth International Conference on Machine Learning (ICML-99)*. Bled, Slovenia, 1999.

49. T. Joachims. Estimating the generalization performance of a SVM efficiently. In *Proceedings of the Seventeenth International Conference on Machine Learning (ICML-2000)*, pages 431–438. Morgan Kaufmann, San Francisco, CA, 2000.

50. T. Joachims. The maximum-margin approach to learning text classifiers: methods, theory, and algorithms. PhD thesis, Artificial Intelligence Unit, Department of Computer Science, University of Dortmund, February 2001.

51. G. John, R. Kohavi, and K. Pfleger. Irrelevant features and the subset selection problem. In *Proceedings of the Eleventh International Conference on Machine Learning (ICML-94)*, pages 121–129. Morgan Kaufmann, San Mateo, CA, 1994.

52. S. Jung. Capacitive CMOS fingerprint sensor with on-chip parallel signal processing. PhD thesis, University of Dortmund, 2000.

53. S. Jung, C. Hierold, T. Scheiter, P. W. von Basse, R. Thewes, K. Goser, and W. Weber. Intelligent CMOS fingerprint sensors. In *Transducers*, pages 966–969. Sendai, Japan, 1999.

54. S. Jung, R. Thewes, T. Scheiter, K. Goser, and W. Weber. A low-power and high-performance CMOS fingerprint sensing and encoding architecture. In *Proceedings of the 24th European Solid-State Circuits Conference*, pages 324–327. The Hague, September 1998.

55. S. Jung, R. Thewes, T. Scheiter, K. Goser, and W. Weber. CMOS fingerprint sensor with automatic local contrast adjustment and pixel-parallel encoding logic. In *VLSI Circuits Symposium*, pages 161–164. Kyoto, 1999.

56. S. Jung, R. Thewes, T. Scheiter, K. F. Goser, and W. Weber. A low-power and high-performance cmos fingerprint sensing and encoding architecture. *IEEE Journal of Solid-State Circuits*, 1999.

57. K. Kaczmarski and D. Antos. Fast finite difference method for solving multi-component adsorption-chromatography models. *Computers & Chemical Engineering*, 20(11):1271–1276, 1996.

58. A. Kanstein. Possibilistic neural networks for process modelling. PhD thesis, University of Dortmund, 1998.

59. A. Kanstein. *Possibilistic Neural Networks for Process Modelling*. VDE Verlag, Berlin, 1999.

60. A. Kanstein, M. Thomas, and K. Goser. Possibilistic reasoning in a computational neural network. In *IEEE-International Conference on Neural Networks*, pages 2541–2546. Houston, TX, June 1997.

61. J.-U. Kietz. Some lower bounds for the computational complexity of inductive logic programming. *Arbeitspapiere der GMD* 718, German National Research Center for Computer Science (GMD), St. Augustin, Germany, December 1992.

62. J.-U. Kietz. Induktive Analyse relationaler Daten. PhD thesis, Technical University of Berlin, October 1996.

63. K. Kira and L. Rendell. The feature selection problem: Traditional methods and a new algorithm. In *Proceedings of the Tenth National Conference on Artificial Intelligence*, pages 129–134. AAAI Press, Menlo Park, CA, 1992.

64. K.-U. Klatt, G. Dünnebier, S. Engell, and F. Hanisch. Model-based optimization and control of chromatographic processes. *Computers & Chemical Engineering*, 48:1119–1126, 2000.

65. R. Klinkenberg. Using labeled and unlabeled data to learn drifting concepts. In M. Kubat and K. Morik, editors, *Workshop notes of the IJCAI-01 Workshop on* Learning from Temporal and Spatial Data, pages 16–24. AAAI Press, Menlo Park, CA, 2001. Held in conjunction with the International Joint Conference on Artificial Intelligence (IJCAI).

66. R. Klinkenberg. Informed parameter setting for support vector machines: Using additional user knowledge in classification tasks. Technical Report CI-126/02, Collaborative Research Center 531, University of Dortmund, 2002.

67. R. Klinkenberg. Transductive learning of drifting concepts. Technical Report CI-125/02, Collaborative Research Center 531, University of Dortmund, 2002.

68. R. Klinkenberg and T. Joachims. Detecting concept drift with support vector machines. In *Proceedings of the Seventeenth International Conference on Machine Learning (ICML-2000)*, pages 487–494. Morgan Kaufmann, San Francisco, CA, 2000.

69. R. Klinkenberg and I. Renz. Adaptive information filtering: Learning in the presence of concept drifts. In M. Sahami, M. Craven, T. Joachims, and A. McCallum, editors, *Workshop Notes of the Workshop* Learning for Text Categorization *held at the* Fifteenth International Conference on Machine Learning (ICML-98), pages 33–40. AAAI Press, Menlo Park, CA, 1998.

70. R. Kohavi and G. H. John. Wrappers for feature subset selection. *Artificial Intelligence Journal, Special Issue on Relevance*, 97(1–2):273–324, 1997.

71. R. Kohavi, D. Sommerfield, and J. Dougherty. Data mining using MLC++: A machine learning library in C++. In *Tools with Artificial Intelligence*, pages 234–245. IEEE Computer Society Press, Los Alamitos, CA, 1996. http://www.sgi.com/tech/mlc/.

72. M. Kubat and K. Morik, editors. *Workshop notes of the IJCAI-01 Workshop on* Learning from Temporal and Spatial Data. AAAI Press, Menlo Park, CA, 2001. Held in conjunction with the International Joint Conference on Artificial Intelligence (IJCAI).

73. A. Kuh, T. Petsche, and R. L. Rivest. Learning time-varying concepts. In *Advances in Neural Information Processing Systems (NIPS)*, volume 3, pages 183–189. Morgan Kaufmann, San Mateo, CA, 1991.

74. C. Lanquillon. Information filtering in changing domains. In T. Joachims, A. McCallum, M. Sahami, and L. Ungar, editors, *Working Notes of the Workshop Machine Learning for Information Filtering* held at the *Sixteenth International Joint Conference on Artificial Intelligence (IJCAI-99)*, pages 41–48. Stockholm, August 1999.

75. C. Lanquillon. Partially supervised text classification: Combining labeled and unlabeled documents using an EM-like scheme. In R. López de Mántaras and E. Plaza, editors, *Proceedings of the 11th Conference on Machine Learning (ECML 2000)*, volume 1810 of LNCS, pages 229–237. Springer, Berlin, 2000.

76. N. Lavrac, D. Gamberger, and P. D. Turney. A relevancy filter for constructive induction. *IEEE Intelligent Systems*, 13(2):50–56, 1998. Special issue on Feature Transformation and Subset Selection.

77. A. Lunts and V. Brailovskiy. Evaluation of attributes obtained in statistical decision rules. *Engineering Cybernetics*, 3:98–109, 1967.

78. C. Matheus. The need for constructive induction. In *Proceedings of the Eighth International Workshop on Machine Learning*, pages 173–177. Morgan Kaufmann, San Mateo, CA, 1991.

79. S. Michaelis. Verkehrslastprognose durch Wissensentdeckung. *Informatik Spektrum*, 24(4):218–227, August 2001.

80. M. Mitchell. *An Introduction to Genetic Algorithms*. MIT Press, Cambridge, MA, 1996.

81. T. Mitchell, R. Caruana, D. Freitag, J. McDermott, and D. Zabowski. Experience with a learning personal assistant. *Communications of the ACM*, 37(7):81–91, July 1994.

82. T. M. Mitchell. *Machine Learning*. McGraw Hill, New York, 1996.

83. K. Morik. The representation race – preprocessing for handling time phenomena. In R. López de Mántaras and E. Plaza, editors, *Proceedings of the 11th European Conference on Machine Learning (ECML 2000)*, volume 1810 of *Lecture Notes in Artificial Intelligence (LNAI)*, pages 4–19. Springer, Berlin, 2000.

84. K. Nigam, A. McCallum, S. Thrun, and T. Mitchell. Text classification from labeled and unlabeled documents using EM. *Machine Learning*, 39(2/3):103–134, 2000.

85. W. Nye, D. C. Riley, A. Sangiovanni-Vincentelli, and A. L. Tits. Delight.Spice: An optimization-based system for the design of integrated circuits. *IEEE Transactions on Computer-Aided Design*, 7(4):501–519, 1998.

86. R. Otte. Selbstorganisierende Merkmalsarten zur multivarianten Datenanalyse komplexer technischer Prozesse. PhD thesis, University of Dortmund, 1998.

87. R. Otte and K. Goser. Neue Formen der Prozessvisualisierung und Prozessanalyse. *atp - Automatisierungstechnische Praxis*, 39:28–39, December 1997.

88. R. Otte and K. Goser. New approaches of process visualisation and analysis in power plants. In *Workshop on Self-Organizing Maps*, pages 44–50. Helsinki, 1997.

89. R. Otte and K. F. Goser. Selbstorganisierende neuronale Karten zur multivarianten Prozessanalyse. *VGB Kraftwerkstechnik*, 78(7):52–57, 1998.

90. C. Pacha, K. Goser, A. Brennemann, and W. Prost. A threshold logic full adder based on resonant tunneling transistors. In *European Solid-State Circuits Conference (ESSCIRC'98)*, pages 427–431. The Hague, September 1998.

91. G. Paliouras. The scalability of machine learning algorithms. Master's thesis, Department of Computer Science, University of Manchester, UK, 1993.

92. G. D. Plotkin. A note on inductive generalization. In B. Meltzer and D. Michie, editors, *Machine Intelligence*, chapter 8, pages 153–163. American Elsevier, New York, NY, 1970.

93. W. Punch, E. Goodman, P. Hovland, and R. Enbody. Further research on feature selection and classification using genetic algorithms. In *Proceedings of the Fifth International Conference on Genetic Algorithms*, pages 557–564. Morgan Kaufmann, San Mateo, CA, 1993.

94. J. R. Quinlan. *C4.5: Programs for Machine Learning*. Machine Learning. Morgan Kaufmann, San Mateo, CA, 1993.

95. O. Ritthoff, R. Klinkenberg, S. Fischer, and I. Mierswa. A hybrid approach to feature selection and generation using an evolutionary algorithm. Technical Report CI-127/02, Collaborative Research Center 531, University of Dortmund, 2002.

96. O. Ritthoff, R. Klinkenberg, S. Fischer, I. Mierswa, and S. Felske. YALE: Yet another machine learning environment. In R. Klinkenberg, S. Rüping, A. Fick, N. Henze, C. Herzog, R. Molitor, and O. Schröder, editors, *LLWA 01 – Tagungsband der GI-Workshop-Woche* Lernen – Lehren – Wissen – Adaptivität, pages 84–92. Technical Report No. 763, Department of Computer Science, University of Dortmund, October 2001.

97. I. Rojas, M. Anguita, F. J. Pelayo, P. Gloesekoetter, and A. Prieto. CMOS implementation of RBF neural networks using adaptive gaussian radial functions. In *Proceedings of the 6th International Conference on Microelectronics for Neural Networks, Evolutionary Algorithms and Fuzzy Systems*, pages 195–200. September, 1997.

98. I. Rojas, J. Ortega, F. J. Pelayo, and A. Prieto. Statistical analysis of the main parameters in the fuzzy inference process. *Fuzzy Sets and Systems Elsevier*, 102:157–173, 1999.

99. I. Rojas, F. Pelayo, J. Ortega, A. Prieto, B. Prieto, and C. Prieto. Compact CMOS fuzzy controllers using the normalized product of adaptive membership functions. *Electronics Letters*, 33(3):221–223, January 1997.

100. I. Rojas, F. J. Pelayo, M. Anguita, J. L. Bernier, and A. Prieto. Implementation of adaptable and hierachical fuzzy T-norm. *Electronics Letters*, 35(24):2150–2152, November 1999.

101. I. Rojas, F. J. Pelayo, M. Anguita, and A. Prieto. Development of a configurable and hierechical multiple-input fuzzy T-norm. *Proceedings of the 6th International Conference on Microelectronics for Neural Networks, Evolutionary Algorithms and Fuzzy Systems*, pages 297–304, September 1997.

102. I. Rojas, H. Pomares, J. González, P. Glösekötter, J. Dienstuhl, and K. Goser. Multideme evolutionary algorithm based approach to the generation of fuzzy systems. *Proceedings of the 10th IEEE International Conference on Fuzzy Systems*, December 2001. (In print).

103. I. Rojas, H. Pomares, J. González, P. Glösekötter, J. Dienstuhl, and K. Goser. A new sequential learning algorithm using pseudo-Gaussian functions for neuro-fuzzy systems. *Proceedings of the 10th IEEE International Conference on Fuzzy Systems*, December 2001. (In print).

104. I. Rojas, H. Pomares, J. Ortega, and A. Prieto. Self-organized fuzzy system generation from training examples. *IEEE Transactions on Fuzzy Systems*, 8(1):23–36, February 2000.

105. I. Rojas, O. Valenzuela, M. Anguita, and A. Prieto. Analysis of the operators involved in the definition of the implication functions and in the fuzzy interference process. *International Journal of Approximate Reasoning*, 19:367–389, 1998.

106. J. P. Ros. Learning Boolean functions with genetic algorithms: A PAC analysis. In L. D. Whitley, editor, *Foundations of Genetic Algorithms 2*, pages 257–275. Morgan Kaufmann, San Mateo, CA, 1993.

107. M. Rossmann. Hardwarearchitekturen für künstliche biomorphe neuronale Netze. PhD thesis, University of Dortmund, 2001.

108. M. Rossmann, A. Buehlmeier, G. Manteuffel, and K. F. Goser. Dynamic hebbian learning strategies for VLSI-systems. *Neurocomputing*, 1(28):157–164, 1999.

109. M. Rossmann, A. Bühlmeier, G. Manteuffel, and K. Goser. Short- and long-term-dynamics in a stochastic pulse stream neuron implemented in FPGA. In *Proceedings of the 7th International Conference on Artificial Neural Networks, Lausanne, Switzerland*, volume 1327 of LNCS, pages 1241–1246. Springer, Berlin, 1997.

110. M. Rossmann, A. Bühlmeier, G. Manteuffel, and K. Goser. Dynamic hebbian learning strategies for VLSI. In *Fourth International Conference on Neural Networks and their Applications*, pages 191–194. Marseille, March 1998.

111. M. Rossmann, A. Bühlmeier, G. Manteuffel, and K. Goser. Endeavour: System for the analysis of dynamic large scale neural networks in real-time. In *BMBF-Fachtagung*, Göttingen, Germany, May 1998.

112. M. Rossmann, A. Bühlmeier, G. Manteuffel, and K. Goser. VLSI chip for high speed computation of large scale artificial neural networks based on the modified Hebbian learning algorithm. In *Proceedings of the 26th Göttingen Neurobiology Conference: New Neuroethology on the Move*, volume 1, 1998.

113. M. Rossmann, C. Burwick, A. Bühlmeier, G. Manteuffel, and K. Goser. Dynamic Hebbian learning in system architecture based on high speed chip. In *Proceedings of the Sixth International Conference on Microelectronics for Neural Networks*, pages 251–256. Dresden, September 1997.

114. M. Rossmann, C. Burwick, A. Bühlmeier, G. Manteuffel, and K. Goser. Neural dynamics in real-time for large scale biomorphic neural networks. In *Proceedings of the 8th International Conference on Artificial Neural Networks*, volume 1, pages 481–486. Skövde, Sweden, September 1998. ISBN 3-540-76263-9.

115. M. Rossmann, T. Jost, K. Goser, A. Bühlmeier, and G. Manteuffel. Exponential Hebbian on-line learning implemented in FPGAs. In *Proceedings of the Sixth International Conference on Microelectronics for Neural Networks and Fuzzy Systems*, volume 1112 of LNCS, pages 767–772. Springer, Berlin, July 1996.

116. S. Rüping. *mySVM-Manual*. Artificial Intelligence Unit, Department of Computer Science, University of Dortmund, 2000. http://www-ai.cs.uni-dortmund.de/SOFTWARE/MYSVM/.

117. D. Ruthven. *Principles of Adsorption and Adsorption Processes*. Wiley, New York, 1989.

118. G. Salton and C. Buckley. Term weighting approaches in automatic text retrieval. *Information Processing and Management*, 24(5):513–523, 1988.

119. R. E. Schapire. A brief introduction to boosting. In *Proceedings of the Sixteenth International Joint Conference on Artificial Intelligence (IJCAI-99)*, pages 1401–1406. Morgan Kaufmann, San Mateo, CA, 1999.

120. A. J. Smola and B. Schölkopf. A tutorial on support vector regression. Neuro-COLT2 Technical Report NC-TR-98-030, Royal Holloway College, University of London, UK, 1998.

121. H. Surmann. *Automatisierter Entwurf von Fuzzy Systemen*, volume 8, no. 452. VDI Verlag, Düsseldorf, 1995.

122. C. Taylor, G. Nakhaeizadeh, and C. Lanquillon. Structural change and classification. In G. Nakhaeizadeh, I. Bruha, and C. Taylor, editors, *Workshop Notes on Dynamically Changing Domains – Theory Revision and Context Dependence Issues* held at the *Ninth European Conference on Machine Learning (ECML-97)*, volume 1224 of LNCS, pages 67–78. Springer, Berlin, April 1997.

123. M. Thomas. Design und Analyse integrierter Schaltungen mit Evolutionären Algorithmen. PhD thesis, University of Dortmund, 2001.

124. M. Thomas, C. Burwick, and K. F. Goser. Circuit analysis and design using evolutionary algorithms. *Transaction of the Collaborative Research Center on Computational Intelligence (DFG-SFB 531, University of Dortmund)*, (CI-58/00), 2000.

125. A. P. Ungering. *Integrationsgerechter Entwurf von Fuzzy-Reglern*. Number 210 in Series 20. VDI Verlag, Düsseldorf, 1996.

126. H. Vafaie and K. A. De Jong. *Genetic Algorithms as a Tool for Feature Selection in Machine Learning*. IEEE Computer Society Press, Los Alamitos, CA, 1992.

127. L. G. Valiant. A theory of the learnable. *Communications of the ACM*, 27(11):1134–1142, 1984.

128. C. van Rijsbergen. A theoretical basis for the use of co-occurrence data in information retrieval. *Journal of Documentation*, 33(2):106–119, June 1977.

129. V. N. Vapnik. *Estimation of Dependencies Based on Empirical Data*. Springer, New York, 1982.

130. V. N. Vapnik. *Statistical Learning Theory*. Wiley, Chichester, UK, 1998.

131. G. Widmer and M. Kubat. Learning in the presence of concept drift and hidden contexts. *Machine Learning*, 23(2):69–101, 1996.

132. I. H. Witten and E. Frank. *Data mining: Practical machine learning tools and techniques with Java implementations*. Morgan Kaufmann, San Francisco, CA, 2000. http://www.cs.waikato.ac.nz/ ml/weka/.

133. J. Wnek and M. S. Michalski. Hypothesis-driven constructive induction in AQ17-HCI: A method and experiments. *Machine Learning*, 14(1):139–168, 1994.

134. J. Yang and V. Honavar. Feature subset selection using a genetic algorithm. In J. R. Koza, K. Deb, M. Dorigo, D. B. Fogel, M. Garzon, H. Iba, and R. L. Riolo, editors, *Genetic Programming 1997: Proceedings of the Second Annual Conference*, pages 380–385. Morgan Kaufmann, San Mateo, CA, 1997.

135. R. Zimmermann and W. Fichtner. Low power logic styles: CMOS versus pass-transistor logic. *IEEE Journal of Solid-State Circuits*, 32(7):1079–1090, July 1997.

136. E. Zitzler. Multiobjective evolutionary algorithms: A comparative case study and the strength pareto approach. *IEEE Transactions on Evolutionary Computation*, 3(4):257–271, 1999.

# Index

## Natural Computing Series

W.M. Spears: Evolutionary Algorithms. The Role of Mutation and Recombination. XIV, 222 pages, 55 figs., 23 tables. 2000

H.-G. Beyer: The Theory of Evolution Strategies. XIX, 380 pages, 52 figs., 9 tables. 2001

L. Kallel, B. Naudts, A. Rogers (Eds.): Theoretical Aspects of Evolutionary Computing. X, 497 pages. 2001

M. Hirvensalo: Quantum Computing. XI, 190 pages. 2001

G. Păun: Membrane Computing. An Introduction. XI, 429 pages, 37 figs., 5 tables. 2002

A.A. Freitas: Data Mining and Knowledge Discovery with Evolutionary Algorithms. XIV, 264 pages, 74 figs., 10 tables. 2002

H.-P. Schwefel, I. Wegener, K. Weinert (Eds.): Advances in Computational Intelligence. VIII, 325 pages. 2003

A. Ghosh, S. Tsutsui (Eds.): Advances in Evolutionary Computing. IX, 1006 pages. 2003

L.F. Landweber, E. Winfree (Eds.): Evolution as Computation. DIMACS Workshop, Princeton, January 1999. XV, 332 pages. 2003

M. Amos: Theoretical and Experimental DNA Computation. Approx. 200 pages. 2003